Operations Research and Discrete Analysis

Mathematics and Its Applications

Volume 391

Operations Research and Discrete Analysis

by

Alekseĭ D. Korshunov
Sobolev Institute of Mathematics,
Siberian Branch of the Russian Academy of Sciences,
Novosibirsk, Russia

SPRINGER SCIENCE+BUSINESS MEDIA, B.V.

A C.I.P. Catalogue record for this book is available from the Library of Congress

ISBN 978-94-010-6395-1 ISBN 978-94-011-5678-3 (eBook)
DOI 10.1007/978-94-011-5678-3

The contributions to this volume have all been translated
from the second volume of the Russian journal
Discrete Analysis and Operations Research,
the Sobolev Institute of Mathematics,
Siberian Branch of the Russian Academy of Sciences,
Novosibirsk, Russia, © 1995.

Printed on acid-free paper

TABLE OF CONTENTS

PREFACE

This book contains translations of papers from the second volume of the new Russian-language journal published at the Sobolev Institute of Mathematics (Siberian Branch of the Russian Academy of Sciences, Novosibirsk) since 1994.

In 1994 the journal was titled **Sibirskiĭ Zhurnal Issledovaniya Operatsiĭ**. Since 1995 this journal has the title **Diskretnyĭ Analiz i Issledovanie Operatsiĭ** (Discrete Analysis and Operations Research).

The aim of this journal is to bring together research papers in different areas of discrete mathematics and computer science. The journal **Diskretnyĭ Analiz i Issledovanie Operatsiĭ** covers the following fields:

- discrete optimization
- discrete structures and extremal problems
- combinatorics
- control and reliability of discrete devices
- mathematical models and methods of decision making
- design and analysis of algorithms
- synthesis and complexity
- of control systems
- automata
- graphs
- game theory and its applications
- coding theory
- scheduling theory
- functional systems theory

Contributions presented to the journal can be original research papers and occasional survey articles of moderate length.

The journal is published in one volume of four issues per year that appear in March, June, September, and December. Each volume contains approximately 400 pages.

I express my sincere gratitude to Professor S. S. Kutateladze for his help in editing the English translation.

Editor-in-Chief
of the Journal

A. D. Korshunov

ON DISCRETE APPROXIMATIONS OF CONTINUOUS FUNCTIONS WITH BOUNDED SECOND DERIVATIVE[†)]

G. G. Amanzhaev

Evaluating a continuous function by a discrete device of limited size (complexity) is necessarily approximate; moreover, the evaluation of approximate values of functions is in fact performed only on some finite subset of the domain of definition, i.e., a function with finite domain and range is evaluated. In addition, each class of continuous functions naturally generates a corresponding class of discrete functions.

In this paper, for the class of functions

$$H_{2,c} = \{f : [0,1) \to [0,1) \mid |f'(x) - f'(y)| \le c|x-y|\}$$

the corresponding classes of discrete functions are described and their properties are investigated; connection is revealed between such informational characteristic of continuous function classes as ε-entropy and some analogous characteristic of discrete functions which is closely related to their computational complexity. The problem of such approximate realization of continuous functions was posed in [1].

§ 1. Defining Classes

We start with the following class of continuous functions:

$$H_{2,c} = \{f : [0,1) \to [0,1) \mid |f'(x) - f'(y)| \le c|x-y|\}.$$

Taking in $I = [0,1)$ the discrete subset $\tilde{I}_n = \left\{\frac{1}{2n}, \frac{3}{2n}, \dots, \frac{2n-1}{2n}\right\}$, for each function $f \in H_{2,c}$ we consider a discrete "approximating" function of the form $g : \tilde{I}_n \to \tilde{I}_n$. In particular, as g it is possible to consider the function

$$g(x) = \frac{1}{2n} + \frac{1}{n}\lfloor nf(x)\rfloor, \quad x \in \tilde{I}_n.$$

It is possible in a one-to-one fashion to pass from the set $\tilde{I}_n$ to the set $I_n = \{0, 1, \dots, n-1\}$ by associating with the value $x \in \tilde{I}_n$ its "ordinal index." In

†) This research is financially supported by the Russian Foundation for Basic Research (Grant 93-01-01527).

A. D. Korshunov (ed.), *Operations Research and Discrete Analysis*, 1–12.

addition, instead of the function g it is possible to consider the integral discrete function $h : I_n \longrightarrow I_n$ defined by the formula

$$h(x) = \lfloor nf(1/2n + x/n) \rfloor, \quad x \in I_n.$$

Denote by $\widehat{H}^n_{2,c}$ the set of all such functions, i.e.,

$$\widehat{H}^n_{2,c} = \{h : I_n \to I_n \mid (\exists f \in H_{2,c})(\forall x \in I_n) h(x) = \lfloor nf(1/2n + x/n) \rfloor\}.$$

It is easy to see that the problem of determining whether a discrete function $h : I_n \longrightarrow I_n$ belongs to the class $\widehat{H}^n_{2,c}$ is reduced essentially to enumerating the infinite set of continuous functions in $H_{2,c}$.

In order to exclude such enumerating, we consider a broader class of discrete functions than $\widehat{H}^n_{2,c}$. To this end, we fix a finite tuple Σ of properties enjoyable by each function of $\widehat{H}^n_{2,c}$ and consider the class $H^n_{2,c}(\Sigma)$ of discrete functions defined as follows:

$$H^n_{2,c}(\Sigma) = \{f : I_n \to I_n \mid f \text{ has all properties from } \Sigma\}.$$

It is clear that $\widehat{H}^n_{2,c} \subset H^n_{2,c}(\Sigma)$.

Take a collection of properties so that the class $H^n_{2,c}(\Sigma)$ be not much differ from the class $\widehat{H}^n_{2,c}$ (the meaning of this requirement will be made clear below), the number of properties in Σ is small, and the properties are easily checked.

Define the set Σ as follows. We determine the class of continuous functions $H_{2,c}$ in equivalent form:

$$H_{2,c} = \{f : [0,1) \to [0,1) | (\forall x, y, z \in [0,1), x < y < z) | \, |\Delta_2(f; x, y, z)| \leq c/2\},$$

where $\Delta_2(f; x, y, z)$ is the second separated difference of the function f, i.e.,

$$\Delta_2(f; x, y, z) = \left(\frac{f(x) - f(y)}{x - y} - \frac{f(y) - f(z)}{y - z} \right) \Big/ (x - z).$$

Using this definition, we estimate $\Delta_2(g; x, y, z)$ for $g \in \widehat{H}^n_{2,c}$. Denoting $x' = 1/2n + x/n$, $y' = 1/2n + y/n$, $z' = 1/2n + z/n$ and taking it into account that $g(x) = \lfloor nf(x/n + 1/2n) \rfloor$, we obtain

$$\begin{aligned} &\Delta_2(g; x, y, z) \\ &= \frac{(\lfloor nf(x') \rfloor - \lfloor nf(y') \rfloor)/(x - y) - (\lfloor nf(y') \rfloor - \lfloor nf(z') \rfloor/(y - z)}{x - z} \\ &= \frac{1}{n} \Delta_2(f; x', y', z') - \frac{(\alpha(x') - \alpha(y'))/(x - y) - (\alpha(y') - \alpha(z'))/(y - z)}{x - z}, \end{aligned}$$

where $\alpha(t) = nf(t) - \lfloor nf(t) \rfloor - 1/2$, i.e., $|\alpha(t)| \le 1/2$. Hence, if $x < y < z$, then

$$\Delta_2(g; x, y, z) < \frac{c}{2n} + \varphi_2(x, y, z), \tag{1}$$

where $\varphi_2(x, y, x) = \frac{1}{(x-y)(y-z)} = \sup \Delta_2(h; x, y, z)$, and the supremum is taken over all the functions $h : \mathbb{R} \to [-1/2, 1/2]$.

The tuple of inequalities of the form (1) is the required tuple of properties Σ. In this case

$$H^n_{2,c} = \{g : I_n \to I_n | (\forall x, y, z \in [0,1), \ x < y < z) |\Delta_2(g; x, y, z)| < \frac{c}{2n} + \varphi_2(x, y, z)\}.$$

§ 2. Estimating the Cardinality of the Classes

Lemma. $2^n \le |\widehat{H}^n_{2,c}| \le |H^n_{2,c}| \le 2^{n(1+o(1))}$.

PROOF. By virtue of the inclusion $\widehat{H}^n_{2,c} \subset H^n_{2,c}$, it is sufficient to show only the first and the last inequalities in this chain.

To begin with, we verify the first inequality. Consider the function

$$d_n(x) = \begin{cases} 0 & \text{if } x \notin \left[-\frac{1}{2n}, \frac{1}{2n}\right], \\ \left(\frac{1}{2n} + x\right)^2 \left(\frac{1}{2n} - x\right)^2 & \text{if } x \in \left[-\frac{1}{2n}, \frac{1}{2n}\right]. \end{cases}$$

It is easy to see that the function $d_n(x)$ is twice differentiable everywhere except the points $x = \pm 1/2n$, moreover, $|d''_n(x)| < 2n^{-2}$. Hence, for sufficiently large n and $a_k \in \{0, 1\}$ the function

$$f(x) = \frac{1}{n} - \sum_{k=0}^{n-1} a_k d_n \left(x - \frac{1}{2n} - \frac{k}{n}\right)$$

belongs to the class $H_{2,c}$. Therefore, the corresponding functions $g(x)$ defined at the integer points by the relation $g(x) = \lfloor nf(x/n + 1/2n) \rfloor$ are in $\widehat{H}^n_{2,c}$. But all the functions $g(x)$ are distinct since $g(k) = 1 - a_k$. Hence,

$$|\widehat{H}^n_{2,c}| \ge 2^n.$$

Now we verify the second inequality. Denote by $B(x_1, y_1, \ldots, x_k, y_k)$ the collection of functions in $H^n_{2,c}$ such that at $x = x_i$ these functions take the value y_i, $1 \leq i \leq k$. Let $x_1 = 0$, $x_k = n - 1$, and the remaining points x_i split the segment I_n into parts whose length is not greater than r, where r is an integer.

Let $r = o(\sqrt{n}\,)$. For $x \in \{x_i, x_i + 1, \ldots, x_{i+1}\}$, we consider the function

$$h(x) = f(x) - \left(\frac{x_{i+1} - x}{x_{i+1} - x_i} f(x_i) + \frac{x - x_i}{x_{i+1} - x_i} f(x_{i+1})\right). \tag{2}$$

It is possible to show that for sufficiently large n, at least one of the following situations takes place:

(α) $|h(x)| < 1$;

(β) $-0,5 < h(x) < 1,5$;

(γ) $-1,5 < h(x) < 0,5$.

Therefore, for $x \in \{x_i, x_i + 1, \ldots, x_{i+1}\}$, the function $f(x)$ can be presented in the form

$$f(x) = \frac{x_{i+1} - x}{x_{i+1} - x_i} f(x_i) + \frac{x - x_i}{x_{i+1} - x_i} f(x_{i+1}) + h(x),$$

where $h(x) = \theta + h_1(x)$ and θ is a constant in $\{-1/2, 0, 1/2\}$, $|h_1(x)| < 1$.

Let $(\theta_1, \ldots, \theta_{k-1}, \chi_0, \chi_1, \ldots, \chi_{n-1})$ be a tuple such that $\theta_i \in \{-1/2, 0, 1/2\}$, $\chi_i \in \{0, 1\}$. Consider a subset in $B(x_1, y_1, \ldots, x_k, y_k)$ consisting of functions such that for $x \in \{x_i, \ldots, x_{i+1}\}$,

$$\theta = \theta_i, \quad \chi_x = \begin{cases} 0 & \text{if} \quad h_1(x) \leq 0; \\ 1 & \text{if} \quad h_1(x) > 0. \end{cases}$$

Since the function f is integer-valued, to the tuple under consideration there corresponds at most one function in $B(x_1, y_1, \ldots, x_k, y_k)$. Hence, the cardinality of B does not exceed $3^{k-1}2^n$. Since $y_i \in I_n$, the number of sets B does not exceed n^k. Therefore, $|H^n_{2,c}| \leq 3^{k-1}2^n n^k$.

It remains to choose the values of x_i. Put $r = \lfloor\sqrt[3]{n}\rfloor$, $x_1 = 0$, $x_2 = r, \ldots, x_{k-1} = (k-2)r$, and $x_k = n - 1$, where $k = \lceil n/r + 1\rceil \sim n^{2/3}$.

Thus,

$$\log|H^n_{2,c}| \lesssim n^{2/3}\log 3 + n + n^{2/3}\log n = n(1 + o(1)).$$

The lemma is proved.

Considering the logarithm of the cardinality of a discrete class as measure of information needed for representation of its elements (the code length by which the function is uniquely restored), we obtain that for a precise representation of the functions in $\widehat{H}^n_{2,c}$ and $H^n_{2,c}$ we need codes of lengths near to n.

Since the meaning of the discrete functions in $\widehat{H}^n_{2,c}$ consists in approximation of continuous functions in $H_{2,c}$, it follows that such amount of information (n bits) contained in discrete functions is redundant. Indeed, we consider the approximations of $f \in H_{2,c}$ at the points of $\tilde{I}_n$ with accuracy $\leq 1/2n$, but even more detailed information (the values on the whole domain of definition $[0,1)$ with the same accuracy) can be obtained having a code whose length is equal in order to $\sqrt{n}$ (more precisely, $\sqrt{cn}$; see bounds on ε-entropy in [2]). The use of codes of such large length for $\widehat{H}^n_{2,c}$ and $H^n_{2,c}$ is explained by the fact that the last digit in the list of approximate values of the function behaves practically like a random variable with uniform discrete distribution. Hence, to represent it we are compelled to indicate all its values; the number of such values is equal to the size of the domain of definition.

§ 3. 1-Approximating Sets

Instead of precise realization of the functions in $\widehat{H}^n_{2,c}$ and $H^n_{2,c}$ we consider 1-approximating sets for these classes.

We call the set M of functions of the form $f : I_n \to I_n$ a 1-*approximating* set for functions in the class K if the condition

$$(\forall g \in K)(\exists f \in M)(\forall x \in I_n)|f(x) - g(x)| \leq 1$$

holds. Denote by Approx(K) the cardinality of the least 1-approximating set for K.

Note that transfer to the 1-approximating sets and the addition of the error are admissible, since the classes of discrete functions under consideration give only an approximation for the original class $H_{2,c}$.

Estimates for $\text{Approx}(H^n_{2,c})$ and $\text{Approx}(\widehat{H}^n_{2,c})$ are given in Theorems 1 and 2 (throughout this paper log 2 denotes the base 2 logarithm).

Theorem 1. $\log \text{Approx}(H^n_{2,c}) \leq c_1\sqrt{cn}$, *where* $c_1 > 0$ *is an absolute constant.*

Theorem 2. $\log \text{Approx}(\widehat{H}^n_{2,c}) \geq c_2\sqrt{cn}$, *where* $c_2 > 0$ *is an absolute constant.*

PROOF OF THEOREM 1. Let r be a natural number. Put $x_1 = 0$, $x_2 = r, \dots, x_{k-1} = (k-2)r$, and $x_k = n-1$, where $k = k(r) = 1 + \lceil \frac{n-1}{r} \rceil$. As above, let the set $B(x_1, y_1, \dots, x_k, y_k)$ consist of the functions in $H^n_{2,c}$ assuming the value y_i at the point $x = x_i$.

Denote by $M(r)$ the maximum (on $y_1, \dots, y_k$) of Approx($B(x_1, y_1, \dots, x_k, y_k)$). It is obvious that for each r the inequality $\text{Approx}(H^n_{2,c}) \leq n^k M(r)$ holds.

We give an upper bound for the $M(r)$. Firstly, we find the maximal r such that $M(r) = 1$.

Let $f \in B(x_1, y_1, \dots, x_k, y_k)$ and $x \in I_n$. If $x = x_i$ for some i, then $f(x) = y_i$. Estimate $f(x)$ in the case when $x_i < x < x_{i+1}$. Consider the function $h(x)$ from (2). Since $h(x_i) = h(x_{i+1}) = 0$ and $\Delta_2(f) = \Delta_2(h)$, and also $x_i < x < x_{i+1}$ and $x_{i+1} \leq x_i + r$, from the inequality $|\Delta_2(h; x_i, x, x_{i+1})| < c/2n + \varphi(x_i, x, x_{i+1})$ we obtain $|h(x)| < \frac{c}{2n}(x_{i+1} - x)(x - x_i) + 1$. Hence, $|h(x)| < cr^2/8n + 1$.

Let r satisfy the inequality $r \leq 2\sqrt{n/c}$. It is clear that in this case the inequality $cr^2/8n + 1 \leq 1.5$ holds. Let $\tilde{f}(x)$ be the integer nearest to

$$\hat{f}(x) = \frac{x_{i+1} - x}{x_{i+1} - x_i} f(x_i) + \frac{x - x_i}{x_{i+1} - x_i} f(x_{i+1}),$$

for example, $\tilde{f}(x) = \lfloor \hat{f}(x) + 1/2 \rfloor$. Then $|\tilde{f}(x) - f(x)| \leq 1$ since $|\tilde{f}(x) - f(x)| < 1.5 + 0.5$ and the functions $\tilde{f}$ and f are integer-valued. Since $\tilde{f}$ is a 1-approximation to each function $f \in B$, the values $f(x_i)$ are equal and $\tilde{f}$ is related with f by means of them only, by a single function $\tilde{f}$ we can 1-approximate every function in B.

Thus, if $r \leq 2\sqrt{n/c}$, then $M(r) = 1$.

Now we establish a relation between $M(r)$ and $M(2r)$. Show that

$$M(2r) \leq M(r)(\lceil 2 + cr^2/n \rceil)^{k(r)-k(2r)}.$$

This inequality follows from the fact that using the values of the function $f \in H^n_{2,c}$ at the points 0, $2r$, $4r, \dots$ and $n-1$, it is possible to estimate its values at the points r, $3r$, $5r, \dots$: the function $f(x)$ for $x = r(2t+1)$ can take at most $\lceil 2 + cr^2/n \rceil$ values, and the number of the points x is equal to $k(r) - k(2r)$. We now prove this statement.

Let $x = r(2t+1)$, and let x' and x'' be the points in $\{n-1, 0, 2r, 4r, \dots\}$ nearest to x from the left and right; in this event, $x' = 2rt$ and $x'' \leq 2r(t+1)$.

Write $f(x)$ in the form $\hat{f}(x) + h(x)$, where

$$\hat{f}(x) = \frac{x'' - x}{x'' - x'} f(x') + \frac{x - x'}{x'' - x'} f(x''),$$

and $h(x') = h(x'') = 0$, $\Delta_2(f) = \Delta_2(h)$. From the inequality $|\Delta_2(f; x', x, x'')| < c/2n + \varphi(x', x, x'')$ it follows that

$$|h(x)| < \frac{c}{2n}(x'' - x)(x - x') + 1 \leq \frac{c}{2n}\left(\frac{x'' - x'}{2}\right)^2 + 1 \leq \frac{cr^2}{2n} + 1.$$

Hence, the values of $f(x)$ belong to the set

$$(\hat{f}(x) - cr^2/2n - 1,\ \hat{f}(x) + cr^2/2n + 1) \cap \mathbb{Z}$$

that contains at most $\lceil 2 + cr^2/n \rceil$ points.

Thus,

$$M(2r) \le M(r)\lceil 2 + cr^2/n \rceil^{k(r)-k(2r)}.$$

Using $\mathrm{Approx}(H^n_{2,c}) \le n^{k(r)}M(r)$ and $M(r) = 1$ for $r \le 2\sqrt{n/c}$, we obtain the bound for $\mathrm{Approx}(H_{2,c})$ in explicit form.

Let $r_t = \lfloor 2\sqrt{n/c} \rfloor 2^t$. Put $m_t = \log M(r_t)$. Then $m_0 = 0$ and

$$m_t \le m_t + (k(r_t) - k(r_{t+1}))\log\lceil 2 + cr^2_{q-1}/n \rceil.$$

Therefore,

$$m_t \le \sum_{q=1}^{t}(k(r_{q-1}) - k(r_q))\log\left\lceil 2 + \frac{cr^2_{q-1}}{n}\right\rceil.$$

Find an upper bound for this sum. First, note that

$$k(r_{q-1}) - k(r_q) = \left\lceil \frac{n-1}{r_{q-1}} \right\rceil - \left\lceil \frac{n-1}{2r_{q-1}} \right\rceil \le 1 + \frac{n-1}{2r_{q-1}} < 1 + \frac{n}{2^q r_0}.$$

Further, for sufficiently large n, we have $cr^2_{q-1}/n \ge cr^2_0/n = \frac{c}{n}\lfloor 2\sqrt{n/c} \rfloor^2$. Therefore,

$$\lceil 2 + cr^2_{q-1}/n \rceil < 3 + cr^2_{q-1}/n < 2cr^2_{q-1}/n.$$

Hence,

$$m_t \le \sum_{q=1}^{t}\left(1 + \frac{n}{2^q r_0}\right)\left(2q + 1 + \log\frac{cr_0^2}{4n}\right).$$

Since $r_0 = \lfloor 2\sqrt{n/c} \rfloor$, we have $\log(cr_0^2/4n) \le 0$ and

$$m_t \le \sum_{q=1}^{t}\left(1 + \frac{n}{2^q r_0}\right)(2q+1) \le t^2 + 2t + \frac{5n}{r_0}.$$

Finally,

$$\log \mathrm{Approx}(H^n_{2,c}) \le m_t + k(r_t)\log n$$

$$= m_t + \left(1 + \left\lceil \frac{n-1}{2^t r_0} \right\rceil\right)\log n \le t^2 + 2t + \frac{5n}{r_0} + 2\log n + \frac{n\log n}{2^t r_0}.$$

Having chosen t such that the relations $t = o(n^{1/4})$ and $\log n = o(2^t)$ hold, we obtain

$$\log \mathrm{Approx}(H^n_{2,c}) \le \frac{5}{2}\sqrt{nc}\,(1 + o(1)).$$

Theorem 1 is proved.

PROOF OF THEOREM 2. Let $d(x) = \frac{1}{2}(\max(0, x(1-x)))^2$. The function $d(x)$ is differentiable, and its derivative equals 1 at $x \notin [0,1]$, and $x - 3x^2 + 2x^3$ at $x \in [0,1]$. Therefore, this function satisfies the Lipschitz condition with constant 1.

Let r be an integral parameter. Consider the set A of functions $f : [0,1) \to [0,1)$ of the form

$$f(x) = \sum_{k=1}^{K} \frac{cr^2 a_k}{n^2} d\left(\frac{nx - 1/2}{r} - k + 1\right),$$

where $K = \lfloor \frac{n-1}{r} \rfloor$ and $a_k \in \{0,1\}$.

The set A is a subset of $H_{2,c}$. Hence, the discrete functions of the form $g(x) = \lfloor n f(x/n + 1/2n) \rfloor$ belong to sets $\widehat{H}^n_{2,c}$ and $H^n_{2,c}$.

Estimate the distance (in the uniform metric) between distinct functions g. To this end, estimate the value $\rho = \max_{x \in I_n} |D_k(x)|$, where

$$D_k(x) = \left\lfloor \frac{cr^2}{n} d\left(\frac{x}{r} + 1 - k\right) \right\rfloor.$$

Since $\max_{x \in I_n} |D_k(x)| \geq |D_k(r(k-1) + \lfloor r/2 \rfloor)|$, we have $\rho \geq \frac{c}{4n}\left(\frac{r}{2} - 1\right)^2 - 1$. Therefore, $\rho \geq 3$ for any $r \geq 2 + 8\sqrt{n/c}$.

Let $g'(x) = \lfloor n f'(x/n + 1/2n) \rfloor$ and $g''(x) = \lfloor n f''(x/n + 1/2n) \rfloor$ be distinct functions, where f' and f'' belong to the set A and are given by the tuples of coefficients a'_i and a''_i, respectively. Then

$$\max_{x \in I_n} |g'(x) - g''(x)| \geq \max_i |a'_i - a''_i| \rho \geq \rho = 3,$$

i.e., the functions g are 3-distinguishable; therefore, $\log \operatorname{Approx} \widehat{H}^n_{2,c} \geq K$.

For $r = \lceil 2 + 8\sqrt{n/c} \rceil$, we have $K \geq \frac{1}{8}\sqrt{nc}\,(1 + o(1))$. Theorem 2 is proved.

Thus, we have

$$\frac{1}{8}\sqrt{nc}\,(1 + o(1)) \leq \log \operatorname{Approx} \widehat{H}^n_{2,c} \leq \log \operatorname{Approx} H^n_{2,c} \leq \frac{5}{2}\sqrt{nc}\,(1 + o(1)).$$

These estimates conform with the relation $\mathcal{H}_{1/n}(H_{2,c}) \asymp \sqrt{nc}$ (see [2]). Here $\mathcal{H}_\varepsilon(K)$ is the ε-entropy of the set K [2]; the expression $f(n) \asymp g(n)$ means that for all sufficiently large n the inequalities $c_1 g(n) \leq f(n) \leq c_2 g(n)$ hold, where c_1 and c_2 are suitable positive constants.

§ 4. Complexity of Precise and Approximate Calculation of Discrete Functions

For $n = 2^r$, the discrete functions $f : I_n \to I_n$ can be considered as Boolean operators $f : \{0,1\}^r \to \{0,1\}^r$ by associating with each number in I_n its binary representation. Such operators can be realized by *schemes of functional elements* which perform some elementary operations (for example, conjunction, disjunction, and negation). Each scheme S has *complexity* $L(S)$ equal to the number of its elements (for necessary definitions see [3]). The complexity $L(f)$ of an operator f is defined as the complexity of schemes with the least number of elements realizing f; for a sequence of classes of operators $\{K_1, K_2, \dots\}$, where K_i consists of the operators $f : \{0,1\}^i \to \{0,1\}^{m_i}$, the function $L(r) = \max_{f \in K_r} L(f)$ is considered.

We introduce the complexity of approximate realization of functions. Consider the operators $f : \{0,1\}^r \to \{0,1\}^r$ as functions $f : I_n \to I_n$. Put $L^{\text{Approx}}(f) = \min L(g)$, where the minimum is taken over all $g : I_n \to I_n$ such that $|g(x) - f(x)| \le 1$, and let $L^{\text{Approx}}(r) = \max_{f \in K_r} L^{\text{Approx}}(f)$.

We need the following known results about complexity of the realization of functions and classes.

1. *Lower bound* [3, the Corollary of Theorem D. 1]: if K_r consists of the operators $f : \{0,1\}^r \to \{0,1\}^r$ with $\frac{r \log\log |K_r|}{\log |K_r|} \to 0$, then

$$L(r) \ge \frac{\log |K_r|}{\log\log |K_r|}(1 + o(1)).$$

2. *Upper bound* [3, Theorem D. 12]: if K_r consists of all operators $f : \{0,1\}^r \to \{0,1\}^{m_r}$, then

$$L(r) \le \frac{m_r 2^r}{r}(1 + o(1)).$$

The following theorem is true.

Theorem 3. *For $K_r = H^r_{2,c}$ and $K_r = \widehat{H}^r_{2,c}$, the following relations hold*

$$L(r) = \frac{2,k}{r}(1 + o(1)), \quad c_3 \frac{2^{r/2}}{r} \le L^{\text{Approx}}(r) \le c_4 \frac{2^{r/2}}{r},$$

where c_3 and c_4 are positive constants.

PROOF. The *lower bounds* may be obtained from the lower bounds for $|K_r|$ and $\text{Approx}(K_r)$, respectively.

The *upper bound* for $L(r)$ is established as follows. As it was shown at prove of the lemma, the function f in $H^n_{2,c}$ can be recovered on the formula

$$f(x) = \left\lfloor \frac{x_{i+1} - x}{x_{i+1} - x_i} y_i + \frac{x - x_i}{x_{i+1} - x_i} y_{i+1} + \theta_i \right\rfloor + \chi_x \tag{3}$$

for the given tuple $(y_1, \dots, y_k, \theta_1, \dots, \theta_{k-1}, \chi_0, \dots, \chi_{n-1})$ and $n = 2^r, k = \lceil n/(\lfloor \sqrt[3]{n} \rfloor) + 1 \rceil = n^{2/3}(1 + o(1))$, $y_i \in I_n$, $\theta_i \in \{-1/2, 0, 1/2\}$, $\chi_i \in \{0, 1\}$, $i = \lfloor x / \lfloor \sqrt[3]{n} \rfloor \rfloor$, $x_i = i \lfloor \sqrt[3]{n} \rfloor$, $x_{i+1} = \min(x_i + \lfloor \sqrt[3]{n} \rfloor, n - 1)$.

It is possible to show that, when calculating $f(x)$ by the formula (3), the r-ary Boolean function χ_x has a maximal complexity. Hence, $L(f) \le \frac{2^r}{r}(1 + o(1))$.

The *upper bound* for $L^{\text{Approx}}(r)$ is established by means of the following construction.

Note that if $f \in H^n_{2,c}$, then $g(x) = f(n - 1 - x)$ is also a function in $H^n_{2,c}$. Therefore, considering f and g as Boolean operators[†)] $\{0,1\}^r \to \{0,1\}$, it is possible to compute f in the form

$$f(x_1, \dots, x_r) = \bar{x}_1 f(0, x_2, \dots, x_r) \vee x_1 g(0, \bar{x}_2, \dots, \bar{x}_r).$$

Hence, it suffices to be able to compute functions in $H^n_{2,c}$ on the first half of the domain of definition.

Introduce the following notation (here $x \in I_n$, $x = |(x_1, \dots, x_r)|$):

$$x|_m = |(x_1, \dots, x_m, \underbrace{0, \dots, 0}_{r-m})|;$$

$$x|^m = x|_m + 2^{r-m} = |(x_1, \dots, x_m, \underbrace{0, \dots, 0}_{r-m})| + |(\underbrace{0, \dots, 0}_{m-1}, 1, \underbrace{0, \dots, 0}_{r-m})|.$$

Note that $x|_m \le x < x|^m$.

Let $\delta_m(x) = f((x|_m + x|^m)/2) - \lfloor \frac{1}{2}(f(x|_m) + f(x|^m)) \rfloor$. Then

$$f(x|_{m+1}) = \begin{cases} f(x|_m) & \text{if} \quad x_{m+1} = 0, \\ \left\lfloor \frac{1}{2} \lfloor (f(x|_m) + f(x|^m)) \right\rfloor + \delta_m(x) & \text{if} \quad x_{m+1} = 1; \end{cases} \tag{4}$$

$$f(x|^{m+1}) = \begin{cases} \left\lfloor \frac{1}{2}(f(x|_m) + f(x|^m)) \right\rfloor + \delta_m(x) & \text{if} \quad x_{m+1} = 0, \\ f(x|^m) & \text{if} \quad x_{m+1} = 1. \end{cases} \tag{5}$$

---

†) We mean the following correspondence between the Boolean collection $\tilde{x} = (x_1, \dots, x_n)$ and the number $x \in I_{2^n}$: $x = 2^{n-1}x_1 + \dots + 2^0 x_n$.

The following scheme is proposed for computing $f(x)$.

STEP 1. Calculating $\delta_m, \delta_{m+1}, \ldots$ by (4) and (5), we successively find $f(x|_{m+1})$ and $f(x|^{m+1})$, $f(x|_{m+2})$ and $f(x|^{m+2}), \ldots, f(x|_{m_1})$ and $f(x|^{m_1})$.

STEP 2. For some $m = m_0$, we calculate $f(x|_m)$ and $f(x|^m)$.

STEP 3. We find an approximate value of $f(x)$ by linear interpolation between $f(x|_{m_1})$ and $f(x|^{m_1})$; i.e., instead of $f(x)$ we calculate

$$\tilde{f}(x) = \left\lfloor \frac{1}{2} + \frac{x - x|_{m_1}}{x|^{m_1} - x|_{m_1}} f(x|^{m_1}) + \frac{x|^{m_1} - x}{x|^{m_1} - x|_{m_1}} f(x|_{m_1}) \right\rfloor .$$

Estimate the complexity of such computation and choose some m_0 and m_1 in order to minimize the complexity and to make the error obtained at step 3 to be not greater than 1.

It follows from the definition of $\delta_m(x)$ that

$$|\delta_m(x)| \le \frac{1}{2} + \frac{1}{2}\left| f(x|^m) - 2f\left(\frac{x|^m + x|_m}{2}\right) + f(x|_m)\right|$$

$$= \frac{1}{2} + \frac{1}{4} 2^{2(r-m)} \left| \Delta_2 \left(f; x|^m, \frac{x|^m + x|_m}{2}, x|_m \right)\right|$$

$$\le \frac{1}{2} + \frac{c2^{2(r-m)}}{82^r} + 1 \le \max\left(3, \frac{c2^r}{42^{2m}}\right).$$

Considering the value δ_m as the Boolean tuple of length r which represents $\delta_m(x)$ in the ring $\mathbb{Z}_n$, it is possible to note that the significant digits of this tuple are the same, i.e., the number of digits which it is necessary to complete independently does not exceed $\min(3, r - 2m + \log c + c_5)$, where c_5 is an absolute constant. Hence, the complexity of the calculation of $\delta_m(x)$ does not exceed

$$\frac{2^m}{m} \min(3, r - 2m + \log c + c_5)(1 + o(1)).$$

When performing calculations in $\mathbb{Z}_n$, it is possible to show that the complexity of the rest computation by (4) and (5) is equal to $O(r)$, and the complexity of computations at step 2 does not exceed

$$O(r^2) + \left(\sum_{m=m_0}^{m_1} \frac{2^m}{m} \min(3, r - 2m + \log c + c_5)\right)(1 + o(1)).$$

The complexity of computation at step 3 does not exceed $O(r^2)$.

The parameter m_1 is chosen such that the inequality $|f - \tilde{f}| \leq 1$ holds. Since f and $\tilde{f}$ are integer-valued, it is enough to verify that $|f - \tilde{f}| < 2$.

We have

$$|f - \tilde{f}| \leq \frac{1}{2} + \left| \frac{x - x|_{m_1}}{x|^{m_1} - x|_{m_1}} f(x|^{m_1}) + \frac{x|^{m_1} - x}{x|^{m_1} - x|_{m_1}} f(x|_{m_1}) - f(x) \right|$$

$$= \frac{1}{2} + (x - x|_{m_1})(x|^{m_1} - x)|\Delta_2(f; x|^{m_1}, x, x|_{m_1})|$$

$$< \frac{1}{2} + (x - x|_{m_1})(x|^{m_1} - x)\left(c/2^r + \frac{1}{(x - x|_{m_1})(x|^{m_1} - x)}\right)$$

$$\leq \frac{3}{2} + \frac{1}{4}\frac{c2^{2(r-m)}}{2^r} = \frac{3}{2} + \frac{c2^r}{4 \cdot 2^{2m_1}}.$$

Hence, for the inequality $|f - \tilde{f}| < 2$ to be true it suffices to have $\frac{c \cdot 2^r}{4 \cdot 2^{2m_1}} < \frac{1}{2}$, i.e., $m_1 > \frac{r-1+\log c}{2}$. Therefore, setting $m_0 = \lfloor \frac{r}{3} \rfloor$ and $m_1 = \lfloor \frac{r+\log c+1}{2} \rfloor$, we obtain that the complexity of the computation does not exceed

$$\left(2r\frac{2^{m_0}}{m_0} + \sum_{m=m_0}^{m_1} \frac{2^m}{m} \min(3, r - 2m + \log c + c_5) + O(r^2)\right)(1 + o(1))$$

$$\leq \text{const}\, 2^{r/2}/r.$$

Theorem 3 is proved.

Acknowledgments. The author is grateful to his teacher O. B. Lupanov for formulating the problems and discussing the results.

References

1. G. G. Amanzhaev (1992) Discrete functions with a given continuity modulus (in Russian), *Vestnik Moskov. Univ. Ser.* I *Mat. Mekh.* No. 5, 86–89.
2. A. N. Kolmogorov and V. M. Tikhomirov (1959) ε-entropy and ε-capacity of sets in function spaces (in Russian), *Uspekhi Mat. Nauk* **14**, No. 2, 3–86.
3. O. B. Lupanov (1965) On a certain approach to the synthesis of control systems — the principle of local coding (in Russian), in: *Problemy Kibernetiki.* Vol. 14, Nauka, Moscow, pp. 31–110.

Faculty of Mechanics and Mathematics
Moscow State University
Vorob'ëvy Gory
Moscow 119899
RUSSIA

TRANSLATED BY E. BODIN

ON A PROPERTY OF PERFECT BINARY CODES[†)]

S. V. Avgustinovich

It is proved that any perfect binary $(n,3)$-code is uniquely reconstructible from the allocation of its code vertices in the two medium layers of the n-dimensional unit cube. A nontrivial upper bound is obtained for the number of such codes.

The problem of describing the perfect binary codes of length n with distance 3 has no satisfactory solution by now since $2^{2^{(n-1)/2+\log(n+1)}}$, well-known (see [1]) to restrict the number of such codes from below, differs considerably from $2^{2^{n-\log n+\log\log(n+1)}}$, being a trivial upper bound

$$\binom{2^n}{2^n/(n-+1)} < 2^{2^{n-\log n+\log\log(n+1)}};$$

throughout this paper log denotes the base 2 logarithm. The last bound can be strengthened due to the reconstructibility of perfect codes from its fragments which is proved in the present paper.

A subset C of vertices of the n-dimensional unit cube E^n is called a *perfect binary* $(n,3)$*-code* if

(1) for all distinct vertices $\alpha,\beta \in C$ we have $\rho(\alpha,\beta) \geq 3$, where ρ is the Hamming distance,

(2) for every vertex $\gamma \neq C$ there is a unique vertex $\theta \in C$ such that $\rho(\gamma,\theta)=1$.

Conditions 1 and 2 say that the unit balls with centers at code vertices are disjoint and cover the cube E^n.

Denote by $P(n)$ the number of perfect binary $(n,3)$-codes. It is known [1] that perfect binary $(n,3)$-codes exist only for $n=2^k-1$, $k\in\mathbb{N}$, and the number of code vertices is $2^{n-k}=2^{n-\log(n+1)}$. Therefore,

$$P(n) < \binom{2^n}{2^{n-\log(n+1)}} < 2^{2^{n-\log n+\log\log(n+1)}}.$$

†) This research was supported by the Russian Foundation for Basic Research (Grant 93–01–01484).

A. D. Korshunov (ed.), Operations Research and Discrete Analysis, 13–15.

A subset $M \subseteq E^n$ is called *basic* if for all distinct codes C_1 and C_2 there is a vertex $\alpha \in M$ such that $\alpha \in C_1$ and $\alpha \notin C_2$.

It is not hard to see that if, for a code C and a basic set M, the set $C \cap M$ is known, then one can reconstruct the whole code C.

Let $n = 2^k - 1$. Define the sets M_1, M_2, and M_3 of vertices of E^n as follows:

vertices in M_1 have weights at most $(n-3)/2$,
vertices in M_2 have weights $(n-1)/2$ and $(n+1)/2$,
vertices in M_3 have weights at least $(n+3)/2$.

Clearly, $E^n = M_1 \cup M_2 \cup M_3$.

Theorem. *The set M_2 is basic.*

PROOF. Suppose that M_2 is not basic. Then there are two distinct codes C_1 and C_2 which are identical on M_2. Introduce the following notation:

$$C_1 \cap M_1 = C_1^1, \quad C_1 \cap M_2 = C_1^2, \quad C_1 \cap M_3 = C_1^3,$$
$$C_2 \cap M_1 = C_2^1, \quad C_2 \cap M_2 = C_2^2, \quad C_2 \cap M_3 = C_2^3.$$

By assumption, $C_1^2 = C_2^2$ but $C_1^1 \neq C_2^1$. It is easy to see that set $C^* = C_2^1 \cup C_1^2 \cup C_1^3$ is a perfect $(n,3)$-code, since the distance between C_2^1 and C_1^3 is at least 3. Consider the code C_1. It follows from [2] that if $\alpha \in C_1$, then $\bar{\alpha} \in C_1$, where $\bar{\alpha}$ is a vertex obtained from α by replacing units with zeros and visa versa. Hence, the set C_1^1 is uniquely determined by set C_1^3. Applying a similar argument to code C^*, we obtain that C_2^1 is determined by C_1^3. Hence, $C_1^1 = C_2^1$, which is a contradiction.

By using Stirling's formula, it is easy to show that

$$|M_2| < 2^{n - \frac{1}{2}\log n}.$$

Moreover, according to [1], it holds

$$|M_2 \cap C| < |M_2|/n.$$

This implies the following

Corollary. *For every $n = 2^k - 1$, $k \in \mathbb{N}$,*

$$P(n) < \binom{2^{n-\frac{1}{2}\log n}}{2^{n-\frac{3}{2}\log n}} < 2^{2^{n-\frac{3}{2}\log n + \log(en)}}.$$

References

1. Yu. L. Vasil'ev (1962) On ungrouped close-packed codes (in Russian), in: *Problemy Kibernetiki*. Vol. 8, Gosudarstv. Izdat. Fiz.-Mat. Lit., Moscow, pp. 337–339.
2. G. S. Shapiro and D. L. Slotnick (1959) On the mathematical theory of error-correcting codes, *JBM J. Res. Develop.* **3**, No. 1, 25–34.

Sobolev Institute of Mathematics
Universitetskiĭ pr., 4
Novosibirsk 630090
RUSSIA

TRANSLATED BY O. V. BORODIN

EDGE NEIGHBORHOODS IN NORMAL MAPS[†)]

S. V. Avgustinovich and O. V. Borodin

Studying Eulerian polytopes, with each edge e of such polytope Lebesgue [1] associated the nondecreasing quadruple of integers (called hereafter the type of an edge) two of which are the degrees of the vertices incident with e while the others are the numbers of edges in the faces incident with e. He proved that every Eulerian polytope has an edge whose type is majorized by one of the following quadruples: (3 3 3 ∞), (3 3 4 11), (3 3 5 7) and (3 4 4 5). Some of the parameters here may be improved.

In the present paper, a similar problem is solved for the normal maps on the torus (Theorem 1), and it is established (Theorem 2) that sufficiently large normal maps on arbitrary orientable surfaces have the same structure as those torroidal.

We use the definitions of [2]. The vertices and faces of a finite map on an orientable surface will be called *elements*, while the degree of a vertex and the number of edges incident with a face — the *range* of the corresponding element. A map is called *normal* if each of its elements has range at least three. By the *type* of an edge we mean the quadruple of ranges of the four incident elements taken in nondecreasing order. We say that the type $(t_1 t_2 t_3 t_4)$ *majorizes* the type $(t'_1 t'_2 t'_3 t'_4)$, if $t_i \geq t'_i$, $i = 1, 2, 3, 4$.

Theorem 1. *Each normal map on the torus has an edge whose type is majorized by one of the following quadruples:*

$$(3\ 3\ 3\ \infty),\quad (3\ 3\ 4\ 10),\quad (3\ 3\ 5\ 7),\quad (3\ 3\ 6\ 6),\quad (3\ 4\ 4\ 6),\quad \textit{and}\quad (4\ 4\ 4\ 4),$$

where each parameter is best possible.

PROOF. Let us explain in which sense all parameters are best possible. The maps (a)–(f) in Fig. 1 satisfy the hypothesis of Theorem 1 (each picture is mapped on the torus in a natural way by identifying its left and right, as well as top and bottom, boundaries). However, if some parameter in some quadruple has been decreased, then one of the maps becomes a counterexample to the corresponding

†) The work of the first author was carried out with the support of the International Science Foundation (Grant NQ4300) and the Russian Foundation for Basic Research (Grant 93–011–1484); the work of the second author was carried out with the support of the International Science Foundation (Grant NQ4300) and the Russian Foundation for Basic Research (Grant 93–011–1486.)

A. D. Korshunov (ed.), Operations Research and Discrete Analysis, 17–22.

statement. Indeed, any labelled edge of the maps in Fig. 1 has type majorizing at least one of the quadruples in Theorem 1. Furthermore, each picture there is a distinguished edge whose type coincides with one of these quadruples. Under each picture one can see the quadruple whose unimprovability is being confirmed.

Let us proceed to the proof of the main statement of Theorem 1. Suppose that a normal map M with the sets of vertices, edges, and faces denoted by V, F and E, respectively, is a counterexample. Euler's formula $|V| - |E| + |F| = 0$ for M may be rewritten as

$$\sum_{x \in V \cup F} (r(x) - 4) = 0, \tag{1}$$

where $r(x)$ is the range of x, i.e. the number of edges incident with x.

To each element of range n, we assign the number $n - 4$ called the *charge* of the element. (By assumption, $n > 2$.) It follows from (1) that the sum of the charges of all elements of M is zero. We distribute the charge of each element of range n equally (i.e. by $(n-4)/n$) among the incident edges. Such a distribution of charges will be called *uniform*. Then each edge acquires a charge that depends only of its type. Observe that if the type of one edge majorizes that of the other, then the first edge acquires a greater charge. The sum of the charges over all edges remains zero. This implies that either all the edges of M have zero charge, or there is an edge with a negative charge. Consider an edge of type (n_1, n_2, n_3, n_4). The charge of such an edge is nonpositive only if

$$(n_1 - 4)/n_1 + (n_2 - 4)/n_2 + (n_3 - 4)/n_3 + (n_4 - 4)/n_4 \leq 0,$$

i.e.

$$1/n_1 + 1/n_2 + 1/n_3 + 1/n_4 \geq 1.$$

Clearly, n_1 may be equal to 3 or 4. If $n_1 = 4$, then $n_1 = n_2 = n_3 = n_4 = 4$. If $n_1 = 3$, then

$$1/n_2 + 1/n_3 + 1/n_4 \geq 2/3,$$

while n_2 may be equal to 3 or 4. If $n_2 = 3$, then

$$1/n_3 + 1/n_4 \geq 1/3,$$

and in this case only the following quadruples are possible:

$$(3\ 3\ 6\ 6),\ (3\ 3\ 5\ 7),\ (3\ 3\ 5\ 6),\ (3\ 3\ 5\ 5),\ (3\ 3\ 4\ 12), (3\ 3\ 4\ 11), \ldots, (3\ 3\ 4\ 4).$$

If $n_2 = 4$, we have

$$1/n_3 + 1/n_4 \geq 5/12,$$

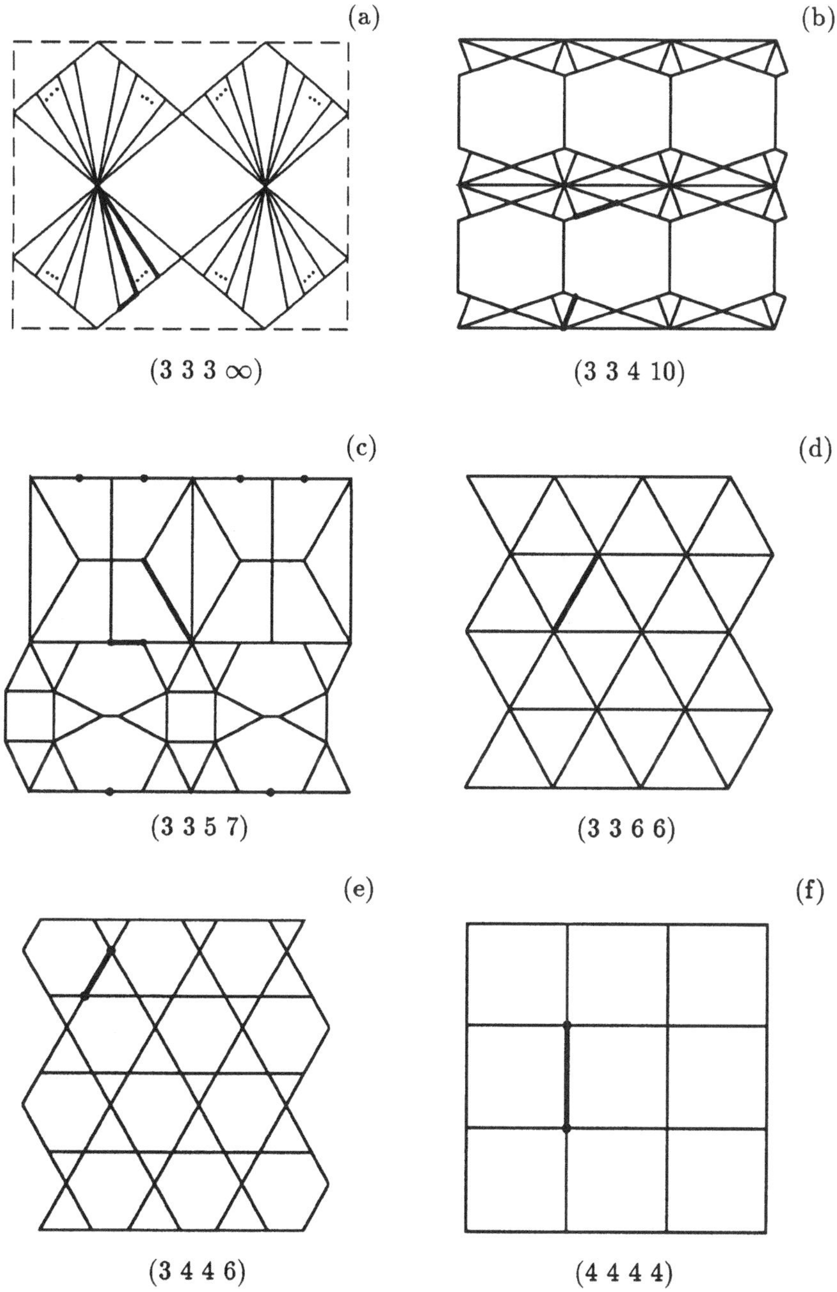

Fig. 1

and in this case only the following quadruples are possible:

$$(3\ 4\ 4\ 6),\ (3\ 4\ 4\ 5),\quad \text{and}\quad (3\ 4\ 4\ 4).$$

Among the quadruples just found, only (3 3 4 11) and (3 3 4 12) are not majorized by those in Theorem 1. Thus, our counterexample M has an element of range 11 or 12. In view of the vertex-face duality, it suffices to consider the case in which this element is a vertex W. We split the edges that are incident with W into 4 subsets:

Subset 1. The edges joining W with vertices of degree less than 5.

Subset 2. The edges joining W with vertices of degree 4 and incident with two faces of range 3, where the vertex of degree 4 is incident with two faces of range at least 6.

Subset 3. The edges joining W with vertices of degree 4 and incident with at least one face of range greater than 3, and also the edges not incident with faces of range 3.

Subset 4. The remaining edges.

The edges in i-subset will be called *i-edges.*

Lemma 1. *Let the map M be a counterexample to Theorem 1 having a vertex W of degree* 11 *or* 12. *Then among any consecutive edges incident with W there is an edge belonging to Subsets* 1, 2 *or* 3.

PROOF. Suppose the contrary, and let r_1, r_2, r_3, r_4, and r_5 be consecutive edges around W, let v_1, v_2, v_3, v_4, and v_5 be the end vertices of these edges, and let F_1, F_2, F_3, and F_4 be the corresponding intermediate faces. We have the following three cases:

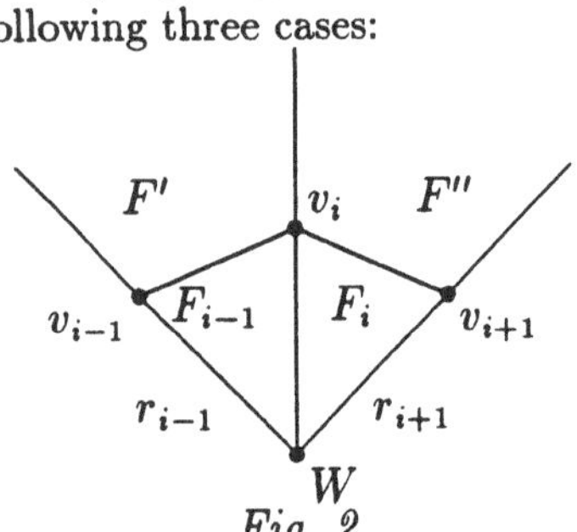

Fig. 2

Case 1. Among the intermediate faces there are two consecutive triangles F_{i-1} and F_i (see Fig. 2), separated by some edge r_i. By assumption, r_{i-1}, r_i, and r_{i+1} are not 1-edges, which implies that v_i, v_{i-1}, and v_{i+1} have degree less than 5. The vertex v_i cannot have degree 3, because otherwise the type of r_i would be majorized by the quadruple $(3\ 3\ 3\ \infty)$. Thus, $d(v_i) = 4$. Faces F' and F'' cannot have range less than 6, because the types of edges $(v_{i-1}v_i)$ and (v_iv_{i+1}) would be majorized by the quadruple (3 4 4 6). Consequently, r_i is 2-edge.

Case 2. Among the intermediate faces there is a triangle F_i surrounded by nontriangles. In this case either v_i or v_{i+1} has degree greater than 3, because otherwise the type of edge (v_iv_{i+1}) would be majorized by the quadruple $(3\ 3\ 3\ \infty)$. Accordingly, either r_i or r_{i+1} is a 3-edge.

Case 3. Among the intermediate edges there is no triangles. By definition, r_2, r_3, and r_4 belong to Subset 3.

Since other cases are impossible, Lemma 1 is proved.

Corollary 1. *In the vicinity of every vertex of degree* 11 *or* 12 *of counterexample* M *there are three edges of Subsets* 1–3.

To continue the proof, we modify the rule for discharging given above in respect to the elements of ranges 11 and 12 and to some of the elements of range 4:

1. Each vertex of degree 4 (face of range 4) incident with a 2-edge sends to it a charge $+1/4$, and sends to the edge separating faces (joining vertices of degree) of range at least 6 a charge $-1/4$. The resulting charge of such a vertex remains zero.

2. Every vertex of degree (face of range) 11 or 12 sends to each surrounding 4-edge a charge equal to $+2/3$, and sends to each other edge a charge equal to $+1/2$.

3. All the other elements distribute their charges equally among the neighbor edges.

It remains to verify that the modified rule of discharging makes the charges of all edges of M nonnegative. Indeed,

(1) every 1-edge obtains a charge at least $-1/3 - 1/3 + 1/5 + 1/2 = 1/30$;

(2) every 2-edge that of at least $-1/3 - 1/3 + 1/2 + 1/4 = 1/12$;

(3) every 3-edge at least $-1/3 + 0 + 0 + 1/2 = 1/6$;

(4) every 4-edge at least $-1/3 - 1/3 + 0 + 2/3 = 0$;

(5) every edge separating two faces of range at least 6, as well as every edge joining two vertices of degree at least 6, obtains a charge at least $-1/3 - 1/4 + 2/6 + 2/6 = 1/12$.

Moreover, due to Corollary 1, this rule leaves at least $7-(8{\cdot}2/3+3{\cdot}1/2) = +1/6$ at each element of range 11 or 12. This contradicts Euler's formula and thus completes the proof of Theorem 1.

Theorem 2. *If a normal map* M *on an orientable surface of genus* P *has more than* $576(P-1)$ *edges, then there is an edge whose type is majorized by one of the following quadruples:* (3 3 3 ∞), (3 3 4 11), (3 3 5 7), (3 3 6 6), (3 4 4 6), *and* (4 4 4 4).

PROOF. As above, we assign to each element of range n of map M a charge $n-4$. Euler's formula $|V| - |E| + |F| = -2P$ for M may be rewritten as

$$\sum_{x \in V \cup F} (r(x) - 4) = 8(p - 1). \qquad (2)$$

Clearly, the sum of the charges of all edges then is $8(P-1)$. Let M be a counterexample to Theorem 2. We prove that the total charge may be distributed among the edges of M so that each edge has the resulting charge of at least $+1/72$.

Lemma 2. *If under the uniform distribution of charges an edge e obtains a charge* $z > 0$, *then* $z > 1/39$.

PROOF. An edge type $(t_1t_2t_3t_4)$ is called *critical*, if under the uniform distribution of charges an edge of such type obtains a strictly positive charge, but the charge becomes nonpositive as soon as any coordinate in $(t_1t_2t_3t_4)$ is decreased by 1. Choosing in lexicographic order all quadruples of integers, it is easy to see that there are precisely the following critical types: (3 3 4 13), (3 3 5 8), (3 3 6 7), (3 4 4 7), (3 4 5 5), and (4 4 4 5). The charges on edges of these types are $+1/39, +1/30, +2/21, +2/21, +1/15$, and $+1/5$ respectively, which completes the proof of Lemma 2.

Now we apply to M the modified rule of distribution of charges. It deals with the edges incident with the elements of range 11 and 12, whose residue charge, as shown in the proof of Theorem 1, is at least $+1/6$. This residue being distributed equally, the charge of each edge becomes at least $+1/72$. It remains to consider an edge incident with an element of range 4 which obtains a charge of $-1/4$. Since such an edge is surrounded by two elements of range at least 6, its charge becomes at least $-1/3 - 1/4 + 2/6 + 2/6 = +1/12$.

The above argument implies that after the distribution of charges, each edges of M acquires a charge at least $+1/72$. If the number of such edges is greater than $576(P-1)$, then the total charge exceeds $8(P-1)$ in (3). This completes the proof of Theorem 2.

In conclusion, we observe that due to the vertical symmetry, all six maps depicted in Fig. 1 can be drawn on the Klein bottle. Therefore, Theorem 1 holds not only for the torus but also for the Klein bottle.

References

1. H. Lebesgue (1940) Quelques consequences simple de la formule d'Euler, *J. Math. Pures Appl.* **9**, 27–43.
2. G. Ringel (1974) *Map Color Theorem*, Springer-Verlag, Berlin, New York, and Heidelberg.

Sobolev Institute of Mathematics
Universitetskiĭ pr., 4
Novosibirsk 630090
RUSSIA

TRANSLATED BY THE SECOND AUTHOR

AN UPPER BOUND FOR THE LENGTH OF A SNAKE IN THE n-DIMENSIONAL UNIT CUBE

P. G. Emel'yanov

By refining some details of the proof in the paper [1], a more precise bound is obtained for the length of an arbitrary snake in the n-dimensional unit cube.

The n-dimensional unit cube Q_n is the *bipartite undirected graph* whose vertex set consists of all possible binary n-tuples and in which two vertices, v_1 and v_2, are adjacent if and only if the vectors v_1 and v_2 differ in precisely one coordinate. We study the length of a maximum cycle S_n (also called a *snake*) in the cube Q_n which has the following property: if two vertices v_1 and v_2 in the cycle S_n are not consecutive, then the Hamming distance $\rho_{Q_n}(v_1, v_2) \geq 2$. For more details on the definitions in use and the history of improving the upper and lower bounds for $|S_n|$, see [1–4] and [5, 6], respectively.

In [1], using the technique of averaging the number of vertices in a snake which are the neighbor vertices in the set $Y = V(Q_n) \setminus V(S_n)$, the bound

$$|S_n| \leq 2^{n-1} - \frac{2^{n-1}}{20n - 41} \quad \text{for} \quad n \geq 12, \tag{1}$$

is obtained which, for $n \geq 39$, improves the following bound in [2]:

$$|S_n| \leq 2^{n-1} - \frac{2^n}{n^2 - n + 2} \quad \text{for} \quad n \geq 7. \tag{2}$$

In this paper, we show that for each $n \geq 19$, the bound

$$|S_n| \leq 2^{n-1} - \frac{2^{n-1}}{6n - 13} \tag{3}$$

holds which improves the bound in [2] for $n \geq 19$. Moreover, we obtain an improvement on upper bounds for $14 \leq n \leq 18$.

The bound (1) follows from the inequality below proved in [1]

$$\frac{\Sigma_{v\in Y}|N(v) \cap V(S_n)|}{|Y|} \leq n - 2.1 \quad \text{for} \quad n \geq 12, \tag{4}$$

A. D. Korshunov (ed.), Operations Research and Discrete Analysis, 23–30.

where $N(v)$ is the set of vertices in Q_n adjacent to the vertex v. We show that, for each $n \geq 19$,

$$\frac{\Sigma_{v \in Y} |N(v) \cap V(S_n)|}{|Y|} \leq n - \frac{7}{3}. \tag{5}$$

We need some definitions and propositions in [1]. Let $Y_m = \{v \in Y \mid |N(v) \cap V(S_n)| = m\}$, i.e., Y_m is the set of vertices each of which does not belong to the snake and has precisely m neighbors in the snake. Put

$$Y_{\leq w} = \bigcup_{i=0}^{w} Y_i.$$

The proof in [1] is based on the following nine lemmas. (The numeration of the lemmas and terminology are taken from [1].)

Lemma 2. $Y_n = \varnothing$.

Lemma 3. *If* $x, y \in Y_{n-1}$, *then* $\rho_Y(x, y) \geq 3$.

Lemma 4. *If* $x \in Y_{n-1}$ *and* $y \in Y_{n-2}$, *then no path of the form* $x - z - y$ *is possible in* Y.

Lemma 5. *If* Y *contains a path of the form* $x - y - z$, *where* $x \in Y_{n-1}$ *and* $y \in Y_{n-2}$, *then* $z \in Y_{\leq n-5}$.

Lemma 6. *If* Y *contains a path of the form* $x_1 - x_2 - x_3$, *then there exists an* $i, 1 \leq i \leq 3$, *such that* $x_i \in Y_{\leq n-3}$.

Lemma 7. *If* Y *contains a configuration of the form* $x_1 - y < {x_2 \atop x_3}$ *and* $x_i \in Y_{n-2}$ *for each* i, $1 \leq i \leq 3$, *then* $y \in Y_{\leq 3}$.

Lemma 8. *If* Y *contains a configuration of the form* $x_1 - y < {x_2 \atop x_3}$ *and* $x_1 \in Y_{n-1}$ *and* $y \in Y_{n-3}$, *then there exists an* i, $2 \leq i \leq 3$, *such that* $x_i \in Y_{\leq (n+1)/2}$.

Lemma 9. *No path of the form* $x - y - z - y' - x'$ *is possible in* Y *if* $x, x' \in Y_{n-1}$ *and* $y, y' \in Y_{n-2}$.

Lemma 10. *No path of the form* $x - y - z - y' - x'$ *is possible in* Y *if* $x, x' \in Y_{n-1}$, $y \in Y_{n-2}$, *and* $y' \in Y_{n-3}$.

Next, in [1], the following notions are introduced.

1. Let $y \in Y_{n-1}$ and s_y be a vertex in Y adjacent to y. A vertex y is called a *type* 1, 2 or 3 vertex if $s_y \in Y_{\leq n-4}$, $s_y \in Y_{n-2}$ or $s_y \in Y_{n-3}$, respectively.

2. The *ally* of a type 2 vertex y is the vertex $a_y \in N(s_y) \cap Y_{\leq n-5}$ (see Lemma 5). The *ally* of a type 3 vertex y is a vertex $a_y \in N(s_y) \cap Y_{\leq (n+1)/2}$ (see Lemma 8).

3. Let a_y be the ally of a type 2 or type 3 vertex y. Define the sets: $T = \{x \in Y_{n-1} \mid a_x = a_y\}$, $R = \{x \in N(a_y) \mid x \in Y_{n-1} \cup Y_{n-2}\}$, and $P = \{x \in Y_{n-2} \cup Y_{n-3} \mid \exists z \in T \text{ such that } x = S_z\}$. The *local* set of a vertex a_y is the set $L = L_{a_y} = \{a_y\} \cup T \cup R \cup P$.

4. A two-element set, consisting of a type 1 vertex and a vertex in Y adjacent to it, is called *special.*

Let **L** and **S** denote the unions of local and special sets, respectively, and $H = Y \setminus (\mathbf{L} \cup \mathbf{S})$. To prove (3), it suffices to show that, for each $\gamma \in \{\mathbf{L}, \mathbf{S}, \mathbf{H}\}$ and each $n \geq 19$,

$$\frac{\sum_{v\in\gamma} |N(v) \cap V(S_n)|}{|\gamma|} \leq n - 2.333.$$

In doing so, (4) will be improved as a result of a more precise estimate of

$$\frac{\sum_{v\in\gamma} |N(v) \cap V(S_n)|}{|\gamma|}$$

for local sets and the set H.

It follows directly from Lemmas 4, 5, 7, 9, and 10 that the local sets of the type 2 vertices have the structures shown in Fig. 1.

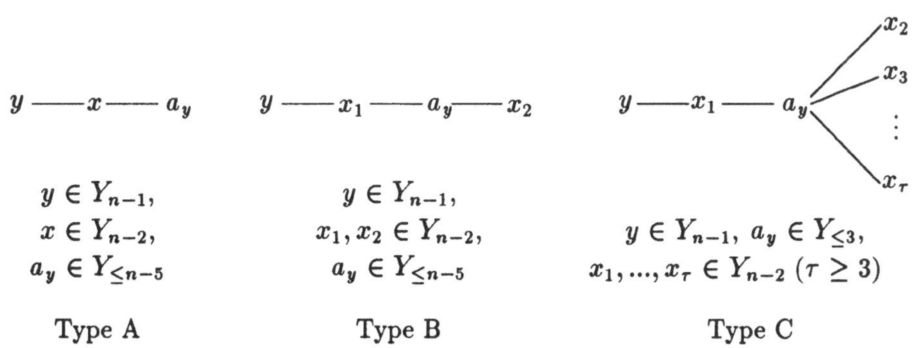

Fig. 1. Local sets for the type 2 vertices.

Obviously, for each n, the bound

$$\frac{\sum_{v\in L} |N(v) \cap V(S_n)|}{|L|} \leq \frac{n-1+n-2+n-5}{3} \leq n - 2.666, \tag{6}$$

holds for a set L of type A and the bound

$$\frac{\sum_{v\in L}|N(v)\cap V(S_n)|}{|L|} \le \frac{n-1+n-5+2(n-2)}{4} \le n-2.5 \qquad (7)$$

for a set L of type B.

For a set L of type C, we have

$$\frac{\sum_{v\in L}|N(v)\cap V(S_n)|}{|L|} \le \max_{0\le\tau\le 3}\frac{n-1+\tau+(n-\tau)(n-2)}{n-\tau+2} \le n-3+\frac{5}{n-1}. \qquad (8)$$

The right-hand side of this inequality is less than $n-2.166$ for $n \ge 7$ and less than $n-2.722$ for $n \ge 19$.

It follows from Lemmas 3, 4, and 10 that the local sets for the type 3 vertices have the structures shown in Fig. 2.

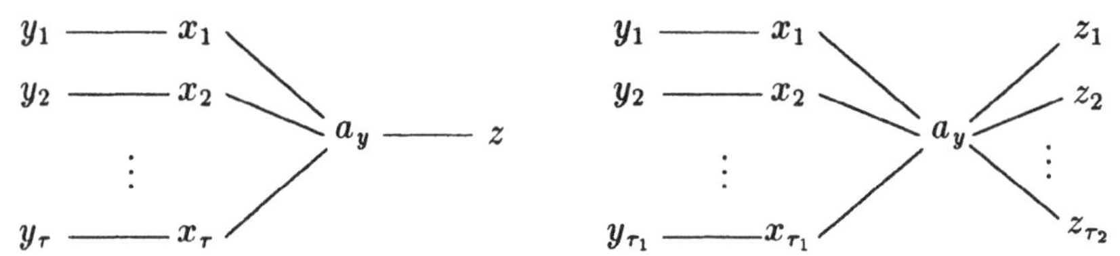

$y_1,\dots,y_\tau, z \in Y_{n-1}$, $x_1,\dots,x_\tau \in Y_{n-3}$, $a_y \in Y_{\le(n+1)/2}$

Type D

$y_1,\dots,y_{\tau_1} \in Y_{n-1}$, $x_1,\dots,x_{\tau_1} \in Y_{n-3}$, $z_1,\dots,z_{\tau_2} \in Y_{n-2}$, $a_y \in Y_{\le(n+1)/2}$

Type E

Fig. 2. Local sets for the type 3 vertices.

If a set L is of type D, then

$$\frac{\sum_{v\in L}|N(v)\cap V(S_n)|}{|L|} \le \max_{0\le\tau\le(n+1)/2}\frac{n-1+\tau+2(n-2)(n-1-\tau)}{2+2(n-1-\tau)}$$
$$= n-2,5+\frac{3}{n-1}, \qquad (9)$$

that is less than $n-2.125$ for $n \ge 9$ and less than $n-2.333$ for $n \ge 19$.

For a set L of type E, we have

$$\frac{\sum\limits_{v\in L}|N(v)\cap V(S_n)|}{|L|} \le \max_{\substack{(n-1)/2\le\tau_1+\tau_2\le n\\ 1\le\tau_1\le n\\ 0\le\tau_2\le n-1}} \frac{(n-2)(2\tau_1+\tau_2)+n-\tau_1-\tau_2}{2\tau_1+\tau_2+1}$$

$$= n - 2,5 + \frac{5}{2n}, \qquad (10)$$

that is less than $n - 2.142$ for $n \ge 7$ and less than $n - 2.368$ for $n \ge 19$.

Clearly, for each n and for each special set S,

$$\frac{\sum\limits_{v\in \mathbf{S}}|N(v)\cap V(S_n)|}{|\mathbf{S}|} \le \frac{n-1+n-4}{2} = n - 2.5. \qquad (11)$$

In [1], some speculations on the properties of the set H are presented and the conclusion is made that (4) is also valid for this set. However, the concrete form of the bound is not given. We obtain this bound. In [1], it is pointed out that

1. H does not contain vertices of the set Y_{n-1} since the vertices in Y_{n-1} lie either in local or in special sets.

2. For each special set $\{y, s_y\}$, it is true that $N(s_y)\cap Y_{n-2} = \varnothing$ (see Lemma 4).

3. It follows from Lemma 4 that, for each local set L, the set $\{v \in L|(N(v)\setminus L)\cap Y_{n-2} \ne \varnothing\}$ may consist only of those vertices which belong to Y_{n-2}.

1–3 directly imply

REMARK. The inclusion $H \subseteq Y_{\le n-2}$ holds and if H contains a vertex v of Y_{n-2}, then, in the set $Y = V(Q_n) \setminus V(S_n)$, the vertex v is adjacent either to vertices in H (by Lemma 6, being adjacent to at most one vertex in Y_{n-2}) or to a vertex in some local set of Y_{n-2} and a vertex in the set $H\cap Y_{\le n-3}$ (see Lemma 6).

Proposition. *For each nonempty set* H,

$$\frac{\sum\limits_{v\in H}|N(v)\cap V(S_n)|}{|H|} \le n - 2.333. \qquad (12)$$

PROOF. We proceed by induction on the cardinality of H.

1. Let $|H| = 1$. Then, by the Remark, $H \subseteq Y_{\le n-3}$ and (12) follows.

2. Let $|H| = 2$. It follows from the Remark that at least one vertex in H belongs to $Y_{\le n-3}$ and hence, for each n,

$$\frac{\sum\limits_{v\in H}|N(v)\cap V(S_n)|}{|H|} \le \frac{n-2+n-3}{2} = n - 2.5.$$

3. Let $|H| = 3$. It follows from the Remark that the estimated value is maximal when H is the path shown in Fig. 3, (a). In this case, for each n,

$$\frac{\sum_{v\in H} |N(v)\cap V(S_n)|}{|H|} \leq \frac{2(n-2)+n-3}{3} = n - 2.333.$$

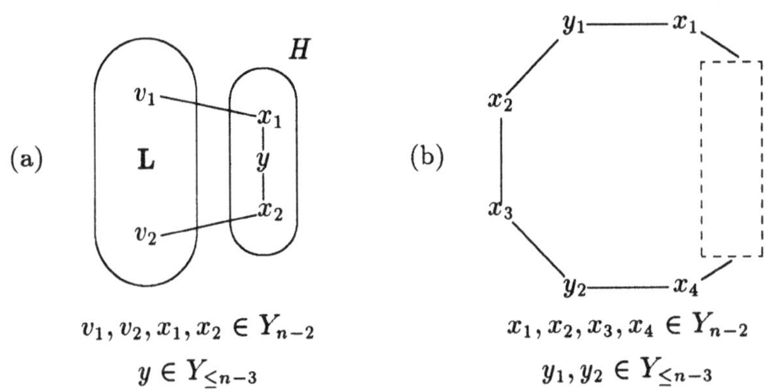

Fig. 3. Configurations of the set H satiated by vertices of Y_{n-2}.

4. Suppose (12) is valid for each set $H = H' \setminus \{v\}$ of cardinality m and prove (12) for the set H' of cardinality $m+1$. The following cases are possible.

Case 1. For the set H, the bound (12) is achieved. Note that this bound is achieved if the cardinality of H is divisible by three, H contains $(2/3)|H|$ vertices of Y_{n-2} and $(1/3)|H|$ vertices of $Y_{\leq n-3}$, and also each vertex in $Y_{\leq n-3}$ being adjacent to precisely two vertices in Y_{n-2}. Therefore, H may consist only of the paths shown in Fig. 3, (a) and the cycles of length $6k$, $k \geq 1$, shown in Fig. 3, (b). (The rectangle means that the vertices x_1 and x_4 are either adjacent or are connected by several copies of the left path $x_1 - y_1 - x_2 - x_3 - y_2 - x_4$.) Then the vertex v cannot belong to Y_{n-2}, since otherwise at least one vertex in H adjacent to v would belong to $Y_{<3}$ (by Lemma 7) which contradicts the assumption that the bound is achieved for H. Thus, $v \in Y_{\leq n-3}$ and (12) follows.

Case 2. The bound (12) is not achieved for the set H. Considering the possible cases $|H| \equiv r \pmod 3$, where $r = 0, 1, 2$, we see that (12) achieves the maximum value for H if and only if H contains a subset H'' of cardinality $m-2$ for which the bound is achieved and two still vertices which cannot belong to Y_{n-2} simultaneously (by the Remark and Lemma 7, this would contradict the fact that (12) is achieved for H''). Adding such vertex of Y_{n-2}, we obtain for H' the inequality

$$\frac{\sum_{v\in H'} |N(v)\cap V(S_n)|}{|H'|} \leq \frac{(|H|-2)(\frac{2}{3}(n-2)+\frac{1}{3}(n-3))+2(n-2)+n-3}{|H|+1}$$
$$= n - 2.333.$$

The proof is complete.

The bound (3) is now immediate from (6)–(12). Note that for $7 \leq n \leq 18$, the bound (9) is maximal among (6)–(12). Choosing it for

$$\frac{\sum_{v\in Y} |N(v)\cap V(S_n)|}{|Y|},$$

we obtain for these n the best upper bound for the length of an arbitrary snake. The comparison of bounds is presented in the table below.

n	Оценка (2)	Оценка (1)	Оценка (3)	$n \leq 18$
7	60	—	—	—
8	122	—	—	—
9	248	—	—	—
10	500	—	—	—
11	1004	—	—	—
12	2016	2036	—	—
13	4044	4076	—	—
14	8102	8156	—	8098
15	16228	16320	—	16200
16	32496	32650	—	32412
17	65056	65316	—	64846
18	130220	130660	—	129732
19	260618	261370	259548	—
20	521542	522826	519388	—
21	1043606	1045808	1039296	—
22	2088112	2091894	2079528	—
23	4177790	4184292	4160748	—
24	8358324	8369498	8324572	—
25	16721476	16740664	16654754	—
26	33451504	33484380	33319784	—
27	66918212	66974376	66658468	—
28	133863590	133959118	133351806	—
29	267775908	267937430	266768154	—
30	535639556	535910498	533656116	—

References

1. H. S. Snevily (1994) The snake-in-the-box problem: A new upper bound, *Discrete Math.* **133**, No. 1-3, 307–314.
2. F. I. Solov'ëva (1987) An upper bound on the length of a cycle in the n-dimensional unit cube (in Russian), *Metody Diskret. Anal.* **45**, 71–76.
3. A. A. Evdokimov and S. A. Malyugin (1987) The "snake-in-the-box" code and paths in a lattice on a torus (in Russian), *Mathematics Today*, Vishcha Shkola, Kiev, pp. 108–116.
4. K. Deimer (1985) A new upper bound for the length of snakes, *Combinatorica* **5**, No. 2, 109–120.
5. H. L. Abbot and M. Katchalski (1991) On the construction of a snake in the box codes, *Utilitas Math.* **40**, 97–116.
6. A. A. Evdokimov (1990) Embedding of chains and cycles in a hypercube. I (in Russian), *Metody Diskret. Anal.* **50**, 10–25.

Institute of Information Systems
Lavrent'ev pr., 6
Novosibirsk 630090
RUSSIA

TRANSLATED BY S. I. SUSLOV

OPERATIONS AND ISOMETRIC EMBEDDINGS OF GRAPHS RELATED TO THE METRIC PROLONGATION PROPERTY[†)]

T. I. Fedoryaeva

In [1] the metric prolongation property (MPP) is introduced for discrete metric spaces, and connections are pointed out of the MPP with metric properties of the locally-isometric embeddings. In the present paper, we continue the study of the MPP for finite graphs [1–5]. Some operations invariant with respect to the MPP are considered whose application simplifies the study of the MPP for certain classes of graphs. A theorem is proved on the isometric embedding preserving the MPP of an arbitrary graph to a graph of given connectivity, and it is proved that the problems of describing the classes of connected, those of connectivity n, and of n-connected graphs with the MPP are equivalent. For graphs of small connectivity, in [3, 5], and also in the present paper, there is obtained a complete characterization of the natural classes of such graphs with the MPP. Conditions are given of the invariance of the MPP for the join operation, and the complete n-partite graphs with the MPP are described. An isometric embedding of an arbitrary graph of diameter 2 to an appropriate graph with the MPP is constructed. Cactii with the MPP are described, which extends the description of trees and unicyclic graphs with the MPP obtained by the author in [3, 4].

Introduction

In the present paper, we consider the metric prolongation property (MPP) for finite simple graphs with the ordinary distance between vertices.

DEFINITION 1. Two vertices of a connected graph G satisfy the MPP if they lie on a diametric path of G. A graph satisfies the MPP if each two of its vertices satisfy the MPP.

In what follows, for a class W of graphs, we denote by W^* the class of all graphs from W which satisfy the MPP.

Let $\kappa(G)$ be the connectivity of G (i.e., the minimal number of vertices whose deletion results in a disconnected or one-vertex graph), and let W_n be the class of graphs G with connectivity $\kappa(G) = n$, where $n \geq 1$. Then $\bigcup_{i\geq 1} W_i$ is the class of all connected graphs, $\bigcup_{i\geq n} W_i$ the class of n-connected graphs, and we have the

†) This research was supported by the Russian Foundation for Basic Research (Grant 93–01–01484).

A. D. Korshunov (ed.), Operations Research and Discrete Analysis, 31–49.

following chain of inclusions (whose strictness follows, for instance, from the fact that $K_{n+1} \in W_n^* \setminus W_{n+1}^*$):

$$\bigcup_{i \geq 1} W_i^* \supset \ldots \supset \bigcup_{i \geq n} W_i^* \supset \bigcup_{i \geq n+1} W_i^* \supset \ldots .$$

The containment of G in W_n^* imposes some metric constraints on G that help in describing graphs with the MPP. For instance, the presence of pendant vertices in G implies that the distance among them is equal to the diameter of G. Moreover, this property is crucial for describing the MPP in the class of trees [4] and is important in describing the unicyclic graphs with the MPP [3]. And in general, the presence of "weak points" — cut-points and bridges in G (i.e., the containment of G in W_1), as will be shown in § 3 of the present paper, gives rise to restrictions on G, in particular, making it possible to describe cactii with the MPP. In case of graphs of class W_2^*, the metric constraints were also employed, for example, to characterize the outerplanar graphs with the MPP (see [5]).

In the present paper, we shall prove that the problems of describing classes

$$\bigcup_{i \geq 1} W_i^*, \quad \bigcup_{i \geq n} W_i^*, \quad W_n^*, \quad n \geq 1,$$

are equivalent in the sense that a description of any of these classes gives a description of all the others (the formal definition is given in § 2). To this end, an operation was introduced in § 2 which constructs for any graph G another graph H of given connectivity $n \geq 1$ preserving the MPP and the metric of G. More exactly, we have the following

Theorem 1. *For every $n \geq 1$, there is an isometric embedding of an arbitrary graph G into an appropriate graph H of connectivity n such that G satisfies* MMP *if and only if H satisfies the* MPP, *where $d(H) = 2d(G) + 1$.*

This theorem implies

Corollary 1. *The problems of describing the classes $\bigcup_{i \geq 1} W_i^*$, $\bigcup_{i \geq n} W_i^*$ and W_n^*, where $n \geq 1$, are equivalent.*

It is known (see, for example, [6]) that almost all graphs have diameter 2. Also, it was proved in [2] that almost all graphs satisfy the MPP and have diameter 2. In § 2 of the present paper, we give a method for rebuilding any diameter 2 graph into a graph satisfying the MPP preserving the original metric and diameter (Theorem 3).

In § 1 we consider the problems of preserving the MPP by the operations of Cartesian product and join of graphs, first employed in the proof of Theorem 1, and second, in the description of some simple operations that produce new graphs preserving the MPP, as well as in the description of complete n-partite graphs with the MPP. In § 2 we give some constructions for isometric embeddings. Theorem 1, Corollary 1, and Theorem 3 are proved. In § 3, for graphs of the class W_1 we define an operation which makes it possible to produce new graphs of this class preserving the MPP and "dismantle" them into simpler graphs in W_1. With the help of this operation we describe the cactii that satisfy the MPP (Theorem 4).

In the paper the ordinary concepts and notations from graph theory are used [6–8]. We deal with finite simple graphs which are connected if the contrary is not stated.

For a graph G denote by
$V(G)$ the set of vertices,
$E(G)$ the set of edges,
$\rho_G(x,y)$ the ordinary distance between vertices x and y,
$d(G)$ the diameter of G,
$e_G(x)$ the eccentricity of a vertex x,
$S_G^1(x)$ the ball of radius 1 with center a vertex x,
xy the edge with endpoints x and y.

As usual, K_n denotes the complete n-vertex graph, $K(p_1,\dots,p_n)$, a complete n-partite graph; and $G \cup H$, a union of graphs G and H. A shortest path P with endpoints u_1 and u_n such that

$$u_2,\dots,u_{n-1} \in V(P), \quad \sum_{i=1}^{n-1} \rho_G(u_i,u_{i+1}) = \rho_G(u_1,u_n)$$

(it is not necessarily presumed that $u_i \neq u_{i+1}$) is denoted by $(u_1,u_2,\dots,u_n)$.

Let G and H be graphs. A mapping $f: V(G) \to V(H)$ is called an *isometric embedding* of G into H if $\rho_G(x,y) = \rho_H(f(x),f(y))$ for all $x,y \in V(G)$. We also say that G is *isometrically embedded* into H. A subgraph G of H is called *isometric* if for all $x,y \in V(G)$ we have $\rho_G(x,y) = \rho_H(x,y)$.

§ 1. Join and Cartesian Product

In this section we study the conditions of invariance of the MPP under the operations of join and Cartesian product over graphs. Let G and H be vertex-disjoint graphs. The *join of the graphs* G and H is defined to be the graph $G+H$ such that

$$V(G+H) = V(G) \cup V(H),$$

$$E(G+H) = E(G) \cup E(H) \cup \{gh \mid g \in V(G), h \in V(H)\}.$$

Clearly, if G and H are arbitrary graphs, not necessarily connected, then $G+H$ is connected.

Observe some metric properties of operation $G+H$. First, introduce a definition.

$B_k(a,b)$, $k \geq 0$

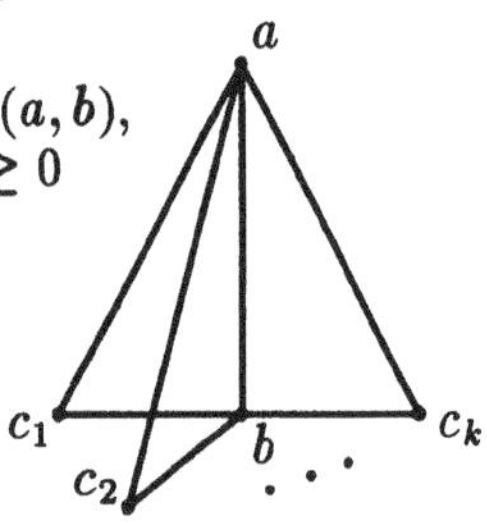

Fig. 1

The graph depicted in Fig. 1 is called the *flounce on vertices* a and b. We shall say that a *graph* G *has a flounce* if in G there is a subgraph $B_k(a,b)$, where $a, b \in V(G)$ and $\deg_G a = \deg_G b = k+1$. In other words, in G the balls of radius 1 centered at vertices a and b coincide.

The definitions imply

REMARK 1. If a graph G of diameter $d(G) \geq 2$ satisfies the MPP, then G has no flounce; for $d(G) = 2$ converse is also true.

Denote by $\eta(g,h)$ the length of the longest shortest path in $G+H$, which contains vertices $g, h \in V(G+H)$. Immediately from the definition of $G+H$ follows

Lemma 1. *Let G and H be graphs without common vertices, and let G_i and H_j be connected components of the graphs G and H, respectively, $1 \leq i \leq n$, $1 \leq j \leq m$. Then*

1. *$d(G+H) \leq 2$, where $d(G+H) = 1$ if and only if G and H are complete graphs.*
2. *If $g_1 \in V(G_i)$ and $g_2 \in V(G_s)$, $i \neq s$, then $\eta(g_1, g_2) = 2$.*
3. *If $g_1, g_2 \in V(G_i)$, $g_1 \neq g_2$, then*

$$\eta(g_1, g_2) = \begin{cases} 1 & \text{if } G_i \text{ has a flounce on vertices } g_1 \text{ and } g_2, \\ 2 & \text{otherwise.} \end{cases}$$

4. *If $g \in V(G_i)$ and $h \in V(H_j)$, then*

$$\eta(g_1, g_2) = \begin{cases} 1 & \text{if } S^1_G(g) = V(G) \text{ and } S^1_H(h) = V(H), \\ 2 & \text{otherwise.} \end{cases}$$

Now let us find out when $G+H$ satisfies the MPP.

Theorem 2. *Let G and H be graphs without common vertices (not necessarily connected). Then $G+H$ satisfies the* MPP *if and only if at least one of the following conditions hold:*

(i) G and H are complete graphs;

(ii) at least one of the graphs G and H has more than one vertex and is not representable as $K_1 + F$ (where F is not necessarily connected graph) and each connected component in G and H has no flounce.

PROOF. Let $G+H$ satisfy the MPP. We assume that $d(G+H) = 2$ (otherwise due to Subsection 1 of Lemma 1, there would hold (i)). Then by Subsection 4 of Lemma 1 at least one of the graphs G and H has more than one vertex and is not representable as $K_1 + F$ and, in virtue of Lemma 1, each connected component of the graphs G and H has no flounce.

Let us show the converse. If (i) holds, then $G+H$ is a complete graph. Assume that (ii) holds. Then by Lemma 1, $d(G+H) = 2$ and $\eta(g_1, g_2) = d(G+H)$ for all vertices $g_1, g_2 \in V(G+H)$, i.e. $G+H$ satisfies the MPP. Theorem 2 is proved.

Now we define a simple operation that transforms graphs satisfying the MPP into a graph also satisfying the MPP.

Corollary 2. *Let G_i and H_j, $1 \le i \le n$, $1 \le j \le m$, $n+m \ge 3$, be graphs of diameter at least two, having no common vertex. If all the graphs G_i and H_j satisfy the* MPP, *then graph $\left(\bigcup_{i=1}^n G_i\right) + \left(\bigcup_{j=1}^m H_j\right)$ satisfies the* MPP.

Corollary 2 follows immediately from Theorem 2 and Remark 1.

REMARK 2. For graphs G_i and H_j of diameter $d \ge 3$ the statement converse to Corollary 2 fails in general.

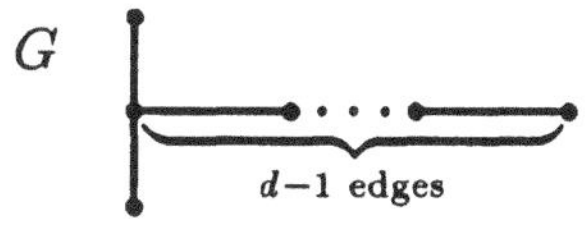

Fig. 2

Indeed the graph G, depicted in Fig. 2, has no flounce, neither does it satisfy the MPP. But the graph $G+(H_1 \cup H_2)$, due to Theorem 2, does satisfy the MPP, where H_1 and H_2 are any graphs of diameter d, having no common vertex and satisfying the MPP (for example, H_i is a path).

However, the operation indicated in Corollary 2 preserves the MPP in the class of graphs of diameter 2. From Theorem 2 and Remark 1 we obtain

Corollary 3. *Let G_i and H_j, $1 \le i \le n$, $1 \le j \le m$, $n+m \ge 3$, be connected graphs of diameter 2 pairwise vertex-disjoint. Then the graph $\left(\bigcup_{i=1}^n G_i\right) + \left(\bigcup_{j=1}^m H_j\right)$ satisfies the* MPP *if and only if all graphs G_i and H_j satisfy the* MPP.

Theorem 2 implies an easy characterization of those complete n-partite graphs $K(p_1, p_2, \ldots, p_n) = \overline{K}_{p_1} + \overline{K}_{p_2} + \cdots + \overline{K}_{p_n}$ that satisfy the MPP.

Corollary 4. *A complete n-partite graph $K(p_1, \ldots, p_n)$ satisfies the* MPP *if and only if at least one of the following conditions holds:*

(1) $p_i = 1$ *for any* $i = 1, 2, \ldots, n$;

(2) $p_i \ne 1$ *for any* $i = 1, 2, \ldots, n$ *and* $n \ge 2$;

(3) $p_i = 1$ *for some* i *and* $p_j \ne 1$ *for* $j \ne i$, $1 \le i, j \le n$.

PROOF. Let $K(p_1, \ldots, p_n)$ satisfy the MPP, and suppose that (1), (2), and (3) are false. Then there are distinct i, j, and k such that $1 \le i, j, k \le n$, $p_i = p_j = 1$,

$p_k \neq 1$. By commutativity and associativity of the join operation, $K(p_1, \dots, p_n) = K_1 + (K_1 + F)$, where F is a graph failing to be complete. This contradicts Theorem 2.

The converse statement follows from Theorem 2, Remark 1, and the fact that $\overline{K}_{p_i}$, $p_i \neq 1$, has no flounce, is not trivial, and is not representable as $K_1 + F$. Corollary 4 is proved.

Consider the operation of the taking of a Cartesian product. The Cartesian product $G \times H$ of graphs G and H is the graph with

$$V(G \times H) = \{\langle x, y\rangle \mid x \in V(G), y \in V(H)\};$$

$$E(G \times H) = \{\langle x_1, y_1\rangle\langle x_2, y_2\rangle \mid (x_1 = x_2 \text{ and } y_1y_2 \in E(H)) \text{ or}$$

$$(y_1 = y_2 \text{ and } x_1x_2 \in E(G))\},$$

where $\rho_{G\times H}(\langle x_1, y_1\rangle, \langle x_2, y_2\rangle) = \rho_G(x_1, x_2) + \rho_H(y_1, y_2)$.

Proposition 1. *The graph $G \times H$ satisfies the* MPP *if and only if the graphs G and H satisfy the* MPP.

PROOF. Let $G \times H$ satisfy the MPP. Suppose, for example, that G does not satisfy the MPP. Then there are $x_1, x_2 \in V(G)$ such that $\rho_G(x_1, x_2) < d(G)$ and there is no shortest path in G containing x_1, x_2 and of length greater than $\rho_G(x_1, x_2)$. Consider $y_1, y_2 \in V(H)$ such that $\rho_H(y_1, y_2) = d(H)$. By the choice of x_1, x_2 and the relation $\rho_{G\times H}(\langle x_1, y_1\rangle, \langle x_2, y_2\rangle) = \rho_G(x_1, x_2) + d(H) < d(G \times H)$, the pair $\langle x_1, y_1\rangle$, $\langle x_2, y_2\rangle$ does not satisfy the MPP. A contradiction. Hence, the graph G (and, similarly, H) satisfies the MPP.

The converse statement follows from [1, Lemma 1.1]. Proposition 1 is proved.

Corollary 5 shows how, preserving the MPP, to embed isometrically any graph G in a graph H with greater connectivity $\kappa(H) \geq \kappa(G)$.

Corollary 5. *For every $n \geq 1$ there is an isometric embedding of any graph G in an appropriate n-connected graph H such that G satisfies the* MPP *if and only if H satisfies the* MPP, *where $d(H) = d(G) + 1$.*

PROOF. Since $\kappa(K_{n+1}) = n$ and $d(K_{n+1}) = 1$, we may assume that G has more than one vertex. Let $H = G \times K_n$. Clearly, G is embedded isomorphically in H. Due to Proposition 1, it remains to show that $\kappa(H) \geq n$. Indeed, using K_n and the nontriviality of G, it is not hard to construct n vertex-disjoint paths in H connecting two arbitrary vertices of H. The well-known criterion of n-connectivity [8, Theorem 5.10] yields $\kappa(H) \geq n$. Corollary 5 is proved.

§ 2. Isometric Embeddings

Let M_1 and M_2 be classes of graphs. We say that the problem of describing the class M_1^* is *reducible* to that of describing the class M_2^* if there exists an operation $\varphi : M_1 \to M_2$ such that for any $G \in M_1$:

- the graph G is isometrically embeddable in $\varphi(G)$,
- $G \in M_1^* \iff \varphi(G) \in M_2^*$.

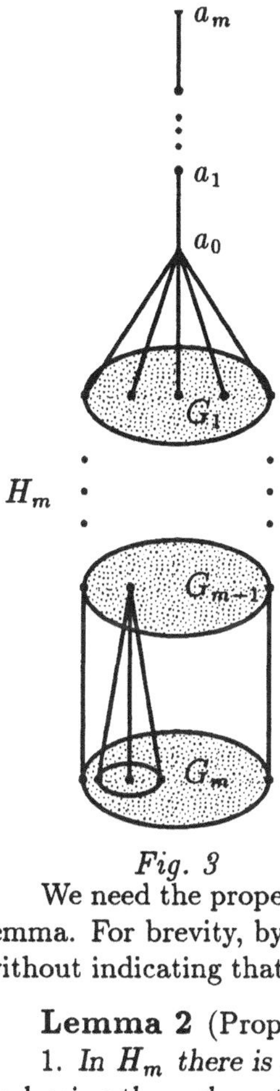

Fig. 3

The problems of describing classes M_1^* and M_2^* are *equivalent* if any one of them is reduced to the other.

To prove equivalence of the classes of graphs

$$\cup_{i\geq 1} W_i^*, \quad \cup_{i\geq n} W_i^*, \quad W_n^*, \quad n \geq 1,$$

we need some auxiliary statements.

Starting from any graph G, we inductively define the graphs $H_1, H_2, \ldots$. Let $G_1, G_2, \ldots$ be vertex-disjoint copies of G. Denote by x^i the vertex of G_i that corresponds to $x \in V(G)$, and put (Fig. 3)

$$V(H_1) = V(G_1) \cup \{a_0, a_1\}, \text{ where } a_0, a_1 \notin V(G_1);$$

$$E(H_1) = E(G_1) \cup \{a_0 a_1\} \cup \{a_0 x \mid x \in V(G_1)\};$$

$$V(H_{i+1}) = V(H_i) \cup V(G_{i+1}) \cup \{a_{i+1}\},$$

where $a_{i+1} \notin V(H_i) \cup V(G_{i+1})$;

$$E(H_{i+1}) = E(H_i) \cup E(G_{i+1}) \cup \{a_i a_{i+1}\}$$

$$\cup \{x^i y^{i+1} \mid x, y \in V(G), y \in S_G^1(x)\}.$$

We need the properties of the graph H_m which are formulated in the following lemma. For brevity, by ρ_m we denote the metric by ρ_{H_m} and use the notation x^i without indicating that $x \in V(G)$.

Lemma 2 (Properties of H_m). *Let $1 \leq j \leq i \leq m$. Then*

1. *In H_m there is a simple path of length $i + j$ connecting vertices x^i and y^j and going through vertices x^1, a_0, and y^1.*

2. *If in the graph G there is a simple path of length n connecting vertices x and y and $i \geq n + 1$, then in H_m there is a simple path of length n connecting vertices x^i and y^{i-n}* (Fig. 4, a).

3. *If* $\rho_G(x,y) \le i-j$, *then in* H_m *there is a simple path* P *of length* $i-j$ *connecting* x^i *and* y^j *and going through* y^{i-s}, *with* $\rho_P(x^i, y^{i-s}) = \rho_G(x,y)$. *Here* $s = \rho_G(x,y)$ (Fig. 4, *b*).

4. *If* $\rho_G(x,y) \ge i-j$, *then in* H_m *there is a simple path* P *of length* $\rho_G(x,y)$ *connecting* x^i *and* y^j *and going through* z^j, *where* $\rho_P(x^i, z^j) = \rho_G(x,z)$ *and* $\rho_P(z^j, y^j) = \rho_G(z,y)$. *Here* (x,z,y) *is any shortest path in* G *and* $\rho_G(x,z) = i-j$ (Fig. 4, *c*).

5. *If in* H_m *there is a shortest path* $(x^i, z_1^{i+1}, z_2^{i+1}, \dots, z_s^{i+1}, y^i)$, *where* $\rho_m(x^i, z_1^{i+1}) = \rho_m(z_s^{i+1}, y^i) = \rho_m(z_r^{i+1}, z_{r+1}^{i+1}) = 1$, $1 \le r \le s-1$, *then in* H_m *there is a shortest path connecting* x^i *and* y^i *and lying in* G_i.

6. *If in* H_m *there is a shortest path* $(x^{i+1}, z_1^i, z_2^i, \dots, z_s^i, y^{i+1})$, *where* $\rho_m(x^{i+1}, z_1^i) = \rho_m(z_s^i, y^{i+1}) = \rho_m(z_r^i, z_{r+1}^i) = 1$, $1 \le r \le s-1$, *then in* H_m *there is a shortest path connecting* x^{i+1} *and* y^{i+1} *and lying in* G_{i+1}.

7. *In* H_m *there is no shortest path* (x^i, y^i, z^{r+1}, z^r) *such that* $\rho_G(x,y) = 1$ *and* $\rho_m(y^i, z^{r+1}) = \rho_G(y,z) = i-(r+1) \ge 0$.

8. *If in* H_m *there is a shortest path* $(x_0^i, \dots, x_s^i, y_0^j, \dots, y_r^j)$, *where* $\rho_m(x_s^i, y_0^j) = \rho_G(x_s, y_0) = i-j$, $\rho_m(x_t^i, x_{t+1}^i) = 1$, $\rho_m(y_l^j, y_{l+1}^j) = 1$, $0 \le t \le s-1$, $0 \le l \le r-1$ (Fig. 4, *d*), *then* $\rho_m(x_0^i, y_r^j) = \rho_G(x_0, y_r)$ *and in* G *there is a shortest path* $(x_0, \dots, x_s, y_0, \dots, y_r)$.

PROOF. Properties 1 and 2 are immediate from the definition of H_m, while Properties 3 and 4 ensue from Property 2 and the construction of H_m.

Show Property 5. By the definition of H_m it follows that $z_1^i \in S^1_{G_i}(x^i)$ and $z_s^i \in S^1_{G_i}(y^i)$. Therefore, the path $P = (x^i, z_1^i, z_2^i, \dots, z_s^i, y^i)$ in G_i is not longer than a shortest path in graph H_m connecting x^i and y^i. Therefore, P is a shortest path.

Property 6 is obtained similarly as Property 5.

Show Property 7. Suppose that in H_m there is an above-mentioned shortest path. Then $\rho_G(x,z) \le i-r$, and by Property 3, $\rho_m(x^i, z^r) \le i-r$. On the other hand, by assumption we have $\rho_m(x^i, z^r) = i-r+1$, a contradiction.

Show Property 8 by induction on $r+s \ge 0$. For $r+s=0$ this property follows by assumption. Now let us suppose that $r>0$ (the case $s>0$ is similar). By the inductive hypothesis, $\rho_G(x_0, y_{r-1}) = \rho_m(x_0^i, y_{r-1}^j)$ and in G there is a shortest path $(x_0, \dots, x_s, y_0, \dots, y_{r-1})$. It remains to show that $\rho_G(x_0, y_r) = \rho_G(x_0, y_{r-1}) + 1$. Suppose not, i.e.,

$$\rho_G(x_0, y_r) \le \rho_m(x_0^i, y_{r-1}^j). \tag{1}$$

Let $\beta = \min\{\rho_G(x_0, y_r), i-j\}$. Choose a vertex $z \in V(G)$ such that

$$\rho_G(x_0, z) + \rho_G(z, y_r) = \rho_G(x_0, y_r), \quad \rho_G(z, y_r) = \beta \le i-j. \tag{2}$$

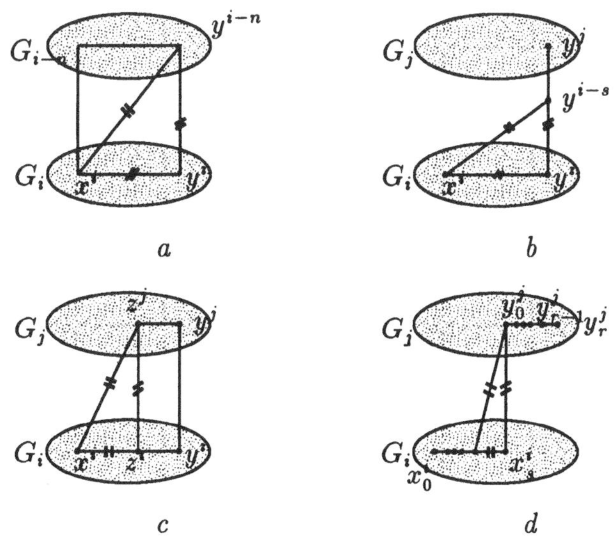

Fig. 4

Using Property 3, we obtain $\rho_m(x_0^i, y_r^j) \leq \rho_m(x_0^i, z^i) + \rho_m(z^i, y_r^j) \leq \rho_G(x_0, z) + i - j = \rho_G(x_0, y_r) + i - j - \beta$. Therefore, if $\beta = i - j$, then, due to (1), $\rho_m(x_0^i, y_r^j) \leq \rho_m(x_0^i, y_{r-1}^j)$, and if $\beta \neq i - j$, then $\rho_m(x_0^i, y_r^j) = i - j = \rho_m(x_s^i, y_0^j) \leq \rho_m(x_0^i, y_{r-1}^j)$. This contradicts the assumption that $(x_0^i, y_{r-1}^j, y_r^j)$ is a shortest path. Lemma 2 is proved.

Lemma 3 (the metric of H_m). *Let $x, y \in V(G)$, $1 \leq j \leq i \leq m$. Then*

$$\rho_m(x^i, y^j) = \begin{cases} i - j & \text{if } \rho_G(x,y) \leq i - j, \\ \rho_G(x,y) & \text{if } i - j < \rho_G(x,y) \leq i + j, \\ i + j & \text{if } i + j < \rho_G(x,y); \end{cases}$$

$$\rho_m(x^i, a_s) = i + s, \quad 0 \leq s \leq m.$$

PROOF. First observe that H_{m-1} is an isometric subgraph of H_m. Indeed, suppose that r and l are such that $1 \leq r \leq l \leq m - 1$, and let P be a shortest path in H_m connecting x^r with y^l. Then $V(P) \subseteq V(H_{m-1}) \cup V(G_m)$. By Property 5, any shortest subpath of the path P of the type $(u^{m-1}, z_1^m, z_2^m, v^{m-1})$ may be replaced by a shortest path connecting u^{m-1} with v^{m-1} and lying in G_{m-1}. Therefore, $\rho_m(x^r, y^l) = \rho_{m-1}(x^r, y^l)$, where $0 \leq k \leq m - 1$.

Proceed by induction on $m \geq 1$. For $m = 1$ the required values of ρ_m follow readily from the definition of H_1.

Assume $m \geq 2$. Since H_{m-1} is an isometric subgraph of H_m, only the following cases should be considered:

(1) $i = j = m$;

(2) $i = m,\ j \leq m - 1$;

(3) $i = m,\ 0 \leq s \leq m$.

Case 1. By Property 1,

$$\rho_m(x^m, y^m) \leq 2m. \tag{3}$$

Therefore we may assume that

$$\rho_m(x^m, y^m) < \rho_G(x, y) \tag{4}$$

(otherwise we have $\rho_m(x^m, y^m) = \rho_G(x, y)$). Then in H_m there is a shortest path $(x^m, u^m, z_1^{m-1}, z_2^{m-1}, v^m, y^m)$ such that $\rho_m(x^m, u^m) = \rho_G(x, u)$, $\rho_m(v^m, y^m) = \rho_G(v, y)$, $u \in S_G^1(z_1)$, $v \in S_G^1(z_2)$. Due to the isometry of H_{m-1} and the inductive hypothesis,

$$\rho_m(z_1^{m-1}, z_2^{m-1}) = \min\{2(m-1),\ \rho_G(z_1, z_2)\}.$$

Then, in view of (4) and Property 6, we have

$$\rho_m(x^m, y^m) \geq \rho_m(z_1^{m-1}, z_2^{m-1}) + 2 = 2m.$$

Therefore, from (3) and (4) it follows that $\rho_m(x^m, y^m) = 2m < \rho_G(x, y)$.

Case 2. In H_m there is a shortest path

$$(x^m, v^m, u^{m-1}, y^j),\ \rho_m(v^m, u^{m-1}) = 1, \quad \rho_m(x^m, v^m) = \rho_G(x, v). \tag{5}$$

By the inductive hypothesis and the isometry of H_{m-1}, the following three subcases may take place:

2.1. $\rho_m(u^{m-1}, y^j) = m-1+j < \rho_G(u, y)$. In this case it follows from (5) that $\rho_m(x^m, y^j) \geq \rho_m(u^{m-1}, y^j)+1 = m+j$, by Property 1 $\rho_m(x^m, y^j) = m+j$, and by Property 3 $\rho_G(x, y) > m-j$. Then by Property 4 $m+j = \rho_m(x^m, y^j) \leq \rho_G(x, y)$.

2.2. $m - 1 - j < \rho_m(u^{m-1}, y^j) = \rho_G(u, y) \leq m - 1 + j$. Then due to Property 4, in H_m there is a shortest path (u^{m-1}, z^j, w^j, y^j), where (u, z, w, y) is a shortest path in G and $\rho_m(u^{m-1}, z^j) = \rho_G(u, z) = m - 1 - j$, $\rho_m(z^j, y^j) = \rho_G(z, y)$, $\rho_G(z, w) = 1$. Then, taking (5) into account, we have

$$\rho_m(v^m, z^j) = \rho_m(v^m, u^{m-1}) + \rho_m(u^{m-1}, z^j) = m - j, \quad \rho_G(v, z) \leq m - j. \tag{6}$$

From (6) and Property 3, in H_m there is a shortest path

$$(v^m, z^{m-s}, z^j),\ \rho_m(v^m, z^{m-s}) = \rho_G(v, z) = s. \tag{7}$$

Thus, in H_m we have a shortest path $(x^m, v^m, z^{m-s}, z^j, w^j, y^j)$. Moreover, $\rho_G(z, w) = 1$ and in H_m there is an edge $z^{j+1}w^j$, therefore $m - s = j$. From (7) and Property 8, we obtain that $\rho_m(x^m, y^j) = \rho_G(x, y)$, and (x, v, z, y) is a shortest path in G. Moreover, from (7) it follows that $\rho_G(x, y) \geq m - j$, and due to Property 1, $\rho_G(x, y) = \rho_m(x^m, y^j) \leq m + j$.

2.3. $\rho_m(u^{m-1}, y^j) = m - 1 - j \geq \rho_G(u, y)$. In this case it follows from (5) that $\rho_m(v^m, y^j) = m - j$ and $\rho_G(v, y) \leq m - j$. By Property 3, in H_m there is a shortest path (v^m, y^{m-s}, y^j), where

$$\rho_m(v^m, y^{m-s}) = \rho_G(v, y) = s. \tag{8}$$

Thus, in H_m there is a shortest path (x^m, v^m, y^{m-s}, y^j). Due to (8) and Property 7, $x = v$ or $m - s = j$. If $x = v$, then $\rho_m(x^m, y^j) = m - j \geq \rho_G(x, y)$. Let now $m - s = j$. From (8) and Property 8 it follows that $\rho_m(x^m, y^j) = \rho_G(x, y)$ and, as in Case 2.2, $m - j \leq \rho_G(x, y) \leq m + j$.

Case 3. Consider a shortest path $(x^m, u^m, v^{m-1}, a_0, a_s)$ in H_m, where $\rho_m(u^m, v^{m-1}) = 1$. From the inductive hypothesis, we have $\rho_m(v^{m-1}, a_s) = m - 1 + s$. Then it is evident that $(x^m, \ldots, x^2, x^1, a_0, a_1, \ldots, a_s)$ is a shortest path of length $m + s$.

Lemma 3 is proved.

Corollary 6. *If $d(G) = 2k$, then $d(H_k) = d(G)$ and G_k is an isometric subgraph of H_k.*

Lemma 4. *If $d(G) = 2k$, then G satisfies the* MPP *if and only if H_k satisfies the* MPP.

PROOF. Let G and, hence, G_k satisfy the MPP, $1 \leq j \leq i \leq k$, $0 \leq s \leq k$, and $x, y \in V(G)$. In view of Lemma 3, any two vertices x^i and a^s satisfy the MPP. Therefore, it remains to prove that if $\rho_k(x^i, y^j) < d(H_k)$, then in H_k there is a shortest path containing x^i and y^j of length greater than $\rho_k(x^i, y^j)$.

If $\rho_G(x, y) > i + j$, then $i < k$ or $j < k$. Let, for example, $i < k$. By Lemma 3, in H_k there is a shortest path (x^{i+1}, x^i, y^j). Therefore, we assume in what follows that $\rho_G(x, y) \leq i + j$.

If $\rho_G(x, y) \leq i - j$, then, due to Lemma 3, in H_k there is a shortest path (x^i, y^j, a_0).

Let now $i - j < \rho_G(x, y) \leq i + j$. By Lemma 3 and Corollary 6, $\rho_G(x^i, y^j) = \rho_G(x, y) < d(G)$. Due to the MPP, in G there is a shortest path (x, y, u) or (v, x, y) such that $\rho_G(y, u) = 1$ and $\rho_G(v, x) = 1$. Let, for example, there is a shortest path

(x, y, u) (the case (v, x, y) is similar). Obviously, $i - s < \rho_G(x, u) \leq i + s$, $1 \leq s \leq k$, where $s = j + 1$ for $i + j < 2k$ and $s = j$ for $i + j = 2k$. By Lemma 3 $\rho_k(x^i, u^s) = \rho_G(x, u)$. Since in H_k there is an edge $y^j u^s$, it follows that $\rho_k(x^i, u^s) = \rho_k(x^i, y^j) + \rho_k(y^j, u^s)$, i.e. (x^i, y^j, u^s) is a sought shortest path. Thus, H_k satisfies the MPP.

Now let H_k satisfy the MPP, $x, y \in V(G)$ and $1 \leq \rho_G(x, y) < d(G)$. Due to the MPP and Corollary 6, in H_k there is a shortest path P containing x^k and y^k, of length greater than $\rho_k(x^k, y^k) = \rho_G(x, y)$. Due to Lemma 3 and the condition $\rho_k(x^k, y^k) \geq 1$, we have $a_0, \ldots, a_m \notin V(P)$. Therefore, in H_k there is a shortest path (x^k, y^k, u^s) such that $\rho_k(y^k, u^s) = 1$, $1 \leq s \leq k$, $u \in V(G)$ (the case (u^s, x^k, y^k) is similar). If $s = k$, then by Corollary 6, in G_k there is a shortest path (x^k, y^k, u^k) and, hence, (x, y, u) is a shortest path in G of length greater than $\rho_G(x, y)$.

Now let $s < k$. By the construction of H_k, we have $s = k-1$, $y^k \in S, 1_{G_k}(u_k)$, and by Property 7 $u \neq y$. From Property 8, in G there is a shortest path (x, y, u). Thus, every two vertices x and y in G satisfy the MPP. Lemma 4 is proved.

In what follows, by $H_k(G)$ we denote the graph H_k that is constructed from G as described above.

PROOF OF THEOREM 1. Let the graph G have more than one vertex (otherwise $H = K_{n+1}$ will do). By Proposition 1, G satisfies the MPP if and only if G^2 satisfies the MPP, where $d(G^2) = 2d(G)$. Denote $H = H_{d(G)}(G^2) \times K_n$. By Corollary 6, $d(H) = 2d(G) + 1$ and G is isometrically embeddable in H. By Lemma 4 and Proposition 1, the graph G^2 satisfies the MPP if and only if H satisfies the MPP. It remains to show that $\kappa(H) = n$. From the proof of Corollary 5, it follows that $\kappa(H) \geq n$. On the other hand, the graph $H_{d(G)}(G^2)$ has a pendant vertex which implies $\delta(H) \leq n$, where $\delta(H)$ is the minimal degree of H. By Whitney [8, Theorem 5.1], $\kappa(H) \leq \delta(H)$. Thus, $\kappa(H) = n$. Theorem 1 is proved.

PROOF OF COROLLARY 1. In Theorem 1, for every $n \geq 1$ the following operation is defined:

$$f_n : \bigcup_{i \geq 1} W_i \to W_n, \; f_n(G) = H_{d(G)}(G^2) \times K_n,$$

where G is isometrically embeddable in $f_n(G)$ and

$$G \in \bigcup_{i \geq 1} W_i^* \Longleftrightarrow f_n(G) \in W_n^*.$$

This yields for every $n \geq 1$ the diagram

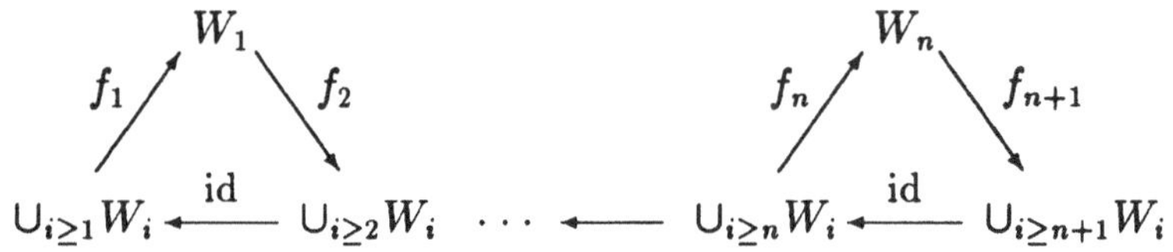

which implies the equivalence of the problems of describing the classes $\bigcup_{i\geq 1} W_i^*$, $\bigcup_{i\geq n} W_i^*$, and W_n^*, $n \geq 1$. Corollary 1 is proved.

Theorem 3. *There is an isometric embedding of every graph G of diameter 2 in an appropriate graph H such that H satisfies the* MPP *and $d(H) = d(G)$.*

Proof. Define a sequence of graphs $H_1 = G, H_2, \dots, H_k$ such that
(a) H_i is an isometric subgraph of H_{i+1},
(b) $d(H_{i+1}) = 2$,
(c) there is a strict inclusion $\Pi(H_{i+1}) \subset \Pi(H_i)$,
(d) $\Pi(H_k) = \varnothing$,
where $\Pi(H_i) = \{xy \in E(H_i) \mid H_i$ has a flounce on vertices x and y.

Suppose that H_i has been defined and $\Pi(H_i) \neq \varnothing$. Let

$$xy \in \Pi(H_i) \tag{9}$$

and $Y = \{y\} \cup \{v \in V(H_i) \mid \rho_{H_i}(y, v) = 1$ and in H_i there is a diametric path with an end vertex y, which contain $v\}$. We set

$$V(H_{i+1}) = V(H_i) \cup \{z\}, \text{ where } z \notin V(H_i),\ E(H_{i+1}) = E(H_i) \cup \{zv \mid v \in Y\}.$$

Let us verify conditions (a)–(d) for the graph H_{i+1}. Since $d(H_i) = 2$ and each new edge is incident with a new vertex z, condition (a) trivially holds. From (a) and the construction of the graph H_{i+1}, we obtain $d(H_{i+1}) = 2$.

Now let us observe that

$$xy, zv \notin \Pi(H_{i+1}), \text{ where } v \in Y. \tag{10}$$

Indeed, $x \notin Y$ due to (9). Therefore, H_{i+1} has no edge zx, but has the edge zy. If $v \in Y \setminus \{y\}$, then in H_i there is a diametric path (y, v, u) (Fig. 5) and, hence, H_{i+1} has no edge uz, but the edge vu. This proves (10).

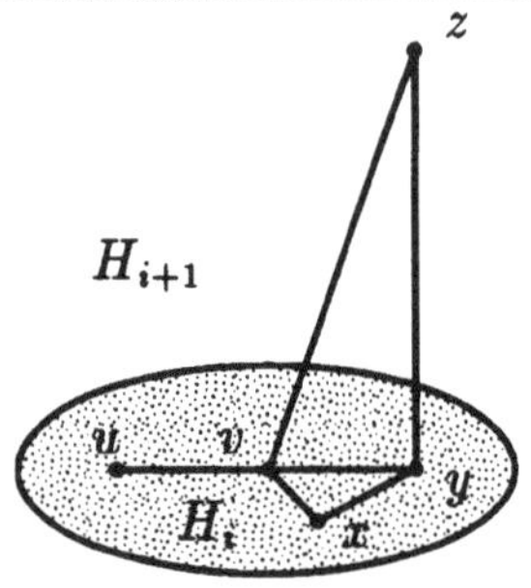

Fig. 5

Let us prove (c). Let $ab \in \Pi(H_{i+1})$. Then $ab \in E(H_{i+1})$. By (10) , $ab \in E(H_i)$ and using (a), we obtain $ab \in \Pi(H_i)$. Hence, using (9) and (10), we have (c).

Thus transition from H_i to H_{i+1} preserves conditions (a)–(c). Continuing this process, we obtain the graph H_k satisfying condition (d). Due to Remark 1, graph H_k possess the MPP. Thus, $H = H_k$ is a required graph. Theorem 3 is proved.

§ 3. Graphs with Cut-Vertices and Cactii

In this section we consider the graphs from the class W_1, i.e., those graph with cut-vertices. Let a be a cut-vertex of a graph G. Using the properties of blocks (see, for example, [6, pp. 140–141]), one can uniquely determine the vertex sets $V_1, \dots, V_{m(a)}$, $m(a) \geq 2$ ($m(a)$ is the number of blocks in the graph G containing vertex a and each V_i contains the only such a block) such that

$$V(G) = \bigcup_{1 \leq i \leq m(a)} V_i, \quad V_i \cap V_j = \{a\}, \quad i \neq j,$$

and any path joining vertices $x \in V_i$ and $y \in V_j$, $i \neq j$, contains a. By G_i we denote the spanned subgraph of G on the vertex set V_i, $1 \leq i \leq m(a)$. In the sequel, connected graphs $G_1, \dots, G_{m(a)}$ will be called the *graphs defined by the cut-vertex a*.

Graphs in which $m(a) = 2$ for each cut-vertex a, and at least one of the graphs defined by a is a path, have the simplest structure among the classes of graphs W_1. In the next proposition, we define a construction for "dismantling" graphs W_1 until arriving at graphs with such simple structure and also for obtaining new graphs in W_1^*.

Proposition 2. *Let $G_1, \dots, G_{m(a)}$ be graphs defined by a cut-vertex a in a graph G. Then the following conditions are equivalent:*

(i) *G satisfies the* MPP;

(ii) *for any distinct $i, j \leq m(a)$ the graph G'_i satisfies the* MPP, *and $d(G'_i) = d(G'_j) = e_i + e_j$.*

Moreover, if (i) *or* (ii) *holds, then $d(G) = d(G'_i)$, and for $m(a) \geq 3$ it holds $e_i = e_j$.*

Here G'_i is the graph obtained from the graph G_i by adding a simple path of length e_{i+1} with the end-vertex a (Fig. 6), *$e_{m(a)+1} = e_1$, and if* (i) *is true, then in* (ii) *e_i is defined as $e_{G_i}(a)$.*

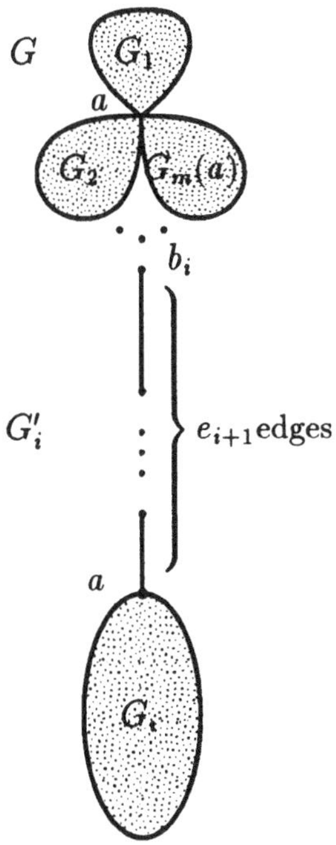

Fig. 6

PROOF. Let us prove the implication (i) $\Rightarrow$ (ii). Due to the MPP, the vertex of G_i that implements the eccentricity of $e_{G_i}(a)$ and the vertex of G_j that implements the eccentricity $e_{G_j}(a)$ belong to a certain diametric path of G. Therefore,

$$d(G) = e_{G_i}(a) + e_{G_j}(a),$$
$$i \neq j,\ 1 \leq i, j \leq m(a). \tag{11}$$

Hence, if $m(a) \geq 3$, then $e_{G_i}(a) = e_{G_j}(a)$. Since the graph G'_i is isometrically embeddable in G and it holds (11), we have $d(G) = d(G'_i)$. It is now easy to see that the graph G'_i satisfies the MPP.

Let prove that (ii) implies (i). As the vertex of the graph G_i which implements the eccentricity $e_{G_i}(a)$ and the vertex b_i (Fig. 6) both belong to a certain diametric path of the graph G'_i, while $d(G'_i) = e_i + e_{i+1}$, then $e_i = e_{G_i}(a)$. Hence, $d(G) = d(G'_i)$ and $e_i = e_j$ for $m(a) \geq 3$. It is now obvious that G satisfies the MPP. Proposition 2 is proved.

In what follows we need the following

REMARK 3. Proposition 2 remains valid if instead of $G_1, \ldots, G_{m(a)}$ we consider the graphs $R_1 = \bigcup_{i=1}^{k_1} G_i$, $R_2 = \bigcup_{i=k_1+1}^{k_2} G_i, \ldots, R_s = \bigcup_{i=k_{s-1}+1}^{k_s} G_i$, where $1 \leq k_1 < k_2 < \cdots < k_s = m(a)$.

Proposition 3. *Let a_1 and a_2 be arbitrary cut-vertices of a graph G satisfying the* MPP. *Then $m(a_i) = 2$ for some i, and at least one of the graphs G_{i1} and G_{i2} defined by a_i is a path.*

PROOF. Suppose the contrary and show that there are distinct cut-vertices b_1 and b_2 such that

$$\deg_G(b_i) - \deg_{G_{i1}}(b_i) \geq 2, \tag{12}$$

where $G_{i1}, \ldots, G_{im(b_i)}$ are the graphs defined by the cut-vertex b_i, and $b_1 \in V(G_{21})$, $b_2 \in V(G_{11})$ (Fig. 7).

Indeed, let us define vertices $b_i^0 = a_i$, $i = 1, 2$. If for b_i^0, inequality (12) does not hold, then $m(b_i^0) = 2$, G_{i2} is not a path (as was supposed above concerning the vertices a_1 and a_2), and in G there is a bridge with end-vertices b_i^0 and b_i^1,

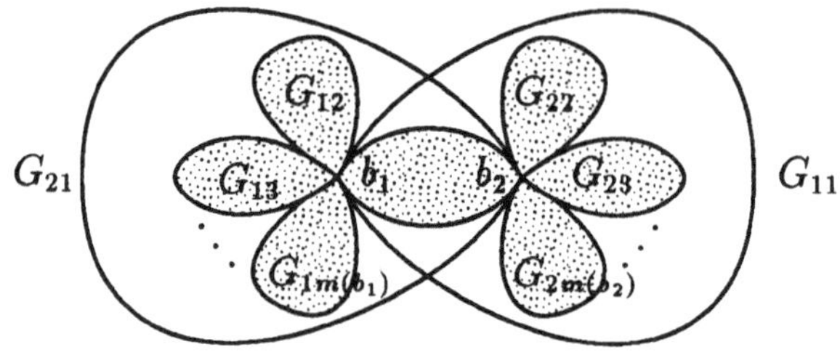

Fig. 7

where $b_i^1 \in V(G_{i2}) \setminus V(G_{i1})$. By applying the same argument to the cut-vertex b_i^1 and going consecutively from b_i^j to b_i^{j+1}, we finally obtain some distinct cut-vertices $b_1 = b_1^k$ and $b_2 = b_2^l$ that satisfy (12). In view of (12), there exist distinct vertices $c_1, c_2 \notin V(G_{11})$ adjacent to b_1. Due to the MPP, the vertices c_1 and c_2 belong to some diametric path of G. Therefore, $e_{G_{1i}}(b_1) \geq d(G)/2$ for a certain $i \geq 2$. Similarly, $e_{G_{2j}}(b_2) \geq d(G)/2$ for certain $j \geq 2$. Hence, $e_{G_{11}}(b_1) > d(G)/2$, $e_{G_{1i}}(b_1) \geq d(G)/2$, $i \geq 2$. On the other hand, $e_{G_{11}}(b_1) + e_{G_{1i}}(b_1) \leq d(G)$. A contradiction. Proposition 3 is proved.

Now, using Propositions 2 and 3, we explicitly describe the cactii that satisfy the MPP.

DEFINITION 2 [9]. A cactus is a connected graph in which no edge belong to more than one simple cycle. Cactii are also called Husimi trees [10].

To describe the cactii that satisfy the MPP, we need

Lemma 5 (cactus properties). *In any cactus:*

(a) *Each vertex of degree at least three is a cut-vertex.*

(b) *Every simple cycle is a block.*

(c) *If L_1 and L_2 are simple cycles without common vertices, then there is a simple path P with end-vertices $a_i \in V(L_i)$, $i = 1, 2$, such that $V(P) \cap V(L_i) = \{a_i\}$, and each path joining a vertex of L_1 with a vertex of L_2 contains a_1 and a_2.*

PROOF.

(a) Let x_1, x_2, and x_3 be distinct vertices adjacent to a which is not a cut-vertex. Due to [8, Theorem 3.1], for any i and j, $1 \leq i, j \leq 3$, $i \neq j$, there is a simple path connecting x_i and x_j and does not containing a. Hence, for some i the edge ax_i belongs to two simple cycles. A contradiction.

Property (b) follows immediately from (a).

(b) By the connectivity of a cactus, there is a simple path P with end-vertices $a_i \in V(L_i)$, $i = 1, 2$, such that $V(P) \cap V(L_i) = \{a_i\}$. Let now P' be any

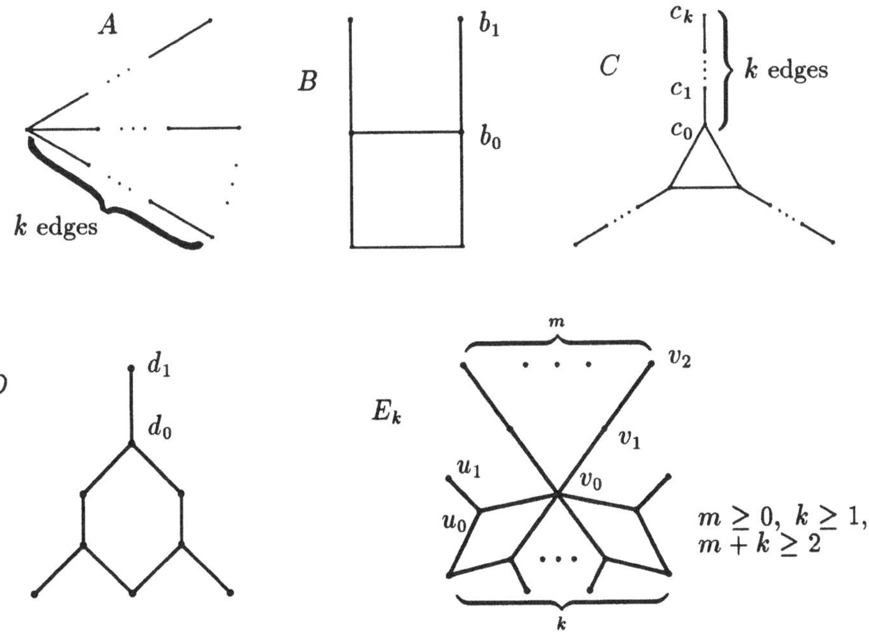

Fig. 8

path joining a vertex in L_1 with L_2, and suppose, for example, that in $a_1 \notin V(P')$. Then there is a subpath P'' of P' with end-vertices $x_i \in V(L_i)$ such that $V(P'') \cap V(L_i) = \{x_i\}$ and $a_1 \notin V(P'')$. Therefore in the simple cycle L_1 there is an edge incident with a_1 which belongs to another simple path arising on the way around the closed route $P_1 \cup P \cup P_2 \cup P''$, where $P_i \subseteq L_i$ is a path with end-vertices a_i and x_i. Lemma 5 is proved.

Theorem 4. *A cactus G satisfies* the MPP *if and only if G is either a path, or a cycle, or else G is one of the graphs in Fig.* 8.

PROOF. Let G be a cactus satisfying the MPP, and let $\mu(G)$ be the number of simple cycles in G. By induction on $\mu(G)$, we show that G has the required type. By the characterization in [3, 4] of the trees and unicyclic graphs satisfying the MPP, for $\mu(G) \leq 1$ the graph G is either a path, or a cycle, or else a graph in A, B, C, D, or E_1.

Let now $\mu(G) \geq 2$. From the above properties (a) and (b) of cactii and Proposition 3, we deduce that every two simple cycles in the cactus G have a common vertex, and hence there is a vertex a that belongs to all simple cycles and is not a cut-vertex. Let $G_1, \ldots, G_{m(a)}$ be the graphs defined by the cut-vertex a (see

Fig. 6). Each such graph contains the only block with a. By property (b), each graph G_i is a tree or a unicyclic graph. After renumbering the graphs G_i , let us suppose that $\mu(G_1) = \mu(G_2) = 1$. Define the cactii $R_1 = G_1$ and $R_2 = \bigcup_{i=2}^{m(a)} G_i$. Then by Proposition 2 and Remark 3, the cactii R'_1 and R'_2 satisfy the MPP, and

$$d(R'_1) = d(R'_2) = d(G) = e_{R_1}(a) + e_{R_2}(a), \tag{13}$$

where the notations R'_1 and R'_2 correspond to those in Proposition 2. Since $1 \leq \mu(R'_i) < \mu(G)$, by inductive assumption R'_1 and R'_2 are graphs of type B, C, D, or E_k. It is clearly seen in Fig. 8 that

- $d(B) = 3$ and $e_{B_1}(b_0) = 2$, where B_1 is the graph obtained from the graph B by deleting the vertex b_1;
- $d(C) = 2k + 1 \geq 3$ and $e_{C_1}(c_0) = k + 1 \geq 2$, where the graph C_1 is obtained from the graph C by consecutive deletion of the vertices $c_1, \dots, c_k$;
- $d(D) = 4$ and $e_{D_1}(d_o) = 3$, where the graph D_1 is obtained from D by deletion of the vertex d_1;
- $d(E_k) = 4$, $e_{E_{k,1}}(u_0) = 3$, and $e_{E_{k,2}}(v_0) = 2$, where the graph $E_{k,1}$ is obtained from E_k by deletion of the vertex u_1, and the graph $E_{k,2}$ is obtained from E_k by deletion of the vertices v_1 and v_2.

Now, using (13), it is not hard to verify that $R'_1 = E_k$, $R'_2 = E_r$ and $a = v_0$ (vertex v_0 is depicted in Fig. 8). Hence, G is a graph of type E_k.

Since the graphs depicted in Fig. 8 satisfy the MPP, Theorem 4 is proved.

References

1. A. A. Evdokimov (1988) Metric properties of embeddings, and the codes preserving distances(in Russian), in: *Modeli i Metody Optimizatsii.* Vol. 10, Trudy Inst. Mat., Nauka, Novosibirsk, pp. 116–132.
2. A. A. Evdokimov (1994) Locally isometric embeddings of graphs and the metric prolongation property, *Sibirsk. Zh. Issled. Oper.* **1**, No. 1, 5–12.
3. T. I. Fedoryaeva (1988) Characterization of a class of graphs with the metric prolongation property (in Russian), *Metody Diskret. Anal.* **47**, 89–93.
4. T. I. Fedoryaeva (1992) Strengthened properties of the metric prolongation (in Russian), *Metody Diskret. Anal.* **52**, 112–118.
5. T. I. Fedoryaeva (1995) Outerplanar graphs satisfying the metric prolongation property (in Russian), Preprint No. 3, Sobolev Institute of Mathematics, Novosibirsk.
6. V. A. Emelichev, O. I. Mel'nikov, V. I. Sarvanov, and R. I. Tyshkevich (1990) *Lectures on Graph Theory* (in Russian), Nauka, Moscow.

7. A. A. Zykov (1987) *Fundamentals of Graph Theory* (in Russian), Nauka, Moscow.

8. F. Harary (1969) *Graph Theory*, Addison-Wesley Publishing Company, Reading, Mass.

9. F. Harary and E. Palmer (1973) *Graphical Enumeration*, Academic Press, New York and London.

10. F. Harary and R. Z. Norman (1953) The dissimilarity characteristic of Husimi trees, *Ann. of Math.* **58**, No. 1, 134–141.

Sobolev Institute of Mathematics
Universitetskiĭ pr., 4
Novosibirsk 630090
RUSSIA

TRANSLATED BY O. V. BORODIN

EFFECTIVE ALGORITHMS FOR SOLVING MULTI-LEVEL PLANT LOCATION PROBLEM[†)]

È. Kh. Gimadi

Two-, three-, and multi-level plant location problems on a chain (MPLPC) are considered provided that the transportation cost of the product unit from a site to a site is equal to the sum of lengths of edges in the subchain connecting these sites. Properties of optimal solutions to the MPLPC such as connectedness of service areas, consistency and centrality of location sites are studied. Two different algorithms for exact solution of the MPLPC are proposed with time complexity $O(pm^p n)$ and $O(pm\, n^3)$, respectively, where p is the number of levels, n is the number of demand sites on the chain and m is the upper bound for the number of possible places of plant location at each level.

Introduction

Recently the researchers exhibit special interest in the multi-level plant location problems (MPLP) [1–5]. This class of problems can be characterized by presence of several levels of production in which raw material is processed before the finished product arrives to a consumer. An example of the two-level production process is the mining and the treatment of natural raw material: petroleum, ore, etc.

In general, this problem is NP-hard (even in the case of the one-level setting [6–7]). In [2–3] for solution of two- and multi-level plant location problems a branch and bound algorithm is suggested which in general is not effective [7]. Therefore, it is worth while to study special classes of the problem which can be solved by polynomial time algorithms [7–18].

In the present paper two-, three-, and multi-level plant location problems on a chain are considered provided that the transportation cost of the product unit from a site to a site is equal to the sum of lengths of edges in the subchain connecting these sites. The author is unaware of any effective algorithm for this case. Note that the polynomial algorithm for solving MPLP in a more general case of a tree-like network which is suggested in [5] is incorrect even in the case of a chain. The paper consists of six sections.

---

†) The research was supported by the Russian Foundation for Basic Research (Grant 93–01–00489) and the International Science Foundation and the Government of the Russian Federation (Grant RPY300).

A. D. Korshunov (ed.), Operations Research and Discrete Analysis, 51–69.

In the first section the problem under study is formulated as a problem of linear integer programming (with using the Boolean variables for the choice of location sites of plants and assigning the chosen plant location sites to the demand sites). As far as research is concerned, another formulation of the problem is more appropriate in which the entries of the assignment matrix are considered as variables.

From the second section we consider the plant location problem on a chain (PLPC). The properties of optimal solutions of the PLPC are studied such as connectedness of service areas and consistency of location sites. These properties appeared to be very useful in constructing exact polynomial time algorithms for the standard (one-level) simple plant location problem on a chain [13–16] and the so-called standardization problem [7–12]. To find an exact solution of the one-level location problem on a tree-like network [13], on a 2-tree [16] and on k-tree [17], some algorithms with time complexity $O(mn)$, $O(m^3n)$ and $O(m^{k+1}n)$ are suggested, where n and m are the number of nodes (demand sites) and the number of possible plant location sites, respectively.

In the third and fourth sections Algorithms A_2 and A_3 for exact solution of the two- and three-level PLPC are described. Their time complexities are equal to $O(m_1m_2n)$ and $O(m_1m_2m_3n)$, respectively, where m_r is the number of possible plant location sites on level r. As regards solution of the MPLPC, the suggested approach gives some Algorithm A_p with time complexity depending linearly on the number of demand sites and exponentially on the number of levels.

In the fifth section some Algorithm $\widetilde{A}_p$ is suggested for exact solution of the MPLPC with time complexity $O(n^3 \sum_{i=1}^{p} m_r)$. The ideas behind Algorithms A_p and $\widetilde{A}_p$ are different.

In the sixth section it is shown that in the MPLPC a central optimal solution may fail to exist at which the chosen plant location sites are situated inside the service areas.

Comparing the algorithms in the concluding remarks we show that for the two- and three-level PLPC it is preferable to use Algorithms A_2 and A_3; whereas the number of levels is greater than 4, Algorithm $\widetilde{A}_p$.

1. Statement of the Problem

The problem is formulated as follows. Let

$N = \{1, \dots, n\}$ be a set of demand sites of the finished product,

$M_r \subset N$ be the set of possible plant location sites on level r, $1 \le r \le p$;

g_i^r be the cost of location of a plant on level r at site i, $i \in M_r$, $g_i^r \ge 0$;

c_{ij} be the transportation cost of the product unit from site i to site j, where $c_{ij} \ge 0$ and $i, j \in N$;

b_j be the size of demand at site j, where $b_j > 0$ and $j \in N$.

It is assumed that each demand site of the finished product and each production site of every level receive the product only from one supplier; thus plant on level r receives the product from a plant on level $r+1$, $1 \le r \le p-1$.

The problem of choosing subsets of location sites for each level (stage) $I^r \subset M_r$, $r = 1, \dots, p$, and assign the chosen plants to the demand sites so as to minimize the total costs of location of all chosen plants and of transportation of the product.

We first present a mathematical formulation of the plant location problem (LP) in the case of two levels with use made of the following Boolean variables of choice and assignment, respectively:

- $x_i = 1\,(y_k = 1)$ if the plant on level 1 (2) is placed at site $i \in M_1$ $(k \in M_2)$, and $x_i = 0$ $(y_k = 0)$ otherwise;
- $x_{kij} = 1$ if demand site j is served by site k on level 2 through site i on level 1, and $x_{kij} = 0$ otherwise.

Using this notation the mathematical formulation of the two-level LP can be written as follows: find the minimum of the function

$$\sum_{i \in M_1} g_i^1 x_i + \sum_{k \in M_2} g_k^2 y_k + \sum_{j \in N} b_j \sum_{k \in M_2} \sum_{i \in M_1} (c_{ki} + c_{ij}) x_{kij} \tag{1}$$

subject to

$$\sum_{k \in M_2} \sum_{i \in M_1} x_{kij} = 1, \; j \in N, \tag{2}$$

$$\sum_{k \in M_2} x_{kij} \le x_i, \; j \in N, \; i \in M_1, \tag{3}$$

$$\sum_{i \in M_1} x_{kij} \le y_k, \; j \in N, \; k \in M_2, \tag{4}$$

$$x_i, y_k, x_{kij} \in \{0, 1\}. \tag{5}$$

Below it will be more convenient to use another equivalent formulation using assignment vectors as variables as it was made for the one-level network location problem [13–15].

We introduce the following notation:

$\pi^r = (\pi_1^r, \dots, \pi_n^r)$ is the plants assignment vector on level r, where π_j^r is the number of sites in M_r in which the plant on level r serving demand site j is placed, $r = 1, 2$ and $1 \le j \le n$;

$\pi = (\pi^1, \pi^2)$ is a pair of assignment vectors;

$I^r(\pi) = \bigcup_{j \in N} \{\pi_j^r\}$ is the set of plants on level r included in a solution $\pi, r = 1, 2$;

$Y_i^r(\pi)$ is the service area of plant i on level r, i.e., the union over all j such that $\pi_j^r = i$, where $i \in M_r$ and $r = 1, 2$.

It is obvious that $I^r(\pi) \subset M_r$ and $\cup Y_i^r(\pi) = N$, where the union is taken over all $i \in I^r(\pi)$, $r = 1, 2$.

The two-level LP in the terms of the variables π_j^r can be written in a more compact form: find the minimum of the function

$$\sum_{i \in I^1(\pi)} g_i^1 + \sum_{k \in I^2(\pi)} g_k^2 + \sum_{j \in N} b_j (c_{\pi_j^2 \pi_j^1} + c_{\pi_j^1 j}). \tag{6}$$

Similarly, but with the help of the entries of the $(p \times n)$-assignment matrix $\pi = (\pi_j^r)$, $1 \le r \le p$, $j \in N$, the MPLP can be represented as follows: find the minimum of the function

$$\sum_{r=1}^{p} \sum_{k \in I^r(\pi)} g_k^r + \sum_{j \in N} b_j \sum_{r=1}^{p} c_{\pi_j^r \pi_j^{r-1}}, \tag{7}$$

where p is the total number of levels; π_j^r is the number of the location sites on level r serving demand site j; $\pi_j^r \in M_r$, $1 \le r \le p$; $\pi_j^0 = j$, $j \in N$; $I^r(\pi) = \bigcup_{j \in N} \{\pi_j^r\}$ is the set of the involved plants on level r, $1 \le r \le p$; $Y_i^r(\pi)$ is the service area of plant i on level r, i.e. the union over all j such that $\pi_j^r = i$, where $i \in M_r$ and $1 \le r \le p$.

2. The Path Location Problem

Below we shall consider the chain location problem in which the demand sites $1, \ldots, n$ are connected sequentially. In this problem the transposition cost c_{ij} of the product unit from site i to site j is the sum of the lengths of the edges in the subchain connecting these sites. It is clear that in this case the function c_{ij} satisfies the triangle property:

$$c_{ij} \le c_{ik} + c_{kj} \quad \text{for all} \quad i, k, j \in N. \tag{8}$$

When constructing economical algorithms for solving some location problems the connectedness property of the service areas turned out to be useful.

• Call a solution π *r-connected* if for every $i \in I^r(\pi)$ the domain $Y_i^r(\pi)$ of demand sites served by plant i on level r is connected (in the case of the location on a chain forms an integer segment).

• A solution π for fixed r, $1 \le r \le p$, is called *r-consistent* if $\pi_1^r \le \cdots \le \pi_n^r$.

It is clear that an r-connected solution can fail to be r-consistent, whereas every r-consistent solution is r-connected. Therefore, to emphasize the importance of the notion of connectedness, further in certain cases we call a consistent solution consistent-connected.

• Call a solution $(1, r)$-*connected* (*consistent*) if it is k-connected (consistent) for every k, $1 \leq k \leq r$.

Let π be an assignment matrix. We introduce the following notation.

$r = r(\pi)$ is the minimal level at which the consistency property does not hold (for a $(1, p)$-consistent solution such parameter does not exist);

$j(\pi)$ is the greatest possible number j such that $\pi_1^r \leq \cdots \leq \pi_j^r$;

$c_j^r(\pi) = c_{\pi_j^p \pi_j^{p-1}} + \cdots + c_{\pi_j^r \pi_j^{r-1}}$, where $j \in N$ and $1 \leq r \leq p$.

We now formulate as remarks the following obvious statements.

REMARK 1. For every assignment matrix π and all $j \in N$ and r, $0 \leq r < p$,

$$c_j^r(\pi) = c_j^{r+1}(\pi) + c_{\pi_j^r \pi_j^{r-1}}.$$

REMARK 2. Let π be an optimal $(p \times n)$-assignment matrix. Also assume that π' is a $(p \times n)$-assignment matrix whose all rows consist only of entries of the corresponding rows of the matrix π. Then

$$c_j^r(\pi) \leq c_j^r(\pi'), \quad j \in N, \quad 1 \leq r \leq p.$$

Theorem 1. *There exists an optimal solution of* the MPLPC *with a set of* $(1, p)$-*consistent service areas.*

PROOF. Suppose that no $(1, p)$-consistent optimal solutions exist. Then for every optimal matrix there is $r(\pi)$, $1 \leq r(\pi) \leq p$. Take an optimal solution π with the greatest possible $r = r(\pi)$ and among such solutions the solution with the maximal $j = j(\pi)$. It is clear that $j < n$, since otherwise it would be possible to increase r.

Put

$$l = \pi_j^{r-1}, \quad k = \pi_j^r, \quad l' = \pi_{j+1}^{r-1}, \quad k' = \pi_{j+1}^r.$$

Due to $(r-1)$-consistency of the chosen solution we have

$$l \leq l', \tag{9}$$

and maximality of $j(\pi)$ implies

$$k' < k. \tag{10}$$

Let $\widehat{\pi}$ be an assignment matrix obtained from π by replacing the elements $\pi_j^r, \ldots, \pi_j^p$ by $\pi_{j+1}^r, \ldots, \pi_{j+1}^p$, respectively; let $\widetilde{\pi}$ be a similar matrix obtained by the reverse replacement.

By Remark 2 we have $c_j^r(\pi) \leq c_j^r(\widehat{\pi})$ and $c_{j+1}^r \leq c_{j+1}^r(\widetilde{\pi})$. By Remark 1 the last inequalities can be rewritten as

$$c_j^{r+1}(\pi) + c_{kl} \leq c_j^{r+1}(\widehat{\pi}) + c_{k'l}, \tag{11}$$

$$c_{j+1}^{r+1}(\pi) + c_{k'l'} \leq c_{j+1}^{r+1}(\widetilde{\pi}) + c_{kl'}. \tag{12}$$

We show that (12) cannot be strict. Indeed, summing (11), (12) and using obvious identities

$$c_j^{r+1}(\pi) = c_{j+1}^{r+1}(\widehat{\pi}), \quad c_{j+1}^{r+1}(\pi) = c_j^{r+1}(\widetilde{\pi}),$$

we come to the strict inequality

$$c_{kl} + c_{k'l'} < c_{k'l} + c_{kl'}. \tag{13}$$

For $k \leq l'$, (9) implies $k' < k \leq l'$. From the equality $c_{k'l'} = c_{k'k} + c_{kl'}$ after cancelling the identical summands we obtain the inequality

$$c_{k'k} + c_{kl} < c_{k'l},$$

which is incompatible with (8).

In the case of $k > l'$, we similarly find the impossible inequality

$$c_{k'l'} + c_{l'l} < c_{k'l}.$$

Thus, it remains to assume that (12) is an equality. This implies that the matrix π can replaced by $\widetilde{\pi}$ not increasing F^*. Since the new optimal matrix satisfies $\pi_j^r = k$, the parameter $j(\pi)$ has increased at least by 1, which contradicts its maximality. Thus, there is no level $r(\pi)$ at which there is inconsistency of the solution. Theorem 1 is proved.

Corollary 1. *There exists an optimal solution to* the MPLPC *with a set of* $(1,p)$*-consistent-connected service areas.*

• We say that the solution π satisfies the *inclusion* property if π has a system of service areas ordered by inclusion, i.e,

$$Y_{\pi_j^r}^r(\pi) \subset Y_{\pi_j^{r+1}}^{r+1}(\pi) \tag{14}$$

for all j and r, where $j \in N$ and $1 \leq r < p$.

Theorem 2. *There exists an* $(1, p)$*-consistent-connected optimal solution* of the MPLPC *satisfying the inclusion property.*

PROOF. Suppose that there is no $(1, p)$-consistent-connected optimal solution to the MPLPC satisfying the inclusion property. Let π be a consistent-connected optimal solution with the maximal possible level r at which condition (14) does not hold. Then there exists a minimal number $j = j(\pi)$, $1 < j \leq n$, such that

$$\pi_j^r = \pi_{j+1}^r = i, \quad \pi_j^{r+1} = k < \pi_{j+1}^{r+1} = k'.$$

Suppose that such solution with the greatest possible parameter $j(\pi)$ has been chosen.

In the case when $c_{j-1}^{r+1}(\pi) < c_j^{r+1}(\pi)$, the replacement of the entries $\pi_j^p, \ldots, \pi_j^{r+1}$ of the assignment matrix by $\pi_{j-1}^p, \ldots, \pi_{j-1}^{r+1}$ decreases the value of the objective function F^*, which contradicts the minimality of F^*.

If $c_{j-1}^{r+1}(\pi) \geq c_j^{r+1}(\pi)$, then replacing the elements $\pi_{j-1}^p, \ldots, \pi_{j-1}^{r+1}$ by the elements $\pi_j^p, \ldots, \pi_j^{r+1}$, in the case of strict inequality we also arrive at a contradiction with the minimality of F^*. In the case of equality, the value of the objective function is not changed, but the value of the parameter $j(\pi)$ will increase at least by 1, which contradicts its maximality. Theorem 2 is proved.

3. Algorithm A_2 for Solving the Two-Level PLPC

Let $\langle M_1, M_2; N\rangle$ be the original two-level PLPC. Consider the family of the following problems:

$$\{\langle M_1^i, M_2^k; [1, j]\rangle \mid i \in M_1, k \in M_2, \quad 1 \leq j \leq n\},$$

where $M_r^s = [1, s] \cap M_r$, $s \in M_r$, and $r = 1, 2$. (The original problem belongs to this family as well.)

Denote by $L_{ik}(j)$ the optimal value of the objective function (the optimum) of each of the defined subproblems, and by $R_{ik}(j)$ and $F_{ik}(j)$, the optima of the subproblems $\langle M_1^i, M_2^k; [1, j] \mid \pi_j^2 = k\rangle$ and $\langle M_1^i, M_2^k; [1, j] \mid \pi_j^1 = i, \pi_j^2 = k\rangle$, respectively. It is clear that $F_{ik}(1) = g_i^1 + g_k^2 + d_{ki1}$, where $d_{kij} = (c_{ki} + c_{ij})b_j$, and it is possible to take $R_{ik}(j)\,(F_{ik}(j))$ equal to ∞ for $k \notin M_2$ (for $i \notin M_1$ or for $k \notin M_2$ respectively).

It is easy to see that the optimum F^* $\langle M_1, M_2; N\rangle$ of the original problem is equal to

$$F^* = \min\{F_{ik}(n) \mid i \in M_1, k \in M_2\}.$$

From the definitions of $F_{ik}(j)$, $R_{ik}(j)$, and $L_{ik}(j)$ we immediately derive

Lemma 1. *For all $i \in M_1$, $k \in M_2$, and $j \in N$ the following recurrent relations*

$$L_{ik}(j) = \min_{k' \in M_2^k} R_{ik'}(j), \tag{15}$$

$$R_{ik}(j) = \min_{i' \in M_1^i} F_{i'k}(j) \tag{16}$$

hold.

Theorem 3. *For all $i \in M_1$ and $k \in M_2$, the values $L_{ik}(j)$, $R_{ik}(j)$, and $F_{ik}(j)$ can be calculated by using the following recurrent relations:*

$$L_{ik}(j) = \min\{L_{i,k-1}(j);\ R_{ik}(j)\},\ j \in N; \tag{17}$$

$$R_{ik}(j) = \min\{R_{i-1,k}(j);\ F_{ik}(j)\},\ j \in N; \tag{18}$$

$$F_{ik}(j) = g_i^1 + g_k^2 + d_{kij} + \min\{F_{ik}(j-1) - g_i^1 - g_k^2;$$
$$R_{i-1,k}(j-1) - g_k^2;\ L_{i-1,k-1}(j-1)\},\ 1 < j \leq n. \tag{19}$$

PROOF. (17) follows from (15):

$$L_{ik}(j) = \min_{k' \in M_2^k} R_{ik'}(j) = \min\{R_{ik}(j),$$

$$\min_{k' \in M_2^{k-1}} R_{ik'}(j)\} = \min\{L_{i,k-1}(j);\ R_{ik}(j)\}.$$

(18) follows from (16):

$$R_{ik}(j) = \min_{i' \in M_1^i} F_{i'k}(j) = \min\{F_{ik}(j);$$

$$\min_{i' \in M_1^{i-1}} F_{i'k}(j)\} = \min\{R_{i-1,k}(j);\ F_{ik}(j)\}.$$

We prove (19).

From existence of a (1,2)-consistent-connected optimal solution it follows that for $j > 1$ in the problem $\langle M_1^i,\ M_2^k;\ [1,j] \mid \pi_j^1 = i,\ \pi_j^2 = k\rangle$ there exists an interval $(j', j]$ such that each demand site is served by the plants at sites i and k (on levels 1 and 2 respectively). In addition, the optimal solution in the interval $[1, j']$ coincides with the optimal solution of at least one of the subproblems $\langle M_1^{i-1},\ M_2^{k-1};\ [1,j']\rangle$ or $\langle M_1^{i-1},\ M_2^k;\ [1,j'] \mid \pi_{j'}^2 = k\rangle$. Thus,

$$F_{ik}(j) = \min_{1 \leq j' < j} \{g_i^1 + g_k^2 + \sum_{t=j'+1}^{j} d_{kit} + \min(L_{i-1,k-1}(j');\ R_{i-1,k}(j') - g_k^2)\}.$$

Represent the minimum over j' in the interval $[1,j)$ as the minimum of two values: the expression in braces over the area $[1,j-1)$ and the same expression for $j'=j-1$:

$$F_{ik}(j) = \min\Big\{ g_i^1 + g_k^2 + d_{kij} + \min(L_{i-1,k-1}(j-1);\ R_{i-1,k}(j-1) - g_k^2);$$

$$\min_{1\le j'<j-1} \Big(g_i^1 + g_k^2 + \sum_{t=j'+1}^{j} d_{kit} + \min(L_{i-1,k-1}(j');R_{i-1,k}(j') - g_k^2)\Big)\Big\}$$

$$= \min\{g_i^1 + g_k^2 + d_{kij} + \min(L_{i-1,k-1}(j-1);\ R_{i-1,k}(j-1) - g_k^2); d_{kij} + F_{ik}(j-1)\}$$
$$= g_i^1 + g_k^2 + d_{kij} + \min\{L_{i-1,k-1}(j-1);\ R_{i-1,k}(j-1) - g_k^2;\ F_{ik}(j-1) - g_i^1 - g_k^2\}.$$

Theorem 3 is proved.

Description of Algorithm A_2 for Solving the Two-Level PLPC

Forward move: the calculation of the functions $F_{ik}(j)$, $R_{ik}(j)$, and $L_{ik}(j)$.

At the preliminary step for each $j \in N$ put the values

$$F_{0k}(j),\ R_{0k}(j),\ L_{0k}(j),\ k \in M_2, \quad \text{and} \quad F_{i0}(j),\ R_{i0}(j),\ L_{i0}(j),\ i \in M_1,$$

equal to ∞.

Step 1. Put

$$F_{ik}(1) = \begin{cases} g_i^1 + g_k^2 + d_{ki1} & \text{for } i \in M_1,\ k \in M_2, \\ \infty & \text{otherwise} \end{cases}$$

and calculate $R_{ik}(1)$ and $L_{ik}(1)$ using (17)–(18).

Step j, $1 < j \le n$. For all $i \in M_1$ and $k \in M_2$ calculate $F_{ik}(j)$, $R_{ik}(j)$, and $L_{ik}(j)$ according to (17)–(19).

Backward move: searching an optimal assignment matrix and location places of the plants on levels 1 and 2.

Preliminary step. Calculate the optimum F^* and the elements π_n^1 and π_n^2 of the optimal assignment vectors in the original problem:

$$F^* = \min_{i\in M_1,\, k\in M_2} F_{ik}(n) = F_{\pi_n^1,\,\pi_n^2}(n).$$

Repeat the following general step for $j = n, \ldots, 2$:

Step j. Put $i = \pi_j^1$; $k = \pi_j^2$; $f_1 = F_{ik}(j-1) - g_i^1 - g_k^2$; $f_2 = R_{i-1,k}(j-1) - g_k^2$; $f_3 = F_{i-1,k-1}(j-1)$; $f_0 = F_{ik}(j) - g_i^1 - g_k^2 - d_{kij}$.

Decreasing j by 1, we put

$$\pi_j^2 = \begin{cases} k & \text{for } f_0 < f_3, \\ k^0 & \text{for } f_0 = f_3, \end{cases}$$

where k^0 is the solution of the equation

$$R_{i-1,k^0}(j) = \min\{R_{i-1,k'}(j) \mid k' \in M_2^{k-1}\}.$$

By making the reassignment $k = \pi_j^2$, put

$$\pi_j^1 = \begin{cases} i & \text{for } f_0 = f_1, \\ i^0 & \text{for } f_0 < f_1, \end{cases}$$

where i^0 is the solution of the equation

$$R_{i^0,k}(j) = \min\{R_{i',k}(j) \mid i' \in M_1^{i-1}\}.$$

Theorem 4. *The time complexity of A_2 does not exceed $O(m_1 m_2 n)$.*

PROOF. At each step of the forward move the calculation of $F_{ik}(j)$, $R_{ik}(j)$, and $L_{ik}(j)$ requires a constant number of operations and the time complexity of making the general step with respect to j does not exceed $O(m_1 m_2)$. To carry out the general step of the backward move, we need $O(m_1 + m_2)$ operations.

Thus, the running time of Algorithm A_2 is mainly determined by realization of the forward move and its time complexity does not exceed $O(m_1 m_2 n)$. Theorem 4 is proved.

4. Algorithms A_3 and A_p for Solving the Three-Level and Multi-Level PLPC

As in the case of the two-level PLPC, we introduce the following notation for the three-level PLPC:

F^* is the optimum of the original problem, and for all $i \in M_1$, $k \in M_2$, $s \in M_3$, and $j \in N$;

$L_{iks}(j)$ is the optimum of the subproblem $\langle M_1^i, M_2^k, M_3^s; [1,j] \rangle$;

$R_{iks}(j)$ is the optimum of the subproblem $\langle M_1^i, M_2^k, M_3^s; [1,j] \mid \pi_j^3 = s \rangle$;

$Q_{iks}(j)$ is the optimum of the subproblem $\langle M_1^i, M_2^k, M_3^s; [1,j] \mid \pi_j^2 = k, \pi_j^3 = s \rangle$;

$F_{iks}(j)$ is the optimum of the subproblem $\langle M_1^i, M_2^k, M_3^s; [1,j] \mid \pi_j^1 = i, \pi_j^2 = k, \pi_j^3 = s \rangle$.

From these definitions we have

Lemma 2. *For all $i \in M_1$, $k \in M_2$, $s \in M_3$, and $j \in N$ the recurrent relations*

$$L_{iks}(j) = \min_{s' \in M_3^s} R_{iks'}(j),$$

$$R_{iks}(j) = \min_{k' \in M_2^k} Q_{ik's}(j),$$

$$Q_{iks}(j) = \min_{i' \in M_1^i} F_{i'ks}(j)$$

hold.

We present the statements similar to Theorems 3 and 4 without proof.

Theorem 5. *For all $i \in M_1$, $k \in M_2$, and $s \in M_3$ the values $L_{iks}(j)$, $R_{iks}(j)$, $Q_{iks}(j)$, and $F_{iks}(j)$ can be calculated by using the following recurrent relations:*
$L_{iks}(j) = \min\{L_{ik,s-1}(j); R_{iks}(j)\}$, $j \in N$;
$R_{iks}(j) = \min\{R_{i,k-1,s}(j); Q_{iks}(j)\}$, $j \in N$;
$F_{iks}(1) = g_i^1 + g_k^2 + g_s^3 + d_{ski1}$;
$F_{iks}(j) = g_i^1 + g_k^2 + g_s^3 + d_{skij} + \min\{L_{i-1,k-1,s-1}(j-1); R_{i-1,ks}(j-1) - g_s^3; Q_{iks}(j-1) - g_k^2 - g_s^3; F_{iks}(j-1) - g_i^1 - g_k^2 - g_s^3\}$, $1 < j \le n$, *where* $d_{skij} = (c_{sk} + c_{ki} + c_{ij})b_j$.

Theorem 6. *The three-level* PLPC *can be solved in $O(m_1 m_2 m_3 n)$ time.*

The idea of construction of an algorithm for solving the two- and three-level PLPC can be extended to the MPLPC. Thus, we obtain some Algorithm A_p for solving the PLPC running in time $O(npm_1m_2 \dots m_p)$ which does not exceed $O(pm^pn)$, where $m = \max\{m_r \mid 1 \le r \le p\}$. Thus, in this way we can construct an algorithm with time complexity depending linearly on the number of the demand sites and exponentially on the number of levels. Moreover, the bounds on time and space complexities coincide.

5. A Polynomial Time Algorithm for Solving the MPLPC

In this section we describe Algorithm $\widetilde{A}_p$ with time complexity $O(n^3 \sum_{r=1}^{p} m_r)$.

The methods of constructing Algorithms A_p and $\widetilde{A}_p$ differ principally from each other. Algorithm $\widetilde{A}_p$ makes essential use of the inclusion property of optimal solutions and of reduction to a special series of the nearest neighbor problems [7–10].

By the nearest neighbor problem we shall mean the following location problem of sites on the integer segment $(0,\ n]$: minimize a function

$$\sum_{s=1}^{m} f(x_{s-1}, x_s) \tag{20}$$

subject to

$$0 = x_0 < \cdots < x_m = n, \tag{21}$$

$$m \geq 1. \tag{22}$$

We denote the nearest neighbor problem (20)–(22) by $\langle 0, n;\ f\rangle$ and the optimal value of the objective function by $S(0, n;\ f)$. The subproblems of the nearest neighbor problem and their optima are denoted by $\langle x, y;\ f\rangle$ and $S(x, y;\ f)$, respectively, $0 \leq x < y \leq n$. The solution of the nearest neighbor problem can be found in time $O(n^2)$ [9].

Let $\langle p;\ N\rangle$ denote the original MPLPC, and let F^* be the optimum of this problem. For all r, x, and y such that $2 \leq r \leq p$, $0 \leq x < y \leq n$, we shall consider the following subproblems of the MPLPC:

$$\langle r;\ (x, y] \mid \pi^r_{x+1} = \cdots = \pi^r_y = k\rangle, \tag{23}$$

$$\langle r;\ (x, y] \mid \pi^r_{x+1} = \cdots = \pi^r_y = k;\ \pi^{r-1}_{x+1} = \cdots = \pi^{r-1}_y;\ g^r_k = 0\rangle. \tag{24}$$

Denote by $F^r_k(x, y)$ and $f^r_k(x, y)$ the minimal values of the objective functions of the subproblems, respectively.

For their computation we shall use the recurrent relations of dynamic programming:

$$f^r_k(x, y) = \min_{i \in M_{r-1}} \{B(x, y)\, c_{ki} + F^{r-1}_i(x, y)\}, \tag{25}$$

$$F^r_k(x, y) = g^r_k + S(x, y;\ f^r_k), \tag{26}$$

where $B(x, y) = \sum_{j=x+1}^{y} b_j$, $0 \leq x < y \leq n$, $k \in M_r$, and $1 < r \leq p$.

Thus, the subproblem (23) on the interval $(x, y]$ splits into a sequence of subproblems of the form (24) on the intervals $(x_{s-1}, x_s]$, $s = 1, \ldots, m$.

Theorem 7. *The optimum F^* of the original problem $\langle p;\ N\rangle$ coincides with the optimum of the nearest neighbor problem $\langle 0, n;\ f\rangle$ with the function*

$$f(x, y) = \min_{k \in M_p} F^p_k(x, y). \tag{27}$$

PROOF. Let π be an consistent-connected optimal solution satisfying the inclusion property. On level p it has the set of plants $I^p(\pi) = \{k_s(\pi) \mid s = 1, \dots, m(\pi)\}$ and the service areas $Y^p_{k_s(\pi)} = (x_{s-1}(\pi), x_s(\pi)], s = 1, \dots, m(\pi),\ 0 = x_0(\pi) < \dots < x_{m(\pi)}(\pi) = n$.

From the inclusion property it follows that the optimal solution of the original problem $\langle p;\ N = [1, n] \rangle$ is the sum of the optimal solutions $m(\pi)$ of the subproblems $\langle x, y;\ f^r_k \rangle$, where $r = p, x = x_{s-1}(\pi),\ y = x_s(\pi),\ k = k_s(\pi)$, and $s = 1, \dots, m(\pi)$, i.e.

$$F^* = \sum_{s=1}^{m(\pi)} F^p_{k_s(\pi)}\big(x_{s-1}(\pi), x_s(\pi)\big). \tag{28}$$

We mark by the tilde the values of the variables at which (20) attains the minimum:

$$\sum_{s=1}^{\tilde{m}} F^p_{\tilde{k}_s}\left(\tilde{x}_{s-1}, \tilde{x}_s\right)$$
$$= \min\Big\{\sum_{s=1}^{m} \min_{k_s \in M_p} F^p_k(x_{s-1}, x_s) \,\big|\, 0 = x_0 < \dots < x_m = n,\ m \geq 1\Big\}. \tag{29}$$

Thus, it suffices to show that the left-hand sides in (28) and (29) are equal. On the one hand, we have the obvious inequality

$$F^* \leq \sum_{s=1}^{\tilde{m}} F^p_{\tilde{k}_s}\left(\tilde{x}_{s-1}, \tilde{x}_s\right). \tag{30}$$

On the other hand, from (29) it follows that for all k_s and x_s, where $1 \leq s \leq m \leq n$, the inequality

$$\sum_{s=1}^{\tilde{m}} F^p_{\tilde{k}_s}\left(\tilde{x}_{s-1}, \tilde{x}_s\right) \leq \sum_{s=1}^{m} F^p_{k_s}\left(x_{s-1}, x_s\right)$$

holds. Since the last inequality is also valid for $m = m(\pi)$, $k_s = k_s(\pi)$, and $x_s = x_s(\pi)$, where $s = 1, \dots, m(\pi)$, we have

$$\sum_{s=1}^{\tilde{m}} F^p_{\tilde{k}_s}\left(\tilde{x}_{s-1}, \tilde{x}_s\right) \leq \sum_{s=1}^{m(\pi)} F^p_{k_s(\pi)}\big(x_{s-1}(\pi), x_s(\pi)\big).$$

Taking (28) and (30) into account, we see that the claim is true. Theorem 7 is proved.

Description of Algorithm $\widetilde{A}_p$ for Solving the MPLPC

Forward move: In the cycle over r from 2 to p repeat Steps 1a and 2a.

Step 1a. For every $k \in M_r$, using (25) compute $f_k^r(x, y)$, $0 \le x < y \le n$.

Step 2a. For every $k \in M_r$, solve the nearest neighbor problem $\langle x, y;\ f_k^r \rangle$, optimizing the number of the intervals by using the following recurrent relations:

$$(x, y;\ f_k^r) = \min_{x \le x' < y} \{S(x, x';\ f_k^r) + f_k^r(x', y)\}, \tag{31}$$

and put $F_k^r(x, y) = g_k^r + S(x, y;\ f_k^r)$, $0 \le x < y \le n$.

Step 3a. Solve the nearest neighbor problem of the upper level

$$F^* = \min\left\{\sum_{s=1}^{m} f(x_{s-1}, x_s) \mid 0 \le x_0 < \cdots < x_m = n,\ m \ge 1\right\}$$

with the function f defined by (27).

Denote by $\widetilde{m}_p$ the found optimal number of plants on level p.

Step 4a. Compute the numbers of plants on level p

$$k_s^p = \arg\min_{k \in M_r} \{F_k^p(x_{s-1}, x_s) \mid \in M_p\}, \quad 1 \le s \le \widetilde{m}_p,$$

where (x_s^p) is the optimal partition of the sets of demand sites by the service areas of the plants on level p, $1 \le s \le \widetilde{m}_p$, obtained at the previous step.

Backward move: in the cycle over $r = p - 1, \dots, 1$ repeat Steps 1b and 2b.

Step 1b. Successively solving the problems $\langle x_{s-1}^{r+1}, x_s^{r+1};\ f_k^{r+1} \rangle$, $k \in I^{r+1}$ (π), determine the optimal number $\widetilde{m}_r$ of plants on level r and the bounds (x_i^r) (increasing orders) of their areas of service, $i = 1, \dots, \widetilde{m}_r$.

Step 2b. For every area $(x_{s-1}^r, x_s^r] \in Y_k^{r+1}$, $k \in I^{k+1}$, find the corresponding optimal plant on level r by the relation

$$k_s^r = \arg\min_{i \in M_r} \{B(x_{s-1}^r, x_s^r)c_{ki} + F_i^r(x_{s-1}^r, x_s^r)\}, \quad 1 \le s \le \widetilde{m}_r.$$

Algorithm $\widetilde{A}_p$ terminates.

Theorem 8. *The time and space complexity of Algorithm $\widetilde{A}_p$ does not exceed $O\left(n^3 \sum_{r=1}^{p} m_r\right)$ and $O\left(n^2 \sum_{r=1}^{p} m_r\right)$, respectively.*

PROOF. In the forward move of the algorithm the most laborious part is the search of the values $S(x, y;\ f_k^r)$ at Step 2a. With given $x \in [0, n)$, $k \in M_r$, and $r, 1 < r \leq p$, the problem $\langle x, y;\ f_k^r \rangle$ with the fixed right-hand endpoint $y = n$ can be solved in time not exceeding $O(n^2)$ [9] with the help of the recurrent relations (31). Simultaneously the whole collection of the values $\{S(x, y;\ f_k^r),\ x < y \leq n\}$ is calculated. Hence, at Step 2a for fixed r, $1 < r \leq p$, the time complexity of finding the family

$$\{S(x, y;\ f_k^r),\ x < y \leq n\}, \quad k \in M_r, \quad 0 \leq x < n,$$

does not exceed $O\big(n^3 m_r\big)$.

The forward move is executed for every $r, 2 \leq r \leq p$, and runs in time $O(n^3 \sum_{r=1}^{p} m_r)$. This bound remains valid as well for the time complexity of Algorithm $\widetilde{A}_p$, since the running time of the backward move is essentially less. The upper bound for the space complexity of the algorithm follows from the necessity of stories the families of functions $\{F_k^r(x, y),\ f_k^r(x, y),\ S(x, y;\ f_k^r)\}$, $0 \leq x < y \leq n$, $k \in M_r$, $1 < r \leq p$. Theorem 8 is proved.

6. On the Centrality Property of a Solution to the MPLPC

In [14] it is shown that the one-level PLPC has optimal solutions with the so-called centrally-connected service areas. In this case the located plants lie inside their service areas and this fact can be used to decrease exhaustion of all possible sites of plants. We give the definition of the centrality property in the case of the MPLPC and show that it is possible to find optimal solutions in which only the plants on level 1 are central. This is not true for plants of higher levels.

- Call a solution π *r-central* if $i \in Y_i^r(\pi)$ for every $i \in I^r(\pi)$.
- Call a solution π *$(1, r)$-central* if it is k-central for every k, $1 \leq k \leq r$.

Proposition 1. *There exist an optimal solution to* the MPLPC *with a set of* 1*-central service areas.*

PROOF. Let π be an consistent optimal solution satisfying the inclusion property, let $I^1(\pi) = \{i_1, \ldots, i_{\widetilde{m}_1}\}$, $i_1 < i_2 < \cdots < i_{\widetilde{m}_1}$, and let s be the maximal number such that $i_1 \in Y_{i_1}^1(\pi), \ldots, i_s \in Y_{i_s}^1(\pi)$.

We assume that a 1-consistent solution with the minimal parameter $\widetilde{m}_1$ is chosen, and if the number of such solutions is more than one, we choose among them the solution with the maximal parameter s, $s \leq \widetilde{m}_1$. To prove the claim, it suffices to show that $s = \widetilde{m}_1$.

Assume to the contrary that $s < \widetilde{m}_1$. Put $i' = i_{s+1}$ and $Y' = Y_{i'}^1(\pi)$. It is clear that $i' \notin Y'$.

By 1-consistency of the solution, between site i' and the area Y' there are no other sites $i \in I^1(\pi)$. Let i be the plant nearest to Y' such that i' lies on the chain from i to Y', and let l be the demand site to Y' in the area $Y_i^1(\pi)$. It is obvious that $\pi_l^1 = i$.

Let j' be a site in Y', $k' = \pi_{j'}^2$, and $i' = \pi_{j'}^1$. Due to the inclusion property, we have $Y' \subset Y_{k'}^2(\pi)$ and $\pi_j^2 = k'$, $j \in Y'$.

Optimality of the solution π and Remark 2 imply

$$c_l^1(\pi) \leq c_l^1(\widetilde{\pi}), \quad c_j^1(\pi) \leq c_j^1(\widehat{\pi}), \tag{32}$$

where $\widetilde{\pi}$ is obtained from π by replacement of the column π_l by the column $\pi_{j'}$, and $\widehat{\pi}$ by replacement of each column π_j, $j \in Y'$, by the column π_l. By Remark 1 the last inequality implies

$$c_l^2(\pi) + c_{il} \leq c_l^2(\widetilde{\pi}) + c_{i'l}, \tag{33}$$

$$c_j^2(\pi) + c_{i'j} \leq c_j^2(\widehat{\pi}) + c_{ij}, \quad j \in Y'. \tag{34}$$

The inequality (34) cannot hold with equality for each $j \in Y'$. Otherwise, it would be possible to proceed to the matrix $\widehat{\pi}$ without increasing the value of the objective function, while decreasing the parameter $\widetilde{m}_1$ at least by 1. Therefore, there exists $j' \in Y'$ for which the inequality (34) is strict. Summing the inequalities (33) and (34) for $j = j'$, we obtain

$$c_{il} + c_{i'j'} < c_{i'l} + c_{ij'}.$$

With regard to the equality $c_{il} = c_{ii'} + c_{i'l}$ which is true due to the location of i' between sites i and l, we arrive at a contradiction with the triangle inequality:

$$c_{ii'} + c_{i'j'} < c_{ij'}.$$

Proposition 1 is proved.

Corollary 2. *There exists an optimal solution to* the MPLPC *with a set of 1-centrally-connected service areas.*

Proposition 2. *An optimal solution* of the MPLPC *with a set of* $(1,p)$-*central service areas may fail to exist.*

PROOF. The validity of the statement follows from the next smallest instance of the two-level PLPC in which the 2-central optimal solution does not exist.

The set of demand sites: $N = \{1,2,3\}$.

The volumes of demand: $b_j = 1$, $j \in N$.

The lengths of edges: 1, 1.

The list of the plants on level 1: $M_1 = \{1,3\}$.
The list of the plants on level 2: $M_2 = \{1,2\}$.
The cost of location of plants: $g_i^1 = \varepsilon$, $i \in M_1$, and $g_k^2 = \varepsilon$, $k \in M_2$, $0 \le \varepsilon < 0,5$.

1		1	(edge weights)
1........................2		3	(chain $N = \{1,2,3\}$)
1,2	2	1	(levels of the plants)

In the given example there is a unique optimal solution with the following assignment matrix:

j	1	2	3
π_j^1	1	1	3
π_j^2	1	1	2

The optimal value of the objective function is $2 + 4\varepsilon$.

It is obvious that 1- and 2-consistency (and hence, connectedness) and 1-centrality hold. The property of 2-centrality is violated, since the plant on level 2 located at site 2 does not belong to its service area $Y_2^2(\pi) = \{3\}$. Every other assignment matrix in the given example gives a larger value of the objective function. Proposition 2 is proved.

7. Concluding Remarks

From the bounds $O(pnm_1m_2 \dots m_p)$ and $O\left(n^3 \sum_{r=1}^{p} m_r\right)$ for time complexity of the suggested algorithms it follows that A_p runs faster than $\widetilde{A}_p$ if the parameters of the problem are such that the inequality

$$\sqrt{\frac{pm_1 \dots m_p}{m_1 + \dots + m_p}} \le n$$

holds. Setting $m = \max\{m_r \mid 1 \le r \le p\}$, for the time complexities of A_p and $\widetilde{A}_p$ we have the upper bounds $O(m^p\, n)$ and $O(p\, m\, n^3)$. From this it is clear that in the case of two- and three-level PLPC, Algorithms A_2 and A_3 are more efficient than the corresponding analogs $\widetilde{A}_2$ and $\widetilde{A}_3$. If the number of levels is greater than 4, Algorithm $\widetilde{A}_p$ is preferable. Moreover, the space complexity of A_p (unlike $\widetilde{A}_p$) grows exponentially with the number of levels.

The question on existence of polynomial time algorithms for solving the MPLPC on trees remains open and requires further research.

References

1. L. Kaufman, M. V. Eede and P. A. Haunsen (1977) A plant and warehouse location problem, *Oper. Res. Quart.* **28**, No. 3, 547–554.
2. H. B. Ro and D. W. Tcha (1984) A branch and bound algorithm for the two-level uncapacitated facility location problem with some side constraints, *European J. Oper. Res.* **18**, No. 3, 349–358.
3. D. W. Tcha and B. I. Lee (1984) A branch and bound algorithm for the multi-level uncapacitated facility location problem, *European J. Oper. Res.* **18**, No. 1, 35–43.
4. V. S. Mikhalevich, V. A. Trubin, and N. Z. Shor (1986) Optimization Problems of Production-Transportation Planning (in Russian), Nauka, Moscow.
5. V. A. Trubin and F. A. Sharifov (1992) The simplest multi-level location problem on a tree-like network (in Russian), *Kibernet. Sistem. Anal.* (Kiev), No. 6, 128–135.
6. M. R. Garey and D. S. Johnson (1979) *Computers and Intractability*, Freeman, San Francisco.
7. V. L. Beresnev, È. Kh. Gimadi, and V. T. Dement′ev (1978) *Extremal Standardization Problems* (in Russian), Nauka, Novosibirsk.
8. È. Kh. Gimadi and N. I. Glebov (1982) *Extremal Problems of Making Solutions* (in Russian), Novosibirsk University, Novosibirsk.
9. È. Kh. Gimadi (1970) Choice of optimal scales in a class of location, unification, and standardization problems (in Russian), *Upravlyaemye Sistemy* **6**, 57-70.
10. È. Kh. Gimadi and V. T. Dement′ev (1973) Certain problems in the selection of optimal parametric series and methods for their solution (standardization problems) (in Russian), in: *Problemy Kibernetiki.* Vol. 27, Nauka, Moscow, pp. 19–32.
11. V. L. Beresnev (1979) Algorithms for the minimization of polynomials of Boolean variables (in Russian), in: *Problemy Kibernetiki.* Vol. 36, Nauka, Moscow, pp. 225–246.
12. È. Kh. Gimadi (1987) A standardization problem with data of arbitrary sign and with connected quasiconvex and almost quasiconvex matrices (in Russian), *Upravlyaemye Sistemy* **27**, 3–11.
13. È. Kh. Gimadi (1983) An efficient algorithm for solution of a distribution problem with servicing areas that are connected in relation to an acyclic network (in Russian), *Upravlyaemye Sistemy* **23**, 12–23.
14. È. Kh. Gimadi (1984) The problem of distribution on a network with centrally connected service areas. (in Russian), *Upravlyaemye Sistemy* **25**, 38–47.

15. È. Kh. Gimadi (1988) Some mathematical models and methods for the planning of large-scale projects (in Russian), in: *Modeli i Metody Optimizatsii*, Trudy Inst. Math. Vol. 10, Nauka, Novosibirsk, pp. 89-115.

16. A. A. Ageev (1990) A polynomial algorithm for solving the location problem on a series-parallel network (in Russian), *Upravlyaemye Sistemy* **30**, 3–16.

17. A. A. Ageev (1992) A criterion of polynomial-time solvability for the network location problem. *Integer Programming and Combinatorial Optimization.* Proc. IPCO II Conf. Campus Printing. Carnegie Mellon University, pp. 237–245.

18. D. Granot and D. Skorin-Kapov (1994) On some optimization problems on k-trees and partial k-trees, *Discrete Appl. Math.* **48**, No. 2, 129–145.

Sobolev Institute of Mathematics
Universitetskiĭ pr., 4
Novosibirsk 630090
RUSSIA

TRANSLATED BY I. V. RENTTS

ON THE COMPLEXITY OF REALIZATION OF BOOLEAN FUNCTIONS BY CIRCUITS OVER AN INFINITE BASIS[†)]

O. M. Kasim-Zade

We study the complexity of realization of Boolean functions by circuits of functional gates over the infinite basis AC consisting of all antichain Boolean functions, i.e., the functions taking the value 1 only at pairwise incomparable tuples. It is shown that as $n \to \infty$ the complexity of realization of a linear function of n arguments over the basis AC grows at least as $(n/\log n)^{1/2}$. It is established that the maximal complexity of realization of Boolean functions of n arguments over the basis AC grows at least as $(n/\log n)^{1/2}$ and at most as n.

A Boolean function taking the value 1 only at pairwise incomparable tuples is an *antichain function*. Denote by AC the set of antichain functions. It is easy to verify that AC is functionally complete and closed under the operations of substituting constants and identifying variables.

Consider the realization of Boolean functions by circuits of functional gates over the infinite basis AC. The *complexity of a circuit* is understood to be the number of gates in the circuit. Introduce the notation:

$L(S)$ is the complexity of a circuit S;

$L(f)$ is the minimal complexity of realization of a function f by circuits over the basis AC;

$L(n) = \max L(f)$, where the maximum is taken over functions of n arguments.

In [1], the following bounds are obtained

$$n^{1/3} \preccurlyeq L(n) \leq n+1, \quad n \to \infty.$$

The lower bound here follows from the lower bound $L(l_n) \succcurlyeq n^{1/3}$ established in [1] on the complexity of realization of the linear function $l_n = x_1 \oplus \ldots \oplus x_n$ of n arguments.

In this paper, an improved lower bound is obtained on the complexity of realization of linear functions over the basis AC.

---

†) This research was supported by the Russian Foundation for Basic Research (Grant 93-01-01527).

A. D. Korshunov (ed.), Operations Research and Discrete Analysis, 71–84.

Theorem A. *As $n \to \infty$,*

$$L(l_n) > (n/\log n)^{1/2}.$$

Combining the lower bound of Theorem A with the upper bound of [1], we arrive at the following theorem.

Theorem B. *As $n \to \infty$,*

$$(n/\log n)^{1/2} \preccurlyeq L(n) \leq n + 1.$$

The purpose of this paper is to prove Theorem A. We apply a variant of the well-known method of substituting constants which is usually used for obtaining lower bounds over finite bases. Consider a circuit S over the infinite basis AC which realizes the linear function l_n, take a positive integer s, $s \leq n$, and proceed as follows.

> If the circuit S contains a gate whose number of inputs connected to the inputs of the circuit is less than s, then substitute the constant 0 for such inputs; in addition, the number of inputs of the circuit decreases by at most $s - 1$. As a result, we obtain a new circuit S_1 which realizes a linear function of $n_1 > n - s + 1$ arguments.

By construction, the number of gates in the circuit S_1, whose inputs are connected to the inputs of the circuit, is less than that for the original circuit S by at least 1.

Apply this procedure to the resulting circuit S_1 and proceed so as long as possible. If it is possible k times, then $L(S) \geq k$. If k is large, then it yields a rather high lower bound on the complexity of the original circuit S. If k is small, then the circuit S_k constructed at the kth step realizes a linear function of $m \geq n - k(s-1)$ arguments, and m being rather large in this case. Moreover, S_k is s-regular (see definition below).

To estimate the complexity of s-regular circuits, we use a different method. We assume that a circuit realizing a function f also realizes every partial function which coincides with f on its domain. Given an s-regular circuit realizing a linear function of m arguments, we determine the partial functions that are realized by this circuit.

In this process, each subsequent reduction of the domain is carried out so that at least 1 of the gates in the circuit which realized a partial function other than a constant on the previous domain realizes a function equal to a constant on the new domain. In view of the specifics of the basis AC and s-regularity of the circuit, the number of such steps is large compared to m. Hence, the complexity of the circuit considered is sufficiently great.

The proof is divided into three parts:

— an applied variant of the method for substituting constants is described;

— combinatorial lemmas and auxiliary propositions on Boolean functions are proved to be used in studying s-regular circuits;

— s-regular circuits are studied and the main result is proved.

1. Circuits over the Basis AC and Substituting Constants

A circuit S is called *reduced* if a function other than a constant is realized at each output of the gate, and different inputs are connected to different vertices of S.

Denote by $P(S)$ the number of the gates in S at least 1 input of which is connected to the inputs of the circuit. Clearly, $L(S) \geq P(S)$ for each circuit S.

Since the basis AC is closed with respect to the operations of substituting constants and identifying variables, we have

Lemma 1.1. *For each circuit S over the basis AC realizing a function f other than a constant, there exists a reduced circuit S' over AC realizing f such that $L(S') \leq L(S)$ and $P(S') \leq P(S)$.*

Let $s \in \mathbb{N}$. A reduced circuit S is *s-regular* if, for each gate e in S, the number of inputs of the circuit connected to the inputs of e either equals 0, or is at least s.

Lemma 1.2. *Let n and s be integers, $n \geq 2$, $s \geq 2$, and assume that a reduced circuit S over the basis AC realizes the linear function l_n of n arguments and is not s-regular. Then there exist a nonnegative integer n_1, $n - (s-1) \leq n_1 \leq n$, and a reduced circuit S_1 over AC realizing the function l_{n_1} such that $L(S_1) \geq L(S)$ and $P(S_1) \leq P(S) - 1$.*

PROOF. Since S is not s-regular and $n \geq 2$, there exists a gate e in S whose inputs are connected to at most $s-1$ inputs of S. Let r be the number of these inputs in the circuit, $1 \leq r \leq s-1$, and let $x_{i_1}, \ldots, x_{i_r}$ be the variables assigned to them. Substitute the constants $x_{i_r} = \ldots = x_{i_r} = 0$ into the function l_n and substitute in S the constant 0 for the inputs to which $x_{i_1}, \ldots, x_{i_r}$ are assigned. As a result, the function l_n transforms into a function l_{n_1} of the remaining $n_1 = n - r$ variables, where $n_1 > n - (s-1)$, and the circuit S transforms into a circuit S' realizing l_{n_1}. Since e now has no inputs connected to the inputs of S', we obtain $P(S') < P(S) - 1$.

Note that the circuit S is not reduced. Using Lemma 1.1, pass from S' to the corresponding reduced circuit S_1 realizing the same function l_{n_1}. It is easy to see that $L(S_1) \leq L(S)$ and $P(S_1) \leq P(S') \leq P(S) - 1$. Lemma 1.2 is proved.

Lemma 1.3. *Let n, n_0, and s be integers such that $n \geq 2$, $n > n_0 \geq 1$, and $s \geq 2$. Then $L(l_n) \geq (n - n_0)/(s - 1)$ or, for some m, $n_0 < m \leq n$, there exists an s-regular circuit S_0 realizing the function l_m such that $L(l_n) \geq L(S_0)$.*

PROOF. Let S be a reduced circuit which realizes the function l_n and is minimal with respect to the number of gates. By virtue of Lemma 1.1, we have $L(S) = L(l_n)$. If S is s-regular, then set $m = n$ and $S_0 = S$.

If S is not s-regular, then, according to Lemma 1.2, construct a reduced circuit S_1 realizing a function l_{n_1} of $n_1 \geq n - (s - 1)$ arguments such that $L(S_1) \leq L(S)$ and $P(S_1) \leq P(S) - 1$.

If $n_1 \leq 1$ or S_1 is s-regular, then the construction is finished.

If $n_1 \geq 2$ and S_1 is not s-regular, then, according to Lemma 1.2, construct a reduced circuit S_2 that realizes a function l_{n_2} of $n_2 \geq n_1 - (s-1) \geq n - 2(s-1)$ arguments such that $L(S_2) \leq L(S_1) \leq L(S)$ and $P(S_2) \leq P(S_1) - 1 \leq P(S) - 2$ and so on.

After k steps, this process results in a reduced circuit S_k realizing a function l_{n_k} of $n_k \geq n - k(s-1)$ arguments such that $L(S_k) \leq L(S)$ and $P(S_k) \leq P(S) - k$, with either $n_k \leq 1$ or the circuit S_k being s-regular.

If $n_k \leq n_0$, then $L(l_n) = L(S) \geq P(S) \geq P(S_k) + k \geq k$ and since $n_k > n - k(s-1)$, we have $k > (n - n_0)/(s-1)$. Hence, $L(l_n) \geq (n - n_0)/(s-1)$.

If $n_k > n_0$, then $n_k \geq 2$ and the circuit S_k is s-regular. Setting $m = n_k$ and $S_0 = S_k$, we obtain $L(l_n) = L(S) \geq L(S_k) = L(S_0)$. Lemma 1.3 is proved.

2. Combinatorial Lemmas

The *n-dimensional unit cube* B_2^n is the set of all n-tuples of 0s and 1s.

A *chain* in the cube B_2^n is a set C, $C \subseteq B_2^n$, in which any two tuples are comparable.

The *length of a chain* is the number of the tuples in the chain.

Any chain C contains

— the *least tuple*, i.e., the tuple $\widetilde{\alpha} \in C$ such that $\widetilde{\alpha} \leq \widetilde{\beta}$ for every $\widetilde{\beta} \in C$,

— the *greatest tuple*, i.e., the tuple $\widetilde{\gamma} \in C$ such that $\widetilde{\beta} \leq \widetilde{\gamma}$ for every $\widetilde{\beta} \in C$.

The least and the greatest tuples of a chain C are denoted by $m(C)$ and $M(C)$, respectively.

We call a chain *dense* if any two consecutive (under the natural ascending order) tuples differ in exactly one component.

The binary n-tuple consisting of 0s is denoted by $\widetilde{0}^n$ and that consisting of 1s only, by $\widetilde{1}^n$. A dense chain containing the tuples $\widetilde{0}^n$ and $\widetilde{1}^n$ is called *maximal*. Each maximal chain in B_2^n has length $n+1$. The number of distinct maximal chains in B_2^n is $n!$.

The *weight of a tuple* $\widetilde{\alpha}$ is the number of its components equal to 1 and is denoted by $\|\widetilde{\alpha}\|$.

Let $\widetilde{\alpha}$ and $\widetilde{\beta}$ be two tuples in B_2^n such that $\widetilde{\alpha} \leq \widetilde{\beta}$. The set of tuples $\widetilde{\gamma} \in B_2^n$ satisfying $\widetilde{\alpha} \leq \widetilde{\gamma} \leq \widetilde{\beta}$ is called an *interval in the cube* B_2^n. The *range of an interval* is the number of coinciding components in the tuples $\widetilde{\alpha}$ and $\widetilde{\beta}$ that define the interval. By the *weight of an interval* we mean the weight of the tuple $\widetilde{\alpha}$. Obviously, the weight of an interval is not greater than its range.

It is well-known that there is a one-to-one correspondence between the intervals of the cube B_2^n and the elementary conjunctions in the variables $x_1, \ldots, x_n$: to each interval I in B_2^n of range r there corresponds an elementary conjunction of the form $x_{i_1}^{\sigma_1} \ldots x_{i_r}^{\sigma_r}$, where $1 \leq i_1 < \ldots < i_r \leq n$ and $\sigma_1, \ldots, \sigma_r \in \{0,1\}$. This conjunction takes the value 1 at those and only those tuples in B_2^n that belong to I. It is easily seen that the weight of the interval I is equal to the weight of the tuple $\widetilde{\sigma} = (\sigma_1, \ldots, \sigma_r)$.

We need several lemmas on the number of maximal chains which have a given size of the intersection with given intervals of the n-dimensional cube.

Lemma 2.1. *Let n, r, k, and a be integers such that $0 \leq k \leq r \leq n$ and $0 \leq a \leq n-r$, and let I be an interval in the cube B_2^n which is of range r and weight k. Then the number of maximal chains in B_2^n having exactly $a+1$ common vertices with I is equal to*

$$\binom{n-a-1}{r-1}\binom{r}{k}^{-1}\binom{n}{r}^{-1} n!. \tag{2.1}$$

PROOF. Let C be a maximal chain having exactly $a+1$ common vertices with I. Clearly, the intersection $C \cap I$ is a dense chain of length $a+1$. Let $\widetilde{\alpha} = m(C \cap I)$ and $\widetilde{\beta} = M(C \cap I)$ be the least and the greatest tuples of this chain, respectively. Then $\widetilde{\alpha}$ and $\widetilde{\beta}$ satisfy the following conditions:

$$\widetilde{\alpha}, \widetilde{\beta} \in I, \quad \widetilde{\alpha} \leq \widetilde{\beta}, \quad \|\widetilde{\beta}\| = \|\widetilde{\beta}\| + a. \tag{2.2}$$

Represent the chain C as the union $C = C_1 \cup C_2 \cup C_3$ of the three dense chains C_1, C_2, and C_3 such that

$$m(C_1) = \widetilde{0}^n, \quad M(C_1) = \widetilde{\alpha}, \quad C_1 \cap I = \{\widetilde{\alpha}\}, \tag{2.3}$$

$$m(C_2) = \widetilde{\alpha}, \quad M(C_2) = \widetilde{\beta}, \tag{2.4}$$

$$m(C_3) = \widetilde{\beta}, \quad M(C_3) = \widetilde{1}^n, \quad C_3 \cap I = \{\widetilde{\beta}\}. \tag{2.5}$$

Obviously, $C_2 = C \cap I$. It is easy to see that to each maximal chain C in B_2^n such that $|C \cap I| = a + 1$ there corresponds a unique tuple $(\widetilde{\alpha}, \widetilde{\beta}, C_1, C_2, C_3)$ satisfying (2.2)–(2.5). On the other hand, to each such tuple there corresponds a unique maximal chain having exactly $a + 1$ common vertices with I. Thus, the desired number of maximal chains equals the number of the tuples $(\widetilde{\alpha}, \widetilde{\beta}, C_1, C_2, C_3)$ satisfying (2.2)–(2.5).

Introduce the additional numerical parameter $t = \|\widetilde{\alpha}\| - k$. It is easy to check that, for each tuple $\widetilde{\alpha}$ satisfying (2.2), the inequalities $k \leq \|\widetilde{\alpha}\| \leq n - r - a + k$ hold. Hence, t takes values in the interval $[0, n - r - a]$. Fix an arbitrary value t in this interval and count the number of the tuples $(\widetilde{\alpha}, \widetilde{\beta}, C_1, C_2, C_3)$ satisfying (2.2)–(2.5) and the additional condition

$$\|\widetilde{\alpha}\| = k + t. \tag{2.6}$$

It is easy to see that the number of pairs $(\widetilde{\alpha}, \widetilde{\beta})$ satisfying (2.2) and (2.6) equals

$$\binom{n-r}{t}\binom{n-r-t}{a}.$$

It is known [2] that, for each two tuples $\widetilde{\gamma}, \widetilde{\delta} \in B_2^n$ satisfying $\widetilde{\gamma} \leq \widetilde{\delta}$ and $\|\widetilde{\delta}\| - \|\widetilde{\gamma}\| = s$, the number of dense chains with the least tuple $\widetilde{\gamma}$ and the greatest tuple $\widetilde{\delta}$ is $s!$. Using this fact, it is easy to verify the following:

the number of chains C_1 satisfying (2.3) is $(k + t - 1)!k$,
the number of chains C_2 satisfying (2.4) is $a!$,
the number of chains C_3 satisfying (2.5) is $(n - k - a - t - 1)!(r - k)$.

Hence, the number of tuples $(\widetilde{\alpha}, \widetilde{\beta}, C_1, C_2, C_3)$ satisfying (2.2)–(2.6) is equal to

$$\binom{n-r}{t}\binom{n-r-t}{a}(k+t-1)!ka!(n-k-a-t-1)!(r-k)$$
$$= \frac{(n-r)!(k+t-1)!(n-k-a-t-1)!k(r-k)}{t!(n-r-t-a)!}.$$

Multiply and divide the last equality by $(k - 1)!(r - k - 1)!r!n!$. Upon rearranging these factors, we obtain

$$\frac{(k-1+t)!}{t!(k-1)!}\,\frac{(n-k-a-1-t)!}{(n-r-a-t)!(r-k-1)!}\,\frac{k!(r-k)!}{r!}\,\frac{(n-r)!r!}{n!}n!$$

$$= \binom{k-1+t}{k-1}\binom{n-k-a-1-t}{r-k-1}\binom{r}{k}^{-1}\binom{n}{r}^{-1} n!. \qquad (2.7)$$

It remains now to sum up (2.7) over all t in the interval $[0, n-r-a]$ and reduce the resulting sum to the form (2.1). To this end, we make use of the combinatorial equality

$$\sum_{t=0}^{s}\binom{p+t}{p}\binom{q+s-t}{q} = \binom{p+q+s+1}{p+q+1}$$

which holds for any nonnegative integers p, q, and s (see [3, p. 268, Problem 2.13(3)]). Substituting $p = k-1$, $q = r-k-1$, and $s = n-r-a$ into the equality, we obtain

$$\sum_{t=0}^{n-r-a}\binom{k-1+t}{k-1}\binom{n-k-a-1-t}{r-k-1} = \binom{n-a-1}{r-1}.$$

Hence, the sum of (2.7) equals (2.1). Lemma 2.1 is proved.

Lemma 2.2. *Under the hypothesis of Lemma 2.1, the number of maximal chains in the cube B_2^n having at least $a+1$ common vertices with the interval I is equal to*

$$\binom{n-a}{r}\binom{r}{k}^{-1}\binom{n}{r}^{-1} n!. \qquad (2.8)$$

PROOF. In the combinatorial identity

$$\sum_{t=p}^{s}\binom{t}{p} \equiv \binom{s+1}{p+1}$$

holding for any nonnegative integers s, p, and $s \geq p$ (see [3, p. 253, Problem 1.15(8)]), put $s = n-a-1$ and $p = r-1$ and find the sum

$$\sum_{b=a}^{n-r}\binom{n-b-1}{r-1} = \sum_{i=r-1}^{n-a-1}\binom{i}{r-1} = \binom{n-a}{r}.$$

Since by Lemma 2.1, the number of maximal chains in B_2^n having exactly $b+1$ common vertices with the interval I is equal to

$$\binom{n-b-1}{r-1}\binom{r}{k}^{-1}\binom{n}{r}^{-1} n!,$$

we obtain (2.8), as desired. Lemma 2 is proved.

Lemma 2.3. *Let n, r, k, and a be integers such that $1 \leq r \leq n$ and $0 \leq k \leq r$, $a \geq 0$, and let I be an interval in the cube B_2^n which is of range r and weight k. Then the number of maximal chains in B_2^n having at least $a+1$ common vertices with I does not exceed*

$$\binom{r}{k}^{-1} e^{-ar/n} n!. \tag{2.9}$$

PROOF. If $a \leq n-r$, then, by Lemma 2.2, the number of maximal chains in B_2^n having at least $a+1$ common vertices with I is given by (2.8). To see that (2.9) is really an upper bound for (2.8), it suffices to use the inequality

$$\binom{n-a}{r}\binom{n}{r}^{-1} \leq e^{-ar/n}$$

that holds for any nonnegative integers n, a, and r such that $r \leq n-a$ (see [3, p. 277, Problem 4.11(3)]). It remains to observe that any chain in B_2^n has at most $n-r+1$ common vertices with I and, hence, if $a > n-r$, then the number of maximal chains in B_2^n having at least $a+1$ common vertices with I equals 0 whereas (2.9) is positive. Lemma 2.3 is proved.

3. Certain Properties of Boolean Functions

Enumerate all the $n!$ maximal chains in the cube B_2^n arbitrarily from 1 to $n!$ and denote by D_i the chain numbered i, $1 \leq i \leq n!$. This numbering of chains will be fixed throughout the remainder of the paper. Let $R_n = \{1, \dots, n!\}$. For each Boolean function f of n essential arguments, denote by N_f the set of vertices in B_2^n at which f takes the value 1.

Lemma 3.1. *Let n, r, and a be integers such that $a \geq 0$ and $1 \leq r \leq n$. Suppose that $g'(x_{i_1}, \dots, x_{i_r})$, $1 \leq i_1 \leq \dots \leq i_r \leq n$, is a function in AC essentially depending on r arguments and $g(x_1, \dots, x_n)$ is the function whose only difference from $g'(x_1, \dots, x_n)$ consists in the presence of $n-r$ inessential arguments. Then*

$$|K| \leq e^{-ar/n} n!,$$

where K is the set of indices i, $i \in R_n$, of the chains D_i in the cube B_2^n such that $|D_i \cap N_g| \geq a+1$.

PROOF. Consider the representation of g' as a full disjunctive normal form (FDNF):

$$g'(y_1, \dots, y_r) = \bigcup_{j=1}^{M} y_1^{\sigma_{j1}} \dots y_r^{\sigma_{jr}},$$

where $M = |N_{g'}|$. Then $N_{g'} = \{\widetilde{\sigma}_1, \ldots, \widetilde{\sigma}_M\}$, where

$$\widetilde{\sigma}_1 = (\widetilde{\sigma}_{11}, \ldots, \widetilde{\sigma}_{1r}), \ldots, \widetilde{\sigma}_M = (\sigma_{M1}, \ldots, \sigma_{Mr}).$$

For each j, $1 \leq j \leq M$, put $k_j = \|\widetilde{\sigma}_j\|$. Since $g' \in AC$, the tuples $\widetilde{\sigma}_1, \ldots, \widetilde{\sigma}_M$ are pairwise incomparable and, by the Lubell inequality [2],

$$\sum_{j=1}^{M} \binom{r}{k_j}^{-1} \leq 1.$$

The function $g(x_1, \ldots, x_n)$ admits an FDNF representation:

$$g(x_1, \ldots, x_n) = \bigcup_{j=1}^{M} x_{i_1}^{\sigma_{j1}} \ldots x_{i_r}^{\sigma_{jr}}.$$

For each j, $1 \leq j \leq M$, the conjunction $x_{i_1}^{\sigma_{j1}} \ldots x_{i_r}^{\sigma_{jr}}$ corresponding to the tuple $\widetilde{\sigma}_j = (\widetilde{\sigma}_{j1}, \ldots, \widetilde{\sigma}_{jr})$ determines an interval I_j in B_2^n. Since the tuples $\widetilde{\sigma}_1, \ldots, \widetilde{\sigma}_M$ are pairwise incomparable, the tuples lying in different intervals $I_1, \ldots, I_M$ are also incomparable. Hence, each of the chains $D_1, \ldots, D_{n!}$ may have common vertices with one of the above intervals only. Therefore, for each chain D_i such that the intersection $D_i \cap N_g$ is nonempty, there exists a unique interval I_j such that $D_i \cap N_g = D_i \cap I_j$. Denote by K_j the set of indices of the chains D_i such that $|D_i \cap I_j| \geq a + 1$. It follows from the above that K is the union of the sets $K_1, \ldots, K_M$. By Lemma 2.3, for each j, we have

$$|K_j| \leq \binom{r}{k_j}^{-1} e^{-ar/n} n!\,.$$

Thus,

$$|K| \leq \sum_{j=1}^{M} \binom{r}{k_j}^{-1} e^{-ar/n} n!\,,$$

from which, by invoking Lubell's inequality, we obtain

$$|K| \leq e^{-ar/n} n!.$$

Lemma 3.1 is proved.

Below, we use partial Boolean functions. Let A be an arbitrary subset of vertices in the cube B_2^n. A *partial Boolean function* f *in* n *arguments with domain* A is a mapping $f : A \to \{0,1\}$; outside the set A the function f is not defined. The set of all partial Boolean functions of n arguments with a domain A is denoted by $P_2^n(A)$. If $A \subseteq B$ and $f \in P_2^n(B)$, then the function $g \in P_2^n(A)$ coinciding with f on all tuples in A is said to be the *restriction of the function f to the set A* and is denoted by $f|_A$. The restrictions to A of the functions equal to the constants 0 and 1 are denoted by $0|_A$ and $1|_A$ and the restriction of the function equal to the variable x_i by $x_i|_A$. Put

$$C^n(A) = \{0|_A, 1|_A\}, \quad U^n(A) = \{0|_A, 1|_A, x_1|_A, \dots, x_n|_A\}.$$

We deal with restrictions of linear functions onto the sets which are chains in the cubes of the corresponding dimension. In what follows, we need a simple property of such restrictions which is stated in the lemma below.

Lemma 3.2. *If C is an arbitrary (not necessarily dense) chain in the cube B_2^n with length greater than $(n+3)/2$, then $l_n|_C \notin U^n(C)$.*

To prove Lemma 3.2, the following trivial lemma is used.

Lemma 3.3. *If C is a dense chain in the cube B_2^n, then the linear function l_n takes each of the values 0 and 1 at no more than $(|C|+1)/2$ tuples of C.*

4. Estimating the Complexity of s-Regular Circuits

Suppose that a circuit S over the basis AC has n inputs, to which the variables $x_1, \dots, x_n$ are assigned, and realizes a total function $g(x_1, \dots, x_n)$. Let A be an arbitrary subset of vertices in the cube B_2^n, and let f be a partial function of n arguments with domain A such that $g|_A = f$. In this case, we say that the *circuit S realizes the partial function f on the set A*.

For each gate e of the circuit S, denote by $g_e(x_1, \dots, x_n)$ the function of the input variables $x_1, \dots, x_n$ that is realized at the output of e. For any set $A \subseteq B_2^n$, denote by $Q(S, A)$ the set of the gates e in S such that $g_e|_A \in C^n(A)$. The number of gates in $Q(S, A)$ is denoted by $q(S, A)$.

Lemma 4.1. *Let n, s, p, and a be integers such that $1 \le s \le n$, $p \ge 1$, and $a \ge 0$, and let A be a nonempty subset of vertices in the cube B_2^n. Suppose that a circuit S over the basis AC has n inputs, is s-regular, and realizes on a set A a partial function f such that $f \notin U^n(A)$. Let J be a nonempty subset of the set R_n such that $|D_j \cap A| \ge p$ for each $j \in J$. Then there are a nonempty subset $A' \subseteq A$ and a subset $J' \subseteq J$ such that*

$$q(S, A') \ge q(S, A) + 1,$$

$$|D_j \cap A'| \geq p - a \quad \text{for every} \quad j \in J',$$

$$|J'| \geq |J| - e^{-as/n} n!.$$

PROOF. Since $f \notin U^n(A)$, there are gates in S that are not contained in the set $Q(S, A)$. Among them, there exists at least 1 gate each input of which is connected to the output of a gate in $Q(S, A)$ or to an input of S. Moreover, at least 1 of its inputs is connected to an input of the circuit. Fix one of such gates and denote it by e.

Let r be the number of inputs in S to which inputs of the gate e are connected. Since at least 1 of the inputs in e is connected to an input of the circuit and the circuit S is s-regular, the inequality $r \geq s$ holds.

Let $x_{i_1}, \ldots, x_{i_r}$ be the variables assigned to those inputs of S to which the inputs of e, $1 \leq i_1 < \ldots < i_r \leq n$, are connected. Consider the function $g_e(x_1, \ldots, x_n)$ that is realized at the output of e. Observe that, by the definition of the set $Q(S, A)$, the inputs of e connected to outputs of the gates in this set receive the functions whose restrictions to the set A are constants. Hence, the function $g_e(x_1, \ldots, x_n)$ coincides on A with a function $\varphi(x_{i_1}, \ldots, x_{i_r})$ which is obtained from the basis function assigned to e by the corresponding substituting constants. Since the basis function assigned to e belongs to AC and the basis AC is closed with respect to the operation of substituting constants, φ also belongs to AC. Since $e \notin Q(S, A)$, the function φ is not a constant and, hence, essentially depends on all the variables $x_{i_1}, \ldots, x_{i_r}$.

Denote by $g(x_1, \ldots, x_n)$ the function of n arguments whose only difference from $\varphi(x_{i_1}, \ldots, x_{i_r})$ consists in the presence of $n - r$ unessential dummy variables. Clearly, $g|_A = g_e|_A$. Put $A' = A \backslash N_g$. Since $g_e|_A \notin C^n(A)$, the set A' is nonempty and, by $A' \subseteq A$, the inclusion $Q(S, A') \supseteq Q(S, A)$ holds. Taking into account the trivial relationships $g_e|_{A'} = g|_{A'} = 0|_{A'}$, we obtain $Q(S, A') \setminus Q(S, A) \supseteq \{e\}$, which yields $q(S, A') \geq q(S, A) + 1$.

Note that

$$|D_j \cap A| = |D_j \cap A'| + |D_j \cap A \cap N_g| \leq |D_j \cap A'| + |D_j \cap N_g|$$

for each j, $j \in R_n$. Hence, $|D_j \cap A'| \geq p - |D_j \cap N_g|$ for each $j \in J$. Denote by K the set of indices j, $j \in R_n$, of the chains D_j such that $|D_j \cap N_g| \geq a + 1$ and put $J' = J \setminus K$. Then $|D_j \cap A'| \geq p - a$ for each $j \in J'$. Using the bound $|K| \leq e^{-ar/n} n!$ of Lemma 3.1 and the inequality $r \geq s$, we arrive at the inequality $|J'| \geq |J| - e^{-as/n} n!$. Lemma 4.1 is proved.

Lemma 4.2. *Let n, s, and a be positive integers with $s \leq n$, let t be an integer such that $t \geq 0$ and $t < (n-1)/(2a)$, $t < e^{as/n}$, and let S be an s-regular* *circuit over the basis AC which realizes the function l_n. Then there exist nonempty sets*

$$A_0, A_1, \dots, A_{t+1} \subseteq B_2^n, \quad A_0 \supseteq A_1 \supseteq \dots \supseteq A_{t+1},$$

and sets

$$J_0, J_1, \dots, J_{t+1} \subseteq R_n, \quad J_0 \supseteq J_1 \supseteq \dots \supseteq J_{t+1},$$

with $J_0, J_1, \dots, J_t$ nonempty, such that, for each i, $0 \le i \le t$, the following conditions are satisfied:

(1) $q(S, A_i) \ge i$,

(2) $|D_j \cap A_i| \ge n + 1 - ia$ for each $j \in J_i$,

(3) $|J_i| \ge (1 - ie^{-as/n})n!$.

PROOF. Set $A_0 = B_2^n$ and $J_0 = R_n$. Then (1)–(3) are satisfied for $i = 0$. Assume that, for some i, $0 \le i \le t$, the sets A_i and J_i satisfying (1)–(3) are constructed and, in addition, A_i is nonempty.

We describe the construction of the sets A_{i+1} and J_{i+1}. Since

$$|J_i| \ge (1 - ie^{-as/n})n! \ge (1 - te^{-as/n})n! > 0$$

the set J_i is nonempty. Therefore, there exists a number $k \in J_i$ such that

$$|D_k \cap A_i| \ge n + 1 - ia \ge n + 1 - ta > (n+3)/2.$$

By Lemma 3.2, we have $l_n|_{D_k \cap A_i} \notin U^n(D_k \cap A_i)$ and, hence, $l_n|_{A_i} \notin U^n(A_i)$. Thus, the circuit S realizes on the set A_i a partial function not belonging to $U^n(A_i)$. By Lemma 4.1, there exist a nonempty set $A_{i+1} \subseteq A_i$ and a set $J_{i+1} \subseteq J_i$ such that

$$q(S, A_{i+1}) \ge q(S, A_i) + 1 \ge i + 1,$$

$$|D_j \cap A_{i+1}| \ge n + 1 - (i+1)a \quad \text{for each} \quad j \in J_{i+1},$$

$$|J_{i+1}| \ge (1 - (i+1)e^{-as/n})n!.$$

The Lemma 4.2 is proved.

Lemma 4.3. *Let n, s, and a be positive integers such that $s \le n$ and let an s-regular circuit S realize the function l_n. Then*

$$L(S) \ge \min(\lfloor (n-1)/(2a) \rfloor, \lfloor e^{as/n} \rfloor).$$

PROOF. Put

$$t = \min(\lfloor (n-1)/(2a) \rfloor, \lfloor e^{as/n} \rfloor) - 1.$$

Clearly,

$$t \le (n-1)/(2a) - 1 < (n-1)/(2a), \quad t \le e^{as/n} - 1 < e^{as/n}.$$

If $t < 0$, the desired lower bound is trivial. If $t \ge 0$, then, by Lemma 4.2, there exists a nonempty set A_{t+1}, $A_{t+1} \subseteq B$, such that $q(S, A_{t+1}) \ge t+1$. It remains to observe that

$$L(S) \ge q(S, A_{t+1}) \ge t+1.$$

Lemma 4.3 is proved.

5. Completing the Proof of the Main Result

The lemma below summarizes the previous considerations.

Lemma 5.1. *Let n, n_0, s, and a be integers such that $2 \le s \le n_0 < n$ and $a \ge 0$. Then*

$$L(l_n) \ge \min((n-n_0)/(s-1), \lfloor n_0/(2a) \rfloor, \lfloor e^{as/n} \rfloor).$$

PROOF. By Lemma 1.3, either $L(l_n) \ge (n-n_0)/(s-1)$ or, for some m, $n_0 < m \le n$, there exists an s-regular circuit S realizing the function l_m such that $L(l_m) \ge L(S_0)$. In the latter case, by Lemma 4.3 and the inequality $s \le n_0 < m$, we have

$$L(l_n) \ge L(S_0) \ge \min(\lfloor (m-1)/(2a) \rfloor, \lfloor e^{as/m} \rfloor) > \min(\lfloor n_0/(2a) \rfloor, \lfloor e^{as/n} \rfloor).$$

Combining the two bounds for $L(l_n)$, we obtain the desired inequality. Lemma 5.1 is proved.

To complete the proof of Theorem A, in Lemma 5.1, set $n_0 = \lfloor n/2 \rfloor$, $s = \lceil (n \log n)^{1/2} \rceil$, and $a = \lceil (n \log n)^{1/2} \rceil$. Obviously, for sufficiently large n, all the hypotheses of Lemma 5.1 are fulfilled. It suffices to observe that

$$(n-n_0)/(s-1) \asymp n_0/(2a) \asymp (n/\log n)^{1/2}$$

as $n \to \infty$ and $as/n \ge (\log n)/2$ for each n, so that $e^{as/n} \ge n^{1/2}$. Applying Lemma 5.1, we obtain the desired bound. Theorem A is proved.

The author is grateful to O. B. Lupanov for his interest in this research.

References

1. O. M. Kasim-Zade (1994) On the complexity of circuits over an infinite basis (in Russian), *Vestnik Moskov. Univ. Ser.* I *Mat. Mekh.* No. 6, 40–44.
2. M. Aigner (1979) *Combinatorial Theory*, Springer-Verlag, Berlin etc.
3. G. P. Gavrilov and A. A. Sapozhenko (1977) *Problem Book in Discrete Mathematics* (in Russian), Nauka, Moscow.

Faculty of Mechanics and Mathematics
Moscow State University
Vorob′ëvy Gory
Moscow 119899
RUSSIA

TRANSLATED BY S. I. SUSLOV

DYNAMIC PROBLEMS OF CHOOSING OPTIMAL COMPOSITION OF A SYSTEM OF TECHNICAL TOOLS

Yu. A. Kochetov and M. G. Pashchenko

Dynamic problems of choosing optimal composition of a system of technical tools are formulated. Algorithms for their solution based on dual decomposition are presented. An exact polynomial time algorithm for solving subproblems is provided. It is proved that the duality gap can be arbitrarily large. A subclass of problems is identified such that the dual problem is polynomially solvable. Generalizations of the original problem are considered in which the production cost depends nonlinearly on the total output and additional restrictions on the composition of the system of technical tools.

Suppose that during a certain planned period it is necessary to fulfill a given set of jobs. For fulfillment of the jobs there is a system of technical tools. The composition of the system can be altered by adding new models and removing a part of the technical tools because of their moral or physical deterioration.

The dynamic problem of choosing optimal composition of the system is to find such variant of the development of the system which would give a minimum to the total costs of development, production and exploitation with technical tools provided that the given list of jobs is fulfilled each year.

The first formulations of the dynamic problems of choosing optimal composition of a system were studied in [1], where in particular an exact solution to such problem was obtained on assuming that at certain times the old system of technical tools can be completely replaced with a new system. These results are presented in the monograph [2].

One of the special cases of the above-stated problem is a dynamic location problem [3] for which both approximate [4] and exact algorithms [5, 6] based on the computing scheme of the branch and bound method were suggested. In a simpler form this problem was considered in [7] where ideas of the branch and bound method were also used. However, the specifics of the problem, namely a nontrivial subproblem of linear programming obtained when the integer-valued variables are fixed, essentially hinder the use of this method.

In the present paper we consider a dynamic problem of choosing optimal composition of a system and some variants of its generalization. The suggested algorithms are based on dual decomposition and used for approximate solution with an a posteriori precision estimation.

A. D. Korshunov (ed.), Operations Research and Discrete Analysis, 85–98.

In § 1 the mathematical formulation of the problem is presented, effective algorithms of finding the lower bound and also the scheme of solving the problem are described.

In § 2 dynamic problems with additional constraints on the nomenclature of technical tools are considered. Two close statements of the problem are formulated and algorithms for their solution are suggested.

In § 3 the dynamic problems in which the production cost depends on the total output are studied.

In § 4 the results of the numerical experiments are presented and some directions of further research are discussed.

§ 1. A Dynamic Problem of Choosing Optimal Composition of a System

1.1. Formulation of the Problem. Consider a planned period during which a system of technical tools must fulfill certain jobs.

Let l be the duration of the planned period (measured in years for definiteness) and $T = \{1, \dots, l\}$. Denote by $I = \{1, \dots, m\}$ a set of the models of technical tools, and by $J = \{1, \dots, n\}$, a set of jobs to be completed during the whole interval of planning. We assume that the set J is partitioned into some disjoint subsets $J_t, t \in T$, and for each $j \in J$ the index $t(j)$ is known such that $j \in J_{t(j)}$.

For the tools of model i, $i \in I$, we consider as given the following parameters:

c_{it}° is the fixed cost of the tool by year t;

c_{it} is the production cost of a tool in year t;

c_{ij} is the fulfillment cost of job j;

p_{ij} is the number of the tools required for fulfilling job j;

ρ_i is the duration of the exploitation of a tool;

v_{it}° is the initial system by year t;

V_{it} is an upper bound of the outputs of tools in year t.

The variables (z_{it}), (v_{it}), and (x_{it}) mean the following:

$$z_{it} = \begin{cases} 1 & \text{if the development of the tools of model } i \text{ is finished in year } t; \\ 0 & \text{otherwise;} \end{cases}$$

$v_{it} \geq 0$ is the output of the tools of model i in year t,

$x_{ij} \geq 0$ is the part of job j fulfilled by the tools of model i.

Using this notation, the dynamic problem of choosing optimal composition of the system can be written as follows.

PROBLEM P. *Find*

$$\min \sum_{t\in T} \sum_{i\in I} \Big\{ c^{\circ}_{it} z_{it} + c_{it} v_{it} + \sum_{j\in J_t} c_{ij} x_{ij} \Big\} \tag{1}$$

subject to

$$\sum_{i\in I} x_{ij} = 1, \qquad j \in J, \tag{2}$$

$$\sum_{j\in J_t} p_{ij} x_{ij} \le v^{\circ}_{it} + \sum_{\tau=t-\rho_i+1}^{t} v_{i\tau}, \qquad i \in I,\ t \in T, \tag{3}$$

$$0 \le v_{it} \le V_{it} \max_{\tau\le t} z_{i\tau}, \qquad i \in I,\ t \in T, \tag{4}$$

$$0 \le x_{ij} \le \max_{\tau\le t(j)} z_{i\tau}, \qquad i \in I,\ j \in J, \tag{5}$$

$$z_{it} \in \{0,1\}, \qquad i \in I,\ t \in T. \tag{6}$$

The objective function of the problem P means the total costs of development, production and exploitation of technical tools during the planned period. (2) guarantees the fulfillment of all jobs. The inequalities (3) restrict the possibilities of fulfilling the jobs by available articles. The conditions (4) establish an upper bound for possible outputs, and the inequalities (5) require additional expenses on developing the new models after their inclusion in the system. This problem belongs to the class of NP-hard problems of discrete optimization since it generalizes the minimum covering problem [8].

1.2. Lower bounds. We present the original problem in the following equivalent formulation.

PROBLEM P_y. *Find*

$$F_p = \min \sum_{i\in I} \sum_{t\in T} \Big\{ c^{\circ}_{it} z_{it} + c_{it} v_{it} + \sum_{j\in J_t} c_{ij} x_{ij} \Big\} \tag{7}$$

subject to

$$\sum_{i\in I} x_{ij} = 1, \qquad j \in J, \tag{8}$$

$$\sum_{j \in J_t} p_{ij} x_{ij} \leq v^{\circ}_{it} y_{it} + \sum_{\tau = t - \rho_i + 1}^{t} v_{i\tau}, \qquad i \in I, t \in T, \tag{9}$$

$$y_{it} \leq \sum_{\tau \leq t} z_{i\tau} \leq 1, \qquad i \in I, t \in T, \tag{10}$$

$$0 \leq v_{it} \leq V_{it} y_{it}, \qquad i \in I, t \in T, \tag{11}$$

$$0 \leq x_{ij} \leq y_{it_{(j)}}, \qquad i \in I, j \in J, \tag{12}$$

$$y_{it}, z_{it} \in \{0, 1\}, \qquad i \in I, t \in T. \tag{13}$$

Denote by $\overline{\mathrm{P}_y}$ the linear relaxation of the problem P_y, and by $\overline{F_{\mathrm{P}}}$, the optimal value of the objective function of the problem $\overline{\mathrm{P}_y}$.

Below the overbar means transition to the linear relaxation. We shall consider the following estimation problem.

PROBLEM LR(λ). *Find*

$$F_{\mathrm{LR}}(\lambda) = \min \sum_{i \in I} \sum_{t \in T} \Big\{ c^{\circ}_{it} z_{it} + c_{it} v_{it} + \sum_{j \in J_t} (c_{ij} - \lambda_j) x_{ij} \Big\} + \sum_{j \in J} \lambda_j$$

subject to (9)–(13), *where* $\lambda_j, j \in J$, *are Lagrange multipliers corresponding to* (8).

A solution $F_{\mathrm{D}} = \max_{\lambda} F_{\mathrm{LR}}(\lambda)$ of the dual problem D yields a lower bound on the optimal value of the objective function (1):

$$F_{\mathrm{P}} \geq F_{\mathrm{D}} \geq F_{\mathrm{LR}}(\lambda).$$

Theorem 1. *The estimation problem* LR(λ) *is polynomially solvable.*

PROOF. At fixed values of λ_j, $j \in J$, the problem LR(λ) decomposes into i independent subproblems S_i of the following form: Find

$$f_i = \max \sum_{t \in T} \Big\{ \sum_{j \in J_t} (\lambda_j - c_j) x_j - c_t v_t - c^{\circ}_t z_t \Big\}$$

subject to

$$\sum_{j\in J_t} p_j x_j \le v_t^\circ y_t + \sum_{\tau=t-\rho+1}^{t} v_\tau, \qquad t \in T,$$

$$y_t \le \sum_{\tau\le t} z_\tau \le 1, \qquad t \in T,$$

$$0 \le v_t \le V_t y_t, \qquad t \in T,$$

$$0 \le x_j \le y_{t(j)}, \qquad j \in J,$$

$$y_t,\ z_t \in \{0,1\}, \qquad t \in T.$$

For the subproblems S_i, the equality

$$F_{LR}(\lambda) = \sum_{j\in J} \lambda_j - \sum_{i\in I} f_i$$

holds.

For each pair (i,θ), $i \in I$ and $\theta \in T$, consider the problem $S_{i\theta}$ of linear programming in the following form: Find

$$f_{i\theta} = \max \sum_{t\ge\theta} \Big\{ \sum_{j\in J_t} (\lambda_j - c_j) x_j - c_t v_t \Big\}$$

subject to

$$\sum_{j\in J_t} p_j x_j \le v_t^\circ + \sum_{\tau=t-\rho+1}^{t} v_\tau, \qquad t \ge \theta,$$

$$0 \le v_t \le V_t, \qquad t \ge \theta,$$

$$0 \le x_j \le 1, \qquad j \in J_t, t \ge \theta.$$

From $z_t \in \{0,1\}$, $t \in T$, and $\sum_{t\in T} z_t \le 1$ it follows that

$$f_i = \max \{0; \max_{\theta\in T}(f_{i\theta} - c_\theta^\circ)\},$$

which proves Theorem 1.

Theorem 2. *If $\rho_i \geq l$, $i \in I$, then the problem* LR(λ) *is solvable in* $O(nm(l + \log n))$ *time.*

PROOF. We show that the problem S_i is solvable in $O(n(l + \log n))$ time. Put the sets J_t, $t \in T$, in nonincreasing order of $(\lambda_j - c_j)/p_j$, $j \in J_t$. Then the values of the variables x_j, $j \in J$, can be uniquely recovered from the values of the variables v_t in linear time. Optimal values of the variables v_t, $t \in T$, can be found with the help of the coordinate descent procedure. At each step of the procedure an index $\tau \in T$ is sought providing a maximum increase in the objective function as the variable v_t increases. Then the length of the step is determined in the direction τ and the values of the corresponding variables are changed. It is easy to see that if at some step it is impossible to increase the objective function along any coordinate, then the solution obtained is optimal.

Ordering the sets J_t, $t \in T$, can be done in $O(n \log n)$ time. Calculating the increments and finding the best index $\tau \in T$ requires $O(l)$ operations. In the same time the length of a step is found. The number of steps does not exceed $l + n$, and since $l \leq n$, the total running time of the algorithm does not exceed $O(nl + n \log n)$. Theorem 2 is proved.

Theorem 3. *If $\rho_i = 1$, $i \in I$, then the dual problem* D *is polynomially solvable.*

PROOF. We show that $\overline{f_i} = f_i$. Denote by (z_t^*), (y_t^*), (v_t^*), and (x_j^*) an optimal solution to the problem $\overline{S_i}$. For each $k \in T$ we construct a solution of the problem S_i by the following rule:

$$z_t(k) = \begin{cases} 1 & \text{if } t = k, \\ 0 & \text{otherwise}, \end{cases} \qquad t \in T,$$

$$y_t(k) = \begin{cases} 1 & \text{if } t \geq k, \\ 0 & \text{otherwise}, \end{cases} \qquad t \in T,$$

$$v_t(k) = \begin{cases} 0 & \text{if } t < k, \\ v_t^*/y_t^* & \text{otherwise}, \end{cases} \qquad t \in T,$$

$$x_j(k) = \begin{cases} 0 & \text{if } t < k, \\ x_j^*/y_{t(j)}^* & \text{otherwise}, \end{cases} \qquad j \in J.$$

If $z_k^* = 0$, then put

$$z_t(k) = y_t(k) = v_t(k) = 0, \quad t \in T, \qquad x_j(k) = 0, \quad j \in J.$$

It is easy to see that for each $k \in T$ the constructed solution is admissible in the problem S_i. Moreover,

$$\begin{aligned}\overline{f_i} &\le \sum_{t\in T}\Big\{\sum_{j\in J_t}(\lambda_j - c_j)\sum_{k\in T} z_k^* x_j(k) - c_t\sum_{k\in T} z_k^* v_t(k) - c_t^\circ \sum_{k\in T} z_k^* z_t(k)\Big\}\\ &= \sum_{k\in T} z_k^* \sum_{t\in T}\Big\{\sum_{j\in J_t}(\lambda_j - c_j)x_j(k) - c_t v_t(k) - c_t^\circ z_t(k)\Big\}\\ &= \sum_{k\in T} z_k^* f_i(k) \le \max_k f_i(k).\end{aligned}$$

Thus, $\overline{f_i} \le f_i$. The validity of the opposite inequality $\overline{f_i} \ge f_i$ is obvious, whence $\overline{f_i} = f_i$ and $F_{\mathrm{LR}}(\lambda) = \overline{F_{\mathrm{LR}}(\lambda)}$, i.e. $F_{\mathrm{D}} = \overline{F_{\mathrm{D}}} = \overline{F_{\mathrm{P}}}$. Therefore, to solve the problem D it suffices to find an optimal solution of the linear programming problem $\overline{\mathrm{P}_y}$ and using the latter to reconstruct a mixed solution of the dual problem. Theorem 3 is proved.

Note that if the hypotheses of Theorem 3 do not hold and $\rho_i > 1$ for some $i \in I$, then the duality gap $\Delta = \overline{f_i} - f_i$ can be considerable.

Theorem 4. *For each $N > 0$ there exist input data of the problem S_i for which $\Delta > N$.*

PROOF. Put

$$T = \{1,2\},\ J = \{1,2\},\ J_1 = \{1\},\ J_2 = \{2\},$$

$$p_j = 1,\ c_j = 0,\ j \in J, \qquad V_t = 2N,\ v_t^\circ = 0,\ t \in T,$$

$$\lambda = (0, N+2),\ c^\circ = (2N, 0),\ c = (0, 2N),\ \rho = 2.$$

It is easy to verify that the zero solution is optimal in the problem S_i, $f_i = 0$, and at a feasible solution

$$z = (1/2N, 1 - 1/2N); \quad y = (1/2N, 1); \quad v = (1,0); \quad x = (0,1)$$

the value of the objective function of the problem $\overline{\mathrm{S}_i}$ is equal to $N+1$. Theorem 4 is proved.

1.3. Framework of the algorithm. The algorithm for solving the original problem is based on polynomial solvability of the estimation problem $\mathrm{LR}(\lambda)$. The algorithm starts with the calculation of the lower bound F_{D}. Let (z_{it}^k), (y_{it}^k), (v_{it}^k), and (x_{ij}^k) be an optimal solution to the problem $\mathrm{LR}(\lambda^k)$ with fixed values of the parameters $\lambda_j = \lambda_j^k$, $j \in J$. If (2) holds, then an exact solution of the original problem is obtained. Otherwise, the vector

$$S_j^k = 1 - \sum_{i \in I} x_{ij}^k, \qquad j \in J,$$

as a subgradient of the function $F_{\mathrm{LR}}(\lambda)$ at the point λ_j^k, $j \in J$, gives the direction of altering the Lagrange multipliers. Take

$$\lambda_j^{k+1} = \lambda_j^k + \beta^k S_j^k, \quad j \in J,$$

and come back to the solution of the problem LR(λ) with new values of the parameters λ [9].

The algorithm of constructing the upper bound F_0 returns the set of the indices

$$I_0 = \left\{ i \in I \mid \sum_k \sum_{t \in T} z_{it}^k > 0 \right\}$$

and for each $i_0 \in I_0$ fulfills the following operations.

STEP 0. Put $I_1 = \{i_0\}$.

STEP 1. Solve the dual problem D under the additional constraints $\sum_{t \in T} z_{it} = 1$, $i \in I_1$. At each iteration of the solution of the problem an optimal solution of the problem LR(λ) is extended to a feasible solution of the original problem. The best solution is stored as F_0.

STEP 2. Find an index $i_1 \in I \setminus I_1$ for which $\gamma = \sum_k \sum_{t \in T} z_{i_1 t}^k$, the frequency of the entering an optimal solution, is a maximum. If $\gamma_{i_1} = 0$, then go to the next element of the set I_0. Otherwise, the algorithm puts $I_1 = I_1 \cup \{i_1\}$ and returns at step 1.

The running time of the algorithm does not exceed $O(m^2 Q)$, where Q is the running time of the dual problem. The quality of the solution obtained is judged from the relative deviation $\varepsilon = (F_0/F_{\mathrm{D}} - 1)$ which together with the relative error of the algorithm $\delta = (F_0/F - 1)$ includes the error due to the duality gap.

§ 2. Dynamic Problems with Restrictions on the Nomenclature of Technical Tools

Suppose that additional constraints on the nomenclature of technical tools are imposed on the development of a system during a planned period. Let b_{it}, $i \in I$, $t \in T$, give the cost of some generalized resource related to introducing new models in the composition of the system, and let B be the total resource allotted to the development of the system in the whole planned period.

The additional constraints related to the nomenclature of technical tools can be written as follows:

$$\sum_{t\in T}\sum_{i\in I} b_{it}z_{it} \leq B. \tag{14}$$

For $b_{it} = 1$, $i \in I$, $t \in T$, inequality (14) can be treated as a restriction on the total number of models used for fulfillment of jobs in the planned period. The problem P with the constraint (14) is denoted by P_B.

Consider another variant of constraints on the nomenclature when the total resource is allocated not at once for the whole planned period, but it is broken in years.

Denote by K_t, $t \in T$, the value of the resource delivered in year t. Then the constraints on the nomenclature of models can be written as follows:

$$\sum_{i\in I} b_{it}z_{it} \leq K_t, \qquad t \in T. \tag{15}$$

The problem P subject to these constraints is denoted by the problem P_K.

The construction of algorithms solving the problems P_B and P_K is performed in the same way as for the original problem P. The solution of the dual problem to the problem P_B is reduced to solving the following problem: Find

$$\max \sum_{\theta\in T}\sum_{i\in I} f_{i\theta}z_{i\theta}$$

subject to

$$\sum_{\theta\in T} z_{i\theta} \leq 1, \qquad i \in I;$$

$$\sum_{\theta\in T}\sum_{i\in I} b_{i\theta}z_{i\theta} \leq B;$$

$$z_{i\theta} \in \{0,1\}, \qquad i \in I, \quad \theta \in T.$$

This problem is known as the multiple-choice knapsack problem [10]. The linear relaxation of the above problem is solved in $O(lm)$ time, and among its optimal solutions there always exists a solution in which not more than two variables have fractional values [11]. For $b_{it} = 1$, $i \in I$, $t \in T$, the problem degenerates, and an optimal solution to it can be found in one step.

The solution of the dual problem to P_K can be reduced to the solution of the following problem: Find

$$\max \sum_{\theta \in T} \sum_{i \in I} f_{i\theta} z_{i\theta}$$

subject to

$$\sum_{\theta \in T} z_{i\theta} \leq 1, \qquad i \in I;$$

$$\sum_{i \in I} b_{i\theta} z_{i\theta} \leq K_\theta, \qquad \theta \in T;$$

$$z_{i\theta} \in \{0, 1\}, \qquad i \in I, \quad \theta \in T.$$

This problem is known as the generalized assignment problem.

For $b_{it} = 1$, $i \in I$, $t \in T$, this problem becomes the transportation problem and is solved exactly. In general, it is an NP-hard problem of discrete optimization, and for its solution some exact and approximate algorithms are developed [10].

§ 3. Dynamic Problems with the Seriality Factor

Consider a generalization of the original problem P when the production cost of the articles c_{it}, $i \in I$, $t \in T$, is not constant and depends on the total output of the articles

$$c_{it} = c_{it}\Big(\sum_{\tau \in T} v_{i\tau}\Big), \qquad i \in I, \ t \in T,$$

i.e. for each $i \in I$, given are the intervals w_i

$$0 < w_{i1} < w_{i2} < \cdots < w_{iq_i} = \sum_{\tau \in T} V_{i\tau},$$

inside each of which the production cost remains constant, while possibly changing arbitrarily in transition from one interval to another (Fig. 1).

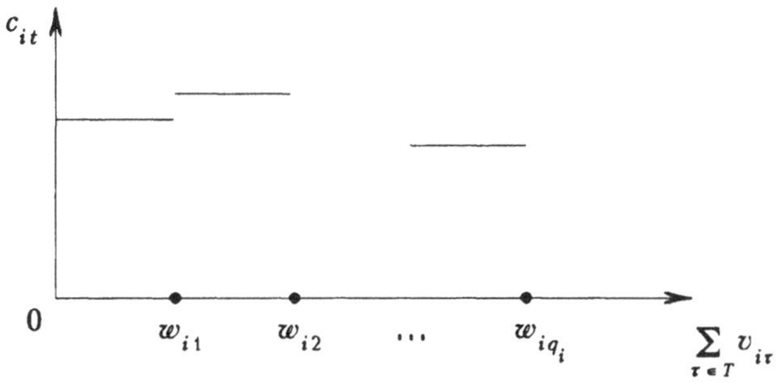

Fig. 1

An algorithm for solving the problem P_w is also based on the polynomial solvability of the relaxed problem LR(λ) in which the problem $S_{i\theta}$ has the following form: Find

$$f_{i\theta} = \max \sum_{t\geq\theta} \left\{ \sum_{j\in J_t} (\lambda_j - c_j)x_j - c_t\Big(\sum_{\tau\in T} v_{i\tau}\Big)v_t \right\}$$

subject to

$$\begin{aligned} &\sum_{j\in J_t} p_j x_j \leq v_t^\circ + \sum_{\tau=t-\rho+1}^{t} v_\tau, && t\geq\theta,\\ &0\leq v_t\leq V_t, && t\geq\theta,\\ &0\leq x_j\leq 1, && j\in J_t,\ t\geq\theta. \end{aligned}$$

The algorithm for solving this problem successively considers all intervals $[w_{k-1}, w_k)$, $1\leq k\leq q_i$, and for each of them solves the following auxiliary problem: Find

$$f_k = \max \sum_{t\geq\theta} \left\{ \sum_{j\in J_t} (\lambda_j - c_j)x_j - c_t(w_{k-1}, w_k)v_t \right\}$$

subject to

$$\sum_{j\in J_t} p_j x_j \le v_t^\circ + \sum_{\tau=t-\rho+1}^{t} v_\tau, \qquad t \ge \theta,$$

$$w_{k-1} \le \sum_{t\ge\theta} v_t < w_k,$$

$$0 \le v_t \le V_t, \qquad t \ge \theta,$$

$$0 \le x_j \le 1, \qquad j \in J_t,\ t \ge \theta,$$

where $c_t(w_{k-1}, w_k)$ is a constant equal to the production cost of model i in year t of consignment volumes in the interval from w_{k-1} to w_k. It is obvious that $f_{i\theta} = \max_k f_k$.

For $\rho_i \ge l$ or $\rho_i = 1$ this problem is simplified and an optimal solution can be obtained in $O(nq_i(l + n\log n))$ time. In general, it is a linear programming problem and can be solved by standard methods.

REMARK. Let the production cost in each seriality interval is given by the linear function $c_t(v) = c_t^\circ(w_{k-1}, w_k) + c_t(w_{k-1}, w_k)v_t$, $t \in T$. In this case the last problem does not change, but $f_{i\theta}$ can be found with the help of the equality $f_{i\theta} = \max_k\big(f_k + \sum_{t\ge\theta} c_t^\circ(w_{k-1}, w_k)\big)$.

§ 4. Computational Results

The above algorithms are coded in PASCAL and were tested on IBM PC-486 (SX). All problems under solution had the same dimension: $m = 50$, $n = 50$, $l = 5$, $n_t = 10$, $t \in T$.

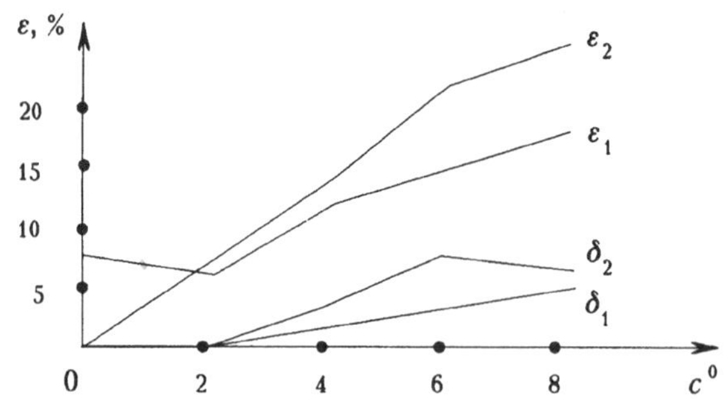

Fig. 2

The files of the initial data were formed with the help of a random number generator with uniform distribution. The values of the parameters were selected as follows: $c_{it}^{\circ} = 300c^{\circ}$, $c_{it} \in [1, 10]$, $c_{ij} \in [1, 1000]$, $p_{ij} \in [1, 10]$, $b_{it} = 1$, and $v_{it}^{\circ} = 0$. The matrices (c_{ij}) and (p_{ij}) were filled in 30% by the number 10^6. The results of computations are represented in Fig. 2. The errors δ_1 and ε_1 were calculated for $\rho_i = 1$, $V_{it} = 100$, and $B = 3$; δ_2 and ε_2 for $\rho_i = 1$, $V_{it} = 100$, and $B = 50$.

Each break point is an average value of the relative errors of 30 test problems. Note that for $c^{\circ} = 0$ the original problem remains a generalization of the NP-hard K-median problem [12]. In this case for $B = 3$ the duality gap is significant (from 7% up to 15%), and the suggested algorithm runs almost without error ($\delta_1 = 0.16$).

For large values of the parameter c°, the original problem degenerates and actually becomes the minimum covering problem. The initial costs dominate in the objective function, and if the algorithm succeeds in finding the minimum nomenclature for fulfillment of all jobs, then an exact solution of the problem is obtained. Otherwise, the deviation from an optimum reaches 10%.

The running time for one problem for $\rho_i = 1$ on the average constitutes 2–3 minutes; for $\rho_i = 5$, from 10 to 15 minutes. For the intermediate values of ρ_i the running time will apparently be more. There is necessity for specialized algorithms to the problem $\mathrm{S}_{i\theta}$ for arbitrary ρ_i, $i \in I$. Furthermore, it is often in practice that a part of technical tools fails annually before the exploitation time terminates. In this case the constraints (3) take the form

$$\sum_{j \in J_t} p_{ij} x_{ij} \le v_{it}^{\circ} + \sum_{\tau = t - \rho_i + 1}^{t} d_{i\tau} v_{i\tau}, \quad i \in I, \quad t \in T,$$

where $d_{i\tau}$ define the share of losses of the technical tools during the exploitation time.

Another important generalization is given by the dynamic problems with total restrictions on the annual outputs or (and) on total composition of the system at each year. Such models require at least an essential modification of the algorithms or pursuing some principally different approaches.

References

1. N. I. Glebov, V. T. Dement'ev, and A. N. Sychëv (1971) On dynamics of development of homogeneous technical systems (in Russian), *Upravlyaemye Sistemy* **8**, 51–67.

2. V. L. Beresnev, È. Kh. Gimadi, and V. T. Dement'ev (1978) *Extremal Standardization Problems* (in Russian), Nauka, Novosibirsk.

3. V. R. Khachaturov and N. D. Astakhov (1976) Dynamic location problems (models and methods for solution) (in Russian), *Ekonom. i Mat. Metody* **12**, No. 1, 93–109.

4. S. K. Jacobsen (1977) Heuristic solutions to dynamic plant location problems, *Advances in Operations Research*: Proc. EURO II. Second European Congress on Operations Research, North-Holland, Amsterdam etc., pp. 207–213.

5. D. Erlenkotter (1981) A comparative study of approaches to dynamic location problems, *European J. Oper. Res.* **6**, No. 2, 133–143.

6. T. J. Van Roy and D. Erlenkotter (1982) A dual-based procedure for dynamic facility location, *Management Sci.* **28**, No. 10, 1091–1105.

7. V. L. Beresnev, G. I. Ibragimov, and Yu. A. Kochetov (1984) Algorithms for solving the problem of optimal choice of a dynamic series of goods (in Russian), *Upravlyaemye Sistemy* **24**, 3–19.

8. M. R. Garey and D. S. Johnson (1979) *Computers and Intractability*, Freeman, San Francisco.

9. M. Held, P. Wolfe, and H. Crowder (1974) Validation of subgradient optimization, *Math. Programming* **6**, No. 1, 62–88.

10. M. E. Dyer (1984) A $O(n)$ algorithm for the multiple-choice knapsack linear program, *Math. Programming* **29**, No. 1, 57–63.

11. S. Martello and P. Toth (1987) Linear assignment problems. *Surveys in Combinatorial Optimization*, North-Holland, Amsterdam and New York, pp. 259–282.

12. S. Ahn, C. Cooper, G. Cornuejols and A. Frieze (1988) Probabilistic analysis of a relaxation for the K-median problem, *Math. Oper. Res.* **13**, No. 1, 1–31.

Sobolev Institute of Mathematics
Universitetskiĭ pr., 4
Novosibirsk 630090
RUSSIA

TRANSLATED BY I. V. RENTTS

LOWER BOUNDS IN THE PROBLEM OF CHOOSING AN OPTIMAL COMPOSITION OF A TWO-LEVEL SYSTEM OF TECHNICAL TOOLS[†)]

Yu. A. Kochetov and M. G. Pashchenko

For the problem of choosing an optimal composition of two-level systems of technical tools, the lower bounds for the objective function are suggested. These bounds are constructed by using Lagrangian relaxations. Two bounds are obtained as a result of relaxing two different groups of restrictions. It is shown that these bounds can be found in polynomial time. For incomparable bounds, domination areas are identified.

Introduction

Assume that for fulfillment of some set of jobs there is a system of technical tools with diverse qualitative and quantitative composition. In order to fulfill the jobs, technical tools are combined into tool-kits which can be considered as new complex technical tools incorporating unified components. The process of fulfilling the jobs splits into stages, a certain part of jobs is fulfilled at the first stage; some, at the second stage and so on. At each stage a known fraction of tool-kits fails and does not participate in further fulfilling the jobs. The problem is to find a composition of the system which would allow us to fulfill all jobs with minimal total expenses.

The first mathematical models of searching a composition of two-level systems of technical tools were considered in [1]. Such systems were studied in the monograph [2] in which, in particular, reduction of the simplest linear problem of searching an optimal composition of two-level systems to a minimization problem of a polynomial in Boolean variables has been established. Most of the articles (see, for example, [3–7]) are devoted to a special case of this problem, to the so-called two-level facility location problem, that generalizes the well-known plant location problem NP-hard in the strong sense [8]. For two-level facility location problems lower bounds and their correlation were studied in [7].

The purpose of the article is to find lower bounds for the objective function in the problems of searching an optimal composition of a two-level system of technical

[†)] This research was supported by the Russian Foundation for Basic Research (Grant 94–01–01326).

A. D. Korshunov (ed.), Operations Research and Discrete Analysis, 99–108.

tools. We consider two lower bounds obtained by relaxing two different groups of restrictions. It is shown that these bounds are incomparable and can be computed in polynomial time. The domination areas are identified.

In § 1 a mathematical statement of the problem is given. In § 2 the main theorems are proved. In § 3 we show that the lower bounds obtained are incomparable. In § 4 the results of some tests are discussed.

§ 1. Formulation of the Problem

Denote by $J = \{1, 2, \dots, n\}$ a set of jobs to be fulfilled. Assume that the set J divides into some disjoint sets J_l, $l \in L$, and the jobs are fulfilled in the following order: the jobs in the set J_1 are fulfilled firstly; the jobs in the set J_2 are fulfilled secondly and so on. We call the set $L = \{1, 2, \dots, r\}$ the set of stages of fulfillment. Denote by $I = \{1, 2, \dots, m\}$ the list of the items of technical tools. To fulfill a job, technical tools unite in tool-kits. Denote by $K = \{1, 2, \dots, q\}$ the collection of the tool-kits of different composition. For $k \in K$ the set $I_k \subseteq I$ defines the composition of the tool-kit of type k, and for $i \in I$ the set $K_i \subseteq K$ defines the list of the tool-kits incorporating some technical tools of the item of type i. For each tool-kit of type k, $k \in K$, the following parameters are designed:

v_k° be the number of tool-kits already incorporated to a system;

c_k° be the cost of the development of the tool-kit;

d_i be the cost of the development of a technical tool of item i;

c_k be the cost of production of one tool-kit including the cost of its technical tools;

V_k be the upper bound of the production volumes of tool-kits;

p_{kj} be the number of tool-kits that are required for fulfilling job j;

c_{kj} be the cost of fulfilling job j by tool-kits;

s_{kl} be the fraction of decrease of tool-kits in fulfilling jobs at the lth stage.

We introduce the following variables:

$$z_k = \begin{cases} 1 & \text{if the tool-kits of type } k \text{ belong to} \\ & \text{the composition of a system,} \\ 0 & \text{otherwise;} \end{cases}$$

$$y_i = \begin{cases} 1 & \text{if the technical tools of item } i \\ & \text{belong to the composition of a system,} \\ 0 & \text{otherwise;} \end{cases}$$

$v_k \geq 0$ be the volume of production of the tool-kits of type k;

$x_{kj} \geq 0$ be the fraction of job j performed by the tool-kits of type k.

Now the problem of searching an optimal composition of a two-level system can be written as a problem P of mixed-integer programming

$$\min_{x,y,z,v}\Big(\sum_{k\in K}\Big(c_k^\circ\ z_k + c_k\ v_k + \sum_{j\in J} c_{kj}\ x_{kj}\Big) + \sum_{i\in I} d_i\ y_i\Big) \tag{1}$$

subject to

$$\sum_{k\in K} x_{kj} = 1, \quad j\in J; \tag{2}$$

$$\sum_{j\in J_l} p_{kj}x_{kj} \le v_k^\circ z_k + v_k - \sum_{l'=1}^{l-1} s_{kl'} \sum_{j\in J_{l'}} p_{kj}x_{kj}, \quad k\in K, \quad l\in L; \tag{3}$$

$$0\le v_k \le V_k z_k, \quad k\in K; \tag{4}$$

$$0\le x_{kj}\le z_k, \quad k\in K, \quad j\in J; \tag{5}$$

$$y_i\ge z_k, \quad k\in K_i, \quad i\in I; \tag{6}$$

$$y_i, z_k\in\{0,1\}, \quad i\in I, \quad k\in K. \tag{7}$$

The objective function (1) means the total costs of development, production and exploitation of technical tools. The equalities (2) provide fulfillment of all jobs. Inequalities (3) show connection between volumes of production of tool-kits and volumes of jobs. For each pair (k,l), the left-hand side of (3) defines the required number of the tool-kits of type k at the lth stage of fulfilling the jobs. The right-hand side defines the number of tool-kits of type k available at the beginning of the lth stage with account taken of the losses made at the previous stages. Inequalities (4) give upper bounds on all possible volumes of production of tool-kits. Inequalities (5) and (6) admit use of tool-kits and technical tools only if they are included in the composition of the system under study.

§ 2. Lower Bounds

We consider two auxiliary problems L_f and L_h which are obtained by introducing into the objective function (1) some restrictions of the problem P. We associate with conditions (2) the dual variables α_j, $j\in J$. By L_f we denote the problem of finding the following quantity:

$$f(\alpha) = \min_{x,y,z,v}\Big\{\sum_{k\in K}\Big(c_k^\circ z_k + c_k v_k + \sum_{j\in J}(c_{kj}-\alpha_j)x_{kj}\Big) + \sum_{i\in I} d_i y_i\Big\}$$

subject to (3)–(7). The corresponding dual problem D_f of finding

$$F = \max_{\alpha}\Big\{f(\alpha) + \sum_{j\in J}\alpha_j\Big\},$$

permits us to find a lower bound for the objective function (1). We replace inequalities (6) in the system of restrictions of the problem P by

$$y_i \geq \sum_{k\in K_i} x_{kj}, \quad i \in I, \quad j \in J. \tag{8}$$

It is easy to check that in this case the feasible domain does not change. Associate with each restriction (8) the dual variables $\beta_{ij} \geq 0$, $i \in I$, $j \in J$, and denote by L_h the problem of finding

$$h(\alpha,\beta) = \min_{x,y,z,v}\Big\{\sum_{k\in K}\Big(c_k^\circ z_k + c_k v_k + \sum_{j\in J}\Big(c_{kj} - \alpha_j + \sum_{i\in I_k}\beta_{ij}\Big)x_{kj}\Big)$$
$$+\sum_{i\in I}\Big(d_i - \sum_{j\in J}\beta_{ij}\Big)y_i\Big\}$$

subject to (3)–(5) and (7). The solution to the dual problem D_h of finding

$$H = \max_{\alpha,\beta}\Big\{h(\alpha,\beta) + \sum_{j\in J}\alpha_j\Big\},$$

also gives a lower bound on the minimal value of the objective function (1). Let $\overline{S}$ denote the linear relaxation of S.

Theorem 1. *The problems D_f and D_h are polynomially solvable.*

PROOF. We show that the problem D_f is polynomially solvable. For each $k \in K$, we consider the auxiliary linear program of finding

$$g_k(\alpha) = \max_{x,v}\Big\{\sum_{j\in J}(\alpha_j - c_{kj})x_{kj} - c_k v_k - c_k^\circ\Big\}$$

subject to (3), (4), and $0 \leq x_{kj} \leq 1$, $k \in K$, $j \in J$. We now verify that the problem $\overline{L}_g$ of finding

$$\overline{f}_g(\alpha) = \min_{y,z}\Big\{\sum_{i\in I} d_i y_i - \sum_{k\in K} g_k(\alpha) z_k\Big\}$$

subject to (6) and $0 \le y_i \le 1$, $0 \le z_k \le 1$, $i \in I$, $k \in K$, has an integer optimal solution.

Let y_i^*, z_k^* be an optimal solution of the problem $\overline{L}_g$. It is clear that if $g_k(\alpha) \le 0$, then $z_k^* = 0$. If $g_k(\alpha) > 0$, then $z_k^* = \min_{i \in I_k} y_i^*$. Put $\delta = \min\{y_i^* > 0, i \in I\}$; $I_\delta = \{i \in I \mid y_i^* = \delta\}$; $K_\delta = \{k \in K \mid z_k^* = \delta\}$; $\varepsilon_1 = \sum_{i \in I_\delta} d_i$; $\varepsilon_2 = \sum_{k \in K_\delta} g_k(\alpha)$. If $\delta = 1$ then the solution y_i^*, z_k^* is an integer. Assume that $\delta < 1$. If $\varepsilon_1 > \varepsilon_2$ then the decreasing of y_i^* and z_k^*, $i \in I_\delta$, $k \in K_\delta$, results in the decreasing of the objective function. If $\varepsilon_1 < \varepsilon_2$ then increasing the values y_i^* and z_k^*, $i \in I_\delta$, $k \in K_\delta$, leads the same result. Hence, $\varepsilon_1 = \varepsilon_2$. Assuming $y_i = 0$ and $z_k = 0$, $i \in I_\delta$, $k \in K_\delta$, we obtain a new optimal solution with a larger parameter δ. Thus, if in the optimal solution all fractional components are nullified, we obtain an integer optimal solution of the problem $\overline{L}_g$.

Note that the problems $\overline{L}_g$ and $\overline{L}_f$ are equivalent and the optimal values of their objective functions coincide. Hence, the problem D_f is equivalent to the problem $\overline{D}_f$, as desired.

Demonstrate solvability in polynomial time of the problem D_h. For each $k \in K$, we consider the auxiliary following linear program

$$g_k(\alpha,\beta) = \max_{x,v}\Big\{\sum_{j\in J}\Big(\alpha_j - c_{kj} + \sum_{i\in I_k}\beta_{ij}\Big)x_{kj} - c_k v_k - c_k^{\circ}\Big\}$$

subject to (3), (4) and $0 \le x_{kj} \le 1$, $k \in K$, $j \in J$. Then the problem L_h is to find

$$h(\alpha,\beta) = \min_{y,z}\Big\{\sum_{i\in I}\Big(d_i - \sum_{j\in J}\beta_{ij}\Big)\,y_i - \sum_{k\in K} g_k(\alpha,\beta)z_k\Big\}$$

subject to (7). Unlike the problem L_f, the variables y_i, $i \in I$, are not already connected to the other variables and their optimal values are determined only by the signs of the corresponding factors. Hence, the problem L_h is a special case of the problem L_f and transition from it to its linear relaxation does not change the minimal value of the objective function. Hence, the problems D_h and $\overline{D}_h$ are equivalent. Theorem 1 is proved.

The linear programs $\overline{D}_f$ and $\overline{D}_h$ have large size, and their solution by standard methods requires significant efforts. For computing the lower bounds F and H it seems expedient to use the method of subgradient optimization [9, 10]. For each iteration of such method the problems L_f and L_h are solved and on using the residuals in (2) and (8), we correct α_j and β_{ij}, $i \in I$, $j \in J$. If the problems L_f and L_h are polynomially solvable, then such approach has certain advantages. The complexity of solving the problem L_h is determined by the complexity of finding the values $g_k(\alpha,\beta)$. For computing the coefficients of the variables x_{kj} not more than $O(qmn)$ operations is required. Finding $g_k(\alpha,\beta)$ requires not more than $O(nq(r+\log q))$ operations [9]. The total running time of the algorithm for solving the problem L_h is $O(qmn + nq(r + \log q))$.

Theorem 2. *The problem L_f can be solved in $O(m^2n+nq(r+\log q))$ time.*

PROOF. Represent the problem L_f as

$$f(\alpha) = \min_{y,z}\Big\{\sum_{i\in I} d_i\, y_i - \sum_{k\in K} g_k(\alpha) z_k\Big\}$$

subject to (6) and (7). Without loss of generality, we may assume that $g_k(\alpha) \geq 0$, $k \in K$. The restrictions (6) can be written as

$$z_k = \prod_{i\in I_k} y_i, \quad k \in K.$$

We may now exclude the variables z_k from consideration. Then the problem L_f can be represented as the minimization problem of the polynomial

$$\Phi(y) = \sum_{i\in I} d_i\, y_i - \sum_{k\in K} g_k(\alpha) \prod_{i\in I_k} y_i$$

in the Boolean variables y_i, $i \in I$. A solution to this problem can be obtained by reducing it to a minimal cut problem on a bipartite network [11]. Let $G = (V, E)$ be a bipartite network with source s, sink t, node set $V = I \cup K \cup \{s,t\}$, and edge set $E = E_1 \cup E_2 \cup E_3$, where I is the node set of the first part, K is the node set of the second part, $E_1 = \{(s,i) \mid i \in I\}$, $E_2 = \{(i,k) \mid i \in I_k, k \in K\}$, and $E_3 = \{(k,t) \mid k \in K\}$. Each edge e has capacity $w(e) \geq 0$ defined by

$$w(e) = \begin{cases} d_i & \text{if } e = (s,i) \in E_1, \\ W & \text{if } e = (i,k) \in E_2 \text{ and } W > \sum_{i\in I} d_i, \\ g_k(\alpha) & \text{if } e = (k,t) \in E_3. \end{cases}$$

Let φ be the size of a minimal cut of G. It is obvious that a minimal cut does not contain edges in E_2. It is clear that for every solution $y_i \in \{0,1\}$ a cut of G can be easily constructed with capacity $\Phi(y) + \sum_{k\in K} g_k(\alpha)$. To this end, it is enough to take all edges (s,i) for which $y_i = 1$, and all edges (k,t) for which $\prod_{i\in I_k} y_i = 0$. We show that the converse statement is true, i.e. for every minimal cut of a network G there exists a solution $y_i \in \{0,1\}$ such that

$$\varphi = \Phi(y) + \sum_{k\in K} g_k(\alpha). \tag{9}$$

Assume that the set of edges $E' = E_1' \cup E_3'$ is a minimal cut of the network G. Put

$$y_i = \begin{cases} 1 & \text{if } e = (s,i) \in E_1', \\ 0 & \text{otherwise.} \end{cases}$$

We check (9). To this end it suffices to show that the set E_3' coincides with the edge set $E_y = \{e = (k,t) \mid \prod_{i \in I_k} y_i = 0\}$.

We first show that $E_3' \supseteq E_y$. Assume that there is an edge $e \in E_y$ such that $e \notin E_3'$, i.e. for some k, $\prod_{i \in I_k} y_i = 0$ holds and $(k,t) \notin E_3'$. In the set I_k take i such that $y_i = 0$. Then $(s,i) \notin E_1'$ and a path from s to t consists of the sequence of the edges (s,i), (i,k), and (k,t) which is impossible.

We show that $E_3' \subseteq E_y$. Assume that there is an edge $e = (k,t) \in E_3'$ such that $e \notin E_y$, i.e. $y_i = 1$, $i \in I_k$. By construction every path from s to k passes only through the nodes of the set I_k. As all edges (s,i), $i \in I_k$, belong to E_1', after deleting the edge (k,t) the set $E' \setminus \{(k,t)\}$ is a cut, which contradicts minimality of E'.

Thus, solving the problem L_f reduces to finding a minimal cut in the bipartite network G which requires $O(m^2 n)$ operations [12]. Calculation of coefficients of $g_k(\alpha)$, $k \in K$, requires $O(nq(r + \log q))$ operations [9]. The total running time of the algorithm for solving the problem L_f does not exceed $O(m^2 n + nq(r + \log q))$. Theorem 2 is proved.

§ 3. Correlation Between the Quantities F and H

We show that the lower bounds F and H are incomparable with one another. For $v_k^\circ \geq \sum_{j \in J} p_{kj}$, $k \in K$, the original problem P is simplified and reduces to finding the quantity

$$\min_{x,y,z} \Big(\sum_{k \in K} \Big(c_k^\circ \, z_k + \sum_{j \in J} c_{kj} \, x_{kj} \Big) + \sum_{i \in I} d_i \, y_i \Big)$$

subject to

$$\sum_{k \in K} x_{kj} = 1, \quad j \in J;$$

$$0 \leq x_{kj} \leq z_k, \quad k \in K, \quad j \in J;$$

$$y_i \geq z_k, \quad k \in K_i, \quad i \in I;$$

$$y_i, z_k \in \{0,1\}, \quad i \in I, \quad k \in K.$$

In [7] a special case of this problem was considered when $|I_k| = 2$ for all $k \in K$. It is shown that in this case $H \geq F$. (The restriction $|I_k| = 2$, $k \in K$, is insignificant, and the statement holds for the arbitrary sets I_k.)

We give an example of input data of the problem P for which the reverse inequality holds; more exactly, $F \geq NH$ for each integer $N > 0$. Put $K = I = \{1, 2, \dots, N\}$, $I_k = \{k\}$, $J = L = \{1\}$, $c_k^\circ = c_k = c_{kj} = V_k = 0$, $p_{kj} = d_i = 1$, and $v_k^\circ = 1/N$, $k \in K$, $j \in J$, $i \in I$. Then the problem D_f can be written as follows: Find

$$\max_{\alpha}\Big\{\alpha + \min_{x,y}\Big\{\sum_{i\in I} y_i - \sum_{k\in K} \alpha\, x_k\Big\}\Big\}$$

subject to

$$x_k \leq v_k^\circ, \quad 0 \leq x_k \leq y_k, \quad y_k \in \{0,1\}, \quad k \in K.$$

This problem decomposes into k independent subproblems of finding

$$g(\alpha) = \min_{x,y}\{y - \alpha x\}$$

subject to

$$Nx \leq 1, \quad 0 \leq x \leq y, \quad y \in \{0,1\}.$$

Hence, $F = \max_\alpha\{\alpha + Ng(\alpha)\}$. Since $g(\alpha) = \min(0, 1 - \alpha/N)$, we have $F = \max_\alpha\{\alpha + \min(0, N - \alpha)\} = N$.

We calculate H from the given input data. With these data, the problem D_h can be written as follows: Find

$$H = \max_{\alpha,\beta}\Big\{\alpha + \min_{x,y}\Big\{\sum_{k\in K}(1 - \beta_k)y_k + \sum_{k\in K}(\beta_k - \alpha)x_k\Big\}\Big\}$$

subject to

$$0 \leq Nx_k \leq 1, \quad y_k \in \{0,1\}, \quad k \in K.$$

The optimal values of the variables β_k is 1 and $H = \max_\alpha\{\alpha + \min(0; 1 - \alpha)\} = 1$.

§ 4. Results of Test Calculations

In the previous section an example of input data is considered for which optimal values of all variables x_{kj} are sufficiently close to zero. Such examples are nontypical. Fractional values of the variable x_{kj} appear under the restrictions (4) or at small values of v_k°, $k \in K$. Below the results of test calculations showing dependence of a parameter $\delta = H/F$ on rigidity of the restrictions (3) and (4), are indicated.

The problems under consideration had dimension $m = 20$, $n = 20$, $q = 30$, $r = 4$. The values of the quantities were chosen as follows: $c_k^\circ = 50$, $c_k = 1$, $c_{kj} = s_{kl} = 0$, $d_i = 100$, $v_k^\circ = 0.25\lambda$, $V_k = \lambda$. The values of p_{kj} and the sets I_k, $k \in K$, were chosen with the help of the generator of random numbers with uniform distribution. The sets I_k with probability 0.2 contain elements of the set I. Thirty percents of the matrix (p_{kj}) are filled by the number 10^6. The results of calculations are presented on the diagram. Numbering of curves corresponds to the following values of p_{kj}:

$$\delta_1 \text{ for } p_{kj} \in [1, 10], \quad \delta_2 \text{ for } p_{kj} \in [2, 4], \quad \delta_3 \text{ for } p_{kj} = 3.$$

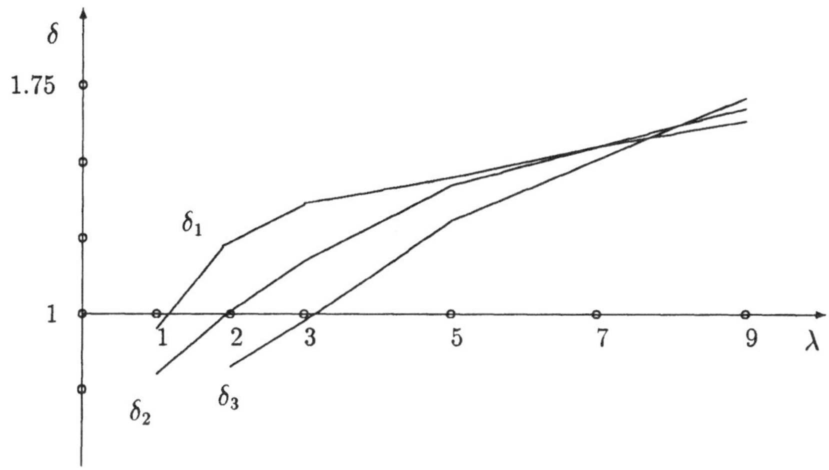

Each break point in the diagram corresponds to the average geometric value of δ for 10 test problems. Note that at $\lambda < 1$ the system of restrictions of the original problem is inconsistent. For $\lambda > 9$ the restrictions (3) and (4) become insignificant and the original problem transforms into the two-level standardization problem (see [2, p. 262]).

References

1. V. L. Beresnev (1977) The problem of optimal choice for series of products and completion nodes. I (in Russian), *Upravlyaemye Sistemy* **16**, 35–46.

2. V. L. Beresnev, È. Kh. Gimadi, and V. T. Dement'ev (1978) *Extremal Standardization Problems* (in Russian), Nauka, Novosibirsk.

3. V. S. Mikhalevich, V. A. Trubin, and N. Z. Shor (1986) Optimization Problems of Production-Transportation Planning (in Russian), Nauka, Moscow.

4. H.-B. Ro and D.-W. Tcha (1984) A branch and bound algorithm for the two-level uncapacitated facility location problem with some side constraints, *European J. Oper. Res.* **18**, No. 3, 349–358.

5. D.-W. Tcha and B.-I. Lee (1984) A branch-and-bound algorithm for the multi-level uncapacitated facility location problem, *European J. Oper. Res.* **18**, No. 1, 35–43.

6. E. N. Goncharov (1994) A mathematical model and a model for solution of a two-level standardization problem (in Russian), in: *Modeli i Metody Optimizatsii*, Trudy Inst. Mat. Vol. 28, Nauka, Novosibirsk, pp. 77–90.

7. A. I. Barros and M. Labbe (1994) A general model for the uncapacitated facility and depot location problem, *Location Science* **2**, No. 3, 173–191.

8. J. Krarup and P. M. Pruzan (1983) The simple plant location problem: survey and synthesis, *European J. Oper. Res.* **12**, No. 1, 36–81.

9. Yu. A. Kochetov and M. G. Pashchenko (1993) Lagrangian relaxations for a choice problem of an optimal system of technical tools (in Russian) *Upravlyaemye Sistemy* **31**, 26–39.

10. M. Held, P. Wolfe, and H. P. Crowder (1974) Validation of subgradient optimization, *Math. Programming* **6**, No. 1, 62–88.

11. A. A. Ageev (1981) Minimization of some polynomials from Boolean variables (in Russian), *Upravlyaemye Sistemy* **21**, 3–5.

12. D. Gusfield, C. Martel, and D. Fernandez-Baca (1987) Fast algorithms for bipartite network flow, *SIAM J. Comput.* **16**, No. 2, 237–251.

Sobolev Institute of Mathematics
Universitetskiĭ pr., 4
Novosibirsk 630090
RUSSIA

TRANSLATED BY I. V. RENTTS

ON SCHEDULES OF A SINGLE MACHINE JOBS WITH PROCESSING TIMES NONLINEAR IN TIME

A. V. Kononov

We study combinatorial complexity of the problem of scheduling jobs of a single machine and some its particular instances. It is established that the problem is NP-hard in the strong sense. The problem remains NP-hard even if we assume that the latest start times are the same for all the jobs, but in this case it can be solved by a pseudopolynomial algorithm.

When examining the problem of scheduling jobs of a single machine we assume that each job has a latest start time and its makespan is the sum of some a fixed processing time and late start penalty. As an optimal criterion we take the length of the schedule obtained; i.e., the makespan of all jobs. The exact formulation of the problem can be found in [1] and will be described below. Note that in [1] some exact algorithms based on the branch and bound method and dynamic programming are suggested together with heuristic polynomial algorithms.

§ 1. Statement of the Problem

Let $I = \{1, 2, \ldots, n\}$ be a set of jobs. Assume that for each job $i \in I$ the following numbers are defined:

$p_i \geq 0$, the fixed processing time,
d_i, the latest start time,
$v_i \geq 0$, the penalty.

The makespan of job i is defined by the function

$$y_i = p_i + \max\{0, v_i(s_i - d_i)\},$$

where s_i is the start time of job i. All jobs are ready to processing at time $t = 0$. Interruption of a job is not permitted. The machine can process only one job at a time. The makespan of all jobs is denoted by $C = \max(s_i + y_i)$, where the maximum is taken over $i \in I$. A schedule minimizing C is called *optimal.*

GP-Problem (of optimal scheduling). *Given* $p_i \geq 0$, $v_i \geq 0$, *and* $d_i \in R$ $(i \in I)$, *find some* s_i *satisfying the conditions*

$$y_i = p_i + \max\{0, v_i(s_i - d_i)\},$$

A. D. Korshunov (ed.), Operations Research and Discrete Analysis, 109–122.

$$y_i + s_i \le s_j \quad \text{if} \quad s_i < s_j,\ j \in I,$$

$$y_i y_j = 0 \quad \text{if} \quad s_i = s_j,\ j \in I,\ i \ne j,$$

$$s_i \ge 0$$

and minimizing the makespan C.

Note that with an optimal schedule the machine jobs without idles. Taking this into account, the objective function can be written as $C = \sum_{i \in I} y_i$, and each schedule is defined entirely by the sequence π of start times.

A *schedule* for the GP-problem is a permutation π determining the order of processing the jobs.

By $C(\pi)$ we denote the makespan defined by the permutation π. A permutation π minimizing $C(\pi)$ is called *optimal*.

Lemma 1 [1]. *Suppose that there is a schedule for the first* k *jobs of the set* I, $\sum_{i=1}^{k} y_i = t$, *such that* $d_j \le t$ *for all* $j > k$. *Then an optimal sequence of the remaining jobs can be obtained by ordering them in ascending order of* $(p_i - d_i v_i)/v_i$.

§ 2. Complexity of the GP-Problem

Consider the following known problem.

SL-Problem (minimizing the total weighted tardiness of a single machine). *Given integer parameters* $\lambda_i \ge 0$, $\nu_i \ge 0$, *and* $\delta_i \ge 0$ $(i \in I)$, *find some* β_i *satisfying the conditions*

$$\xi_i = \max\{0, \beta_i + \lambda_i - \delta_i\},$$

$$\beta_i + \lambda_i \le \beta_j \quad \text{if} \quad \beta_i < \beta_j,\ j \in I,$$

$$\lambda_i \lambda_j = 0 \quad \text{if} \quad \beta_i = \beta_j,\ j \in I,\ i \ne j,$$

$$\beta_i \ge 0$$

and minimizing the quantity $L = \sum_{i \in I} \nu_i \xi_i$.

The parameters λ_i, δ_i, and ν_i characterize the makespan, the due date, and penalty weight of job i, respectively; β_i is the start time of job i, and ξ_i is the tardiness of job i.

In [2] it was shown that the SL-problem is NP-hard in the strong sense. Note that for the SL-problem a schedule is defined by a permutation π describing the order of the jobs start times.

An input of the SL-problem with some $\varepsilon > 0$ gives an input of the GP_ε-problem. We use the following notation: $\nu = \max_{i\in I} \nu_i$ and $\lambda = \max_{i\in I} \lambda_i$. Put

$$p_i = \lambda_i, \quad d_i = \delta_i - \lambda_i, \quad v_i = \nu_i\varepsilon, \quad i \in I, \quad \varepsilon > 0.$$

Let $\Lambda = \sum_{i\in I} \lambda_i$ be the makespan, i.e., the schedule length in the SL-problem, $r_i = s_i - d_i$. Since $\delta_i \geq 0$, we have $\xi_i \leq \Lambda$ for all $i \in I$.

For a permutation π of jobs in the set I, let $S_\pi(GP_\varepsilon) = \{s_i | \ i \in I\}$ and $S_\pi(SL) = \{\beta_i | \ i \in I\}$ be the schedules defined by the permutation π in the GP_ε-problem and the SL-problem. The sets of jobs initialized with tardinesses in the schedules $S_\pi(GP_\varepsilon)$ and $S_\pi(SL)$ are denoted by $J_\varepsilon(\pi) = \{i \in I | \ s_i \geq d_i\}$ and $J(\pi) = \{i \in I | \ \beta_i + \lambda_i \geq \delta_i\}$, respectively.

Let $\{j_1, \dots, j_k\}$ be the jobs in the set $J_\varepsilon(\pi)$ in the order of their processing. Then the value $\rho_{j_q} = \sum_{i=1}^{q} r_{j_i} v_{j_i}$ defines the increase of the completion time of job j_q in the schedule $S_\pi(GP_\varepsilon)$ in comparison with the schedule $S_\pi(SL)$.

Lemma 2. *For every job $j_q \in J_\varepsilon(\pi)$,*

$$\rho_{j_q} \leq \Lambda((1 + \nu\varepsilon)^q - 1). \tag{1}$$

PROOF. By the definition of r_j, d_j, ρ_j, and ξ_j for every j $(1 \leq j \leq k)$

$$r_{j_i} = s_{j_i} - d_{j_i} = \beta_{j_i} + \rho_{j_{i-1}} + \lambda_{j_i} - \delta_{j_i} = \xi_{j_i} + \rho_{j_{i-1}}, \tag{2}$$

where $\rho_{j_0} = 0$. We prove (1) for the jobs $j_q \in J_\varepsilon(\pi)$ by induction on q $(0 \leq q \leq k)$. For $q = 0$, the bound (1) holds since $\rho_{j_0} = 0$. Assume that (1) holds for $q = i - 1$. Then for $q = i$ from (2) it follows that

$$\rho_{j_i} = \rho_{j_{i-1}} + r_{j_i} v_{j_i} \leq \rho_{j_{i-1}} + (\Lambda + \rho_{j_{i-1}})\nu\varepsilon$$

$$= \rho_{j_{i-1}}(1 + \nu\varepsilon) + \Lambda\nu\varepsilon \leq \Lambda((1 + \nu\varepsilon)^i - (1 + \nu\varepsilon) + \nu\varepsilon) = \Lambda((1 + \nu\varepsilon)^i - 1).$$

Lemma 2 is proved.

Lemma 3. *There exist $\varepsilon_0 > 0$ such that for every $\varepsilon < \varepsilon_0$ and every permutation π the equality $J(\pi) = J_\varepsilon(\pi)$ holds.*

PROOF. Since the processing times of jobs in the GP_ε-problem are not less than the processing times of the corresponding jobs in the SL-problem, each job processed with tardiness in the schedule $S_\pi(SL)$ is also processed with tardiness in the schedule $S_\pi(GP_\varepsilon)$. Therefore, $J(\pi) \subseteq J_\varepsilon(\pi)$.

We prove that for some sufficiently small ε the reverse inclusion holds. Indeed, using (2), the inequalities $r_{j_i} \geq 0$ for $j_i \in J_\varepsilon(\pi)$, the integrality of ξ_i and the inequalities

$$\rho_{j_{i-1}} < 1, \quad 1 \leq i \leq k, \tag{3}$$

which can be derived from (1) for small ε for example

$$\varepsilon < \varepsilon_0 = 1/\nu((1/\Lambda + 1)^{1/n-1} - 1),$$

we have $j_q \in J(\pi)$. Lemma 3 is proved.

Theorem 1. *The* GP*-problem is* NP*-hard in the strong sense.*

PROOF. It suffices to prove that some standard NP-hard (in the strong sense) problem can be reduced to an instance of the GP-problem. To this end we take the SL-problem. We show that solving the SL-problem reduces to solving an instance of the GP_ε-problem. Moreover, an optimal permutation π^* for the GP_ε-problem is an optimal permutation of jobs in the SL-problem for sufficiently small ε.

Let π^* be an optimal permutation of jobs of the GP_ε-problem for some $\varepsilon \geq 0$. Assume that it is not optimal for the SL-problem. Since the functional $L(\pi)$ in the SL-problem takes integer values, by assumption we have

$$L(\pi^*) - L(\overline{\pi}) \geq 1, \tag{4}$$

where $\overline{\pi}$ is an optimal permutation for the SL-problem. Demonstrate that (4) implies the inequality

$$C(\pi^*) - C(\overline{\pi}) > 0, \tag{5}$$

contradicting optimality of π^* for the GP_ε-problem. This proves optimality of π^* for the SL-problem. Since Lemma 3 is true for $\varepsilon < \varepsilon_0$, using the lemma and (2) for every permutation π and every job $j_i \in J_\varepsilon(\pi)$ we obtain

$$r_{j_i} = \xi_{j_i} + \rho_{j_{i-1}}.$$

By the definition of the functional C for every permutation π we have

$$C(\pi) = \sum_{i \in I} y_i = \sum_{i \in I} (p_i + v_i(s_i - d_i)^+)$$

$$= \sum_{i \in I} p_i + \sum_{i \in J_\varepsilon(\pi)} v_i(s_i - d_i) = \sum_{i \in I} p_i + \sum_{j_i \in J_\varepsilon(\pi)} v_{j_i} r_{j_i}$$

$$= \sum_{i \in I} p_i + \sum_{j_i \in J_\varepsilon(\pi)} \nu_{j_i} \varepsilon(\xi_{j_i} + \rho_{j_{i-1}}) = \sum_{i \in I} p_i + \varepsilon \sum_{j_i \in J_\varepsilon(\pi)} \nu_{j_i}(\xi_{j_i} + \rho_{j_{i-1}}).$$

By (4) and the definition of the functional L, we obtain

$$C(\pi^*) - C(\overline{\pi})$$

$$= \varepsilon \Bigg[\sum_{j \in J(\pi^*)} \nu_j \xi_j - \sum_{j \in J(\overline{\pi})} \nu_j \xi_j \Bigg] + \varepsilon \Bigg[\sum_{j_i \in J(\pi^*)} \nu_{j_i} \rho_{j_{i-1}} - \sum_{j_i \in J(\overline{\pi})} \nu_{j_i} \rho_{j_{i-1}} \Bigg]$$

$$\geq \varepsilon \Bigg[L(\pi^*) - L(\overline{\pi}) - \sum_{j_i \in J(\overline{\pi})} \nu_{j_i} \rho_{j_{i-1}} \Bigg] \geq \varepsilon - \varepsilon \nu \sum_{j_i \in J(\overline{\pi})} \rho_{j_{i-1}}. \tag{6}$$

We now establish an upper bound for $\varepsilon\nu\sum_{j_i\in J(\overline{\pi})}\rho_{j_{i-1}}$. By (1) for each permutation π we have

$$\varepsilon\nu\sum_{j_i\in J(\overline{\pi})}\rho_{j_{i-1}}\leq\varepsilon\nu\Lambda\left[\sum_{q=1}^{n}(1+\nu\varepsilon)^q-n\right]\leq\varepsilon\nu\Lambda\left[\frac{(1+\nu\varepsilon)^{n+1}-1-\nu\varepsilon}{\nu\varepsilon}-n\right]$$
$$=\Lambda((1+\nu\varepsilon)^{n+1}-1-\nu\varepsilon(n+1))=\Lambda((1+\nu\varepsilon)^{n+1}-1-\nu\varepsilon-n\nu\varepsilon)$$
$$=\Lambda\left[\sum_{i=0}^{n+1}\binom{n+1}{i}(\nu\varepsilon)^i-1-\nu\varepsilon-n\nu\varepsilon\right]=\Lambda\sum_{i=2}^{n+1}\binom{n+1}{i}(\nu\varepsilon)^i$$
$$<\Lambda\sum_{i=2}^{n+1}\frac{(n+1)^i(\nu\varepsilon)^i}{i!}=\Lambda(\nu\varepsilon(n+1))^2\sum_{i=2}^{n+1}\frac{(n+1)^{i-2}(\nu\varepsilon)^{i-2}}{i!}.$$

Take the rational value $\varepsilon=1/(\Lambda\nu^2(n+1)^2)$. Note that $\varepsilon<\varepsilon_0$, and the encoding length of ε does not exceed $\log_2(\Lambda\nu^2(n+1)^2)$ and bounded by a polynomial in the length of the input of the SL-problem. We have

$$\Lambda(\nu\varepsilon(n+1))^2\sum_{i=2}^{n+1}\frac{(n+1)^{i-2}(\nu\varepsilon)^{i-2}}{i!}\leq\varepsilon\sum_{i=2}^{n+1}\frac{1}{i!}\leq(e-2)\varepsilon<0.8\varepsilon,$$

which together with (6) implies (5). Theorem 1 is proved.

PV-Problem. *Solve* the GP-problem *provided that the equality* $v_i=v$ *holds for all* $i\in I$.

Theorem 2. *The* PV-*problem is* NP-*hard.*

The SL-problem with $\nu_i=1$ is known as the problem of minimizing the total tardiness. In [3] it is proved that the problem is NP-hard. Repeating the proof of Theorem 1 we can verify that the problem of minimizing the total tardiness is reducible to the PV-problem.

§ 3. Problems with Common Latest Start Times

Below it is supposed that the latest start times of all jobs are the same.

PD-Problem. *Solve* the GP-problem *provided that the equality* $d_i=D$ *holds for all* $i\in I$.

Consider some special cases of the PD-problem.

(PD≥)-Problem. *Solve* the PD-problem *provided that* $p_i/v_i\geq p_j/v_j$ *for all* i *and* j $(i,j\in I)$ *such that* $p_i>p_j$ *and* $v_i>0, v_j>0$.

(PD≤)-Problem. *Solve* the PD-problem *provided that* $p_i/v_i \leq p_j/v_j$ for all i and j (i, $j \in I$) such that $p_i > p_j$ and $v_i > 0, v_j > 0$.

PDP-Problem. *Solve* the PD-problem *provided that there exists a span of time* $[t_0, t_1]$ ($t_0 < t_1$) *when the machine is repaired; i.e. it does not process any job, and* $D \in [t_0, t_1]$.

In this section we show that the (PD≥)-problem is polynomially solvable and construct some "pseudopolynomial" algorithms for the remaining problems. Note that by Lemma 1 the following statement is true: if $D \leq 0$, then an optimal sequence of jobs in the PD-problem (and, consequently, in all its particular cases) results from scheduling the jobs in ascending order of p_i/v_i.

Suppose that the jobs in the set I are numbered as follows:

$$\begin{gathered} \text{if } p_i < p_j, \text{ then } i < j, \\ \text{if } p_i = p_j \text{ and } p_i/v_i < p_j/v_j, \text{ then } i < j. \end{gathered} \tag{7}$$

Lemma 4. *The permutation* π *indicating the order of processing jobs in accordance with* (7) *defines an optimal schedule.*

PROOF. We prove the lemma by way of contradiction. Let π be an optimal schedule containing some jobs i and j such that $j < i$. Assume that job i is processed before job j. Without loss of generality, we may assume that job j is processed immediately after i. If this is not the case, then there exist jobs i_1 and j_1 which satisfy the condition on the processing order and $j_1 < i_1$.

If $D \leq s_i < s_j$, then jobs i and j belong to the set Q of jobs which are not ordered by time D. Order jobs in Q in accordance with the order of (7). Then, by the definition of (PD≥)-problem these jobs are scheduled in ascending order of p_i/v_i, which by Lemma 1 implies optimality of the schedule.

Let $t_0 = s_i < D < s_j$. Denote by $\bar{\pi}$ the schedule obtained from π by interchanging jobs i and j. Let $t_{ij} = y_i(\pi) + y_j(\pi)$ and $t_{ji} = y_j(\bar{\pi}) + y_i(\bar{\pi})$. Then

$$t_{ij} = p_i + p_j + (t_0 + p_i - D)v_j, \quad t_{ji} = p_i + p_j + (t_0 + p_j - D)^+ v_i.$$

We show that $t_{ij} \geq t_{ji}$. Indeed, this inequality holds if $(t_0 + p_j - D)^+ = 0$. If $(t_0 + p_j - D)^+ > 0$, then

$$\begin{aligned} t_{ij} - t_{ji} &= (t_0 + p_i - D)v_j - (t_0 + p_j - D)v_i \\ &= v_i v_j (p_i/v_i - (D - t_0)/v_i - p_j/v_j + (D - t_0)/v_j) \\ &= v_i v_j (p_i/v_i(1 - (D - t_0)/p_i) - p_j/v_j(1 - (D - t_0)/p_j)) \geq 0. \end{aligned}$$

Hence, interchanging jobs i and j in π does not increase the makespan. Finally, if $s_i < s_j \leq D$, then an alternation of the order of processing jobs i and j in the permutation π does not affect the makespan.

In each of the treated cases interchanging jobs i and j does not worsen the value of the functional C. Hence, there exists an optimal schedule π in which the jobs are scheduled in ascending order of p_i. Lemma 4 is proved.

Theorem 3. *There exists an $O(n \log n)$ algorithm for solving the* (PD $\geq$)-*problem.*

PROOF. The proof follows immediately from Lemma 4.

Lemma 5. *Let π be an optimal schedule in the* PD-*problem such that the performance of job i is finished by the latest time D. Then there exists an optimal schedule π^* in which job i is the first to be processed.*

Lemma 6. *Let π be an optimal schedule in the* PD-*problem such that $s_i(\pi) \geq D$ for every job i and $p_i/v_i \geq p_j/v_j$ for every job $j \in I$ such that $s_j(\pi) \geq D$. Then there exists an optimal schedule π^* in which job i is the last to be processed.*

Lemma 7. *Suppose that in the* PD-*problem job i in an optimal schedule π satisfies the following inequalities: $s_i \leq D$ and $p_i \leq p_j$ for every j such that $s_j < s_i$. Then there exists an optimal schedule π^* in which job i is the first to be processed.*

The claims of Lemmas 5 and 7 are obvious, and Lemma 6 follows immediately from Lemma 1.

Now we proceed to constructing algorithms. Up to the end of this section we suppose that the set I of jobs is indexed as follows:

$$\begin{gathered} \text{if } p_i/v_i < p_j/v_j, \text{ then } i < j, \\ \text{if } p_i/v_i = p_j/v_j \text{ and } p_i > p_j, \text{ then } i < j. \end{gathered} \tag{8}$$

By $\langle k, d, l_0, l_1 \rangle$ we denote the special instance of the PDP-problem in which the set of jobs consists of the first k jobs of the set I, the latest start time is equal to d for all the jobs, and $[l_0, l_1]$ is the span of repairing. Let $\langle n, D, t_0, t_1 \rangle$ be the original PDP-problem, $0 \leq t_0 \leq D \leq t_1$. Consider the following family of instances of the PDP-problem: $\{\langle k, \Delta \rangle \mid k = 1, \dots, n; \Delta = 0, \dots, t_0\}$, where $\langle k, \Delta \rangle = \langle k, D - \Delta, t_0 - \Delta, t_1 - \Delta \rangle$. By $C^*(k, \Delta)$ we denote the optimal value of the objective function C in the problem $\langle k, \Delta \rangle$. To simplify computations we suppose that the function $C^*(k, \Delta)$ is defined for all values of the parameters k and Δ, where $C^*(k, \Delta) = \infty$ for $k \notin [1, n]$ or $\Delta \notin [0, t_0]$.

Lemma 8. *For the* PDP-*problem the following recurrence relations of dynamic programming hold*:

$$C^*(1, \Delta) = \begin{cases} p_1 & \text{for } p_1 \leq t_0 - \Delta, \\ t_1 - \Delta + p_1 + (t_1 - D)v_1 & \text{for } p_1 > t_0 - \Delta, \end{cases} \tag{9}$$

$$C^*(k, \Delta) = \min\{C_1, C_2\}, \tag{10}$$

where

$$C_1 = p_k + C^*(k-1, \Delta + p_k),$$
$$C_2 = C' + p_k + (C' - D + \Delta)v_k,$$
$$C' = \max\{t_1 - \Delta, C^*(k-1, \Delta)\}.$$

PROOF. Relation (9) makes it possible to compute the length of an optimal schedule if we need to perform one job.

To prove (10) we suppose that all the values on the right side of the equality (10) are known. Then by Lemmas 5 and 6, for job k there exists an optimal schedule in which this job is processed either in the first position before repairing or in the last position after repairing.

In the first case an optimal schedule of the remaining jobs is obtained by solving the $\langle k-1, \Delta + p_k \rangle$-problem. Consequently, $C^*(k, \Delta) = C_1$.

In the second case job k is initialized either at time $t_1 - \Delta$ if the preceding jobs were finished by time $t_0 - \Delta$ or otherwise at time $C^*(k-1, \Delta)$. Therefore, $C^*(k, \Delta) = C_2$. Lemma 8 is proved.

Put $D^+ = \max\{0, D\}$.

Theorem 4. *The* PDP*-problem with integer input data, n jobs, and a common latest start time D is solvable in $O(nD^+ + n \log n)$ time.*

PROOF. Indeed, by (9) and (10), to obtain an optimal schedule for the $\langle n, 0\rangle$-problem it suffices to consider $n(t_0 + 1)$ instances of the $\langle k, \Delta \rangle$-problem. Moreover, to obtain the original indexing of the jobs one needs to perform $O(n \log n)$ operations. Hence, solution of the PDP-problem requires $O(nD^+ + n \log n)$ operations. Theorem 4 is proved.

By $\langle k, d\rangle'$ we denote the instance of the PD-problem in which the set of jobs consists of the first k jobs of the set I and the latest start time of all jobs is equal to d. Let $\langle n, D\rangle'$ be the original PD-problem. By $C^*(k, d)$ we denote the optimal value of the objective function C in the $\langle k, d\rangle$-problem. To simplify the computations we suppose that the function $C^*(k, d)$ is defined for all values of the parameters k and d, where $C^*(k, d) = \infty$ if $k \notin [1, n]$ or $d \notin [0, D]$.

Lemma 9. *For the* PD*-problem the following recurrence relations of dynamic programming hold:*

$$C^*(1, d) = p_1, \tag{11}$$

$$C^*(k, d) = \min \begin{cases} C^*(k-1, d-p_k) + p_k, \\ C^*(k-1, d) + p_k + v_k(C^*(k-1, d) - d)^+, \\ \min\limits_{i \in (0, p_k)} \max\{d + i, C^*(k-1, d, d - p_k + i, d + i)\}, \end{cases} \tag{12}$$

where $C^*(k-1,d,d-p_k+i,d+i)$ *is an optimal solution of the instance of the* PDP $\langle k-1,d,d-p_k+i,d+i\rangle$*-problem.*

PROOF. Relation (11) permits us to calculate the length of an optimal schedule if it is necessary to perform only one job.

To prove (12) we suppose that all values on the right side of (12) are known. Then if job k in an optimal schedule is processed by the latest start time, then by Lemma 5 it is the first to be processed; an optimal schedule for the remaining jobs is obtained by solving the $\langle k-1,d-p_k\rangle'$-problem. Hence, $C^*(k,d) = C^*(k-1,d-p_k)+p_k$. If job k in an optimal schedule is processed after the latets start time, then by Lemma 6 it is the last to be processed. Then

$$C^*(k,d) = C^*(k-1,d) + p_k + v_k(C^*(k-1,d)-d)^+.$$

Finally, if job k in an optimal schedule is initialized by the latets start time and is finished after it, then an optimal schedule for the remaining jobs is obtained as a solution of the $\langle k-1,d,d-p_k+i,d+i\rangle$-problem for some $i \in (0,p_k)$. Hence,

$$C^*(k,d) = \min_{i\in(0,p_k)} \max\{d+i, C^*(k-1,d,d-p_k+i,d+i)\}.$$

Lemma 9 is proved.

Theorem 5. *The* PD*-problem with integer input data, n jobs, and a common latets start time D is solvable in $O(nD^+p^2+n\log n)$ time, where $p=\max_{i\in I} p_i$.*

PROOF. Note that for fixed values of d and k there are at most p^2 distinct $\langle k-1,d,d-p_k+i,d+i\rangle$-problems. Therefore, by (11) and (12) to obtain an optimal schedule for the $\langle n,D\rangle'$-problem it is sufficient to solve at most nD^+p^2 $\langle k-1,d,d-p_k+i,d+i\rangle$-problems and at most nD^+ $\langle k,d\rangle'$-problems. Moreover, to obtain the original indexing of the jobs one needs to perform $O(n\log n)$ operations. Hence, a solution of the PD-problem is obtained by performing $O(nD^+p^2+n\log n)$ operations. Theorem 5 is proved.

Lemma 10. *For the* (PD $\leq$)*-problem the following recurrence relations of dynamic programming hold:*

$$C^*(1,d) = p_1, \tag{13}$$

$$C^*(k,d) = \min\begin{cases} C^*(k-1,d-p_k)+p_k, \\ C^*(k-1,d)+p_k+v_k(C^*(k-1,d)-d)^+. \end{cases} \tag{14}$$

PROOF. Relation (13) permits us to calculate the length of an optimal schedule if it is necessary to perform only one job.

To prove (14) we suppose that all values on the right side of (14) are known. Note that since the jobs are indexed in the order stated in (8), for the (PD$\leq$)-problem these jobs are indexed in descending order of p_i. Then if job k in an optimal schedule starts by the latest start time, then, by Lemma 7, it is the first to be processed; an optimal schedule for the remaining jobs can be obtained by solving the $\langle k-1, d-p_k\rangle'$-problem. Hence, $C^*(k,d) = C^*(k-1, d-p_k)+p_k$. If the job k in an optimal schedule is processed after the latets start time, then, by Lemma 6, it is the last to be processed. Then $C^*(k,d) = C^*(k-1,d)+p_k+v_k(C^*(k-1,d)-d)^+$. Lemma 10 is proved.

Theorem 6. *The* (PD $\leq$)*-problem with integer input data, n jobs, and a common latest start time D is solvable in $O(nD^+ + n\log n)$ time.*

PROOF. Indeed, taking (13) and (14) into account, to obtain an optimal schedule for the $\langle n, D\rangle'$-problem it is sufficient to solve nD^+ $\langle k,d\rangle'$-problems. Moreover, to obtain the original indexing of the jobs we need to perform $O(n\log n)$ operations. Hence, a solution for the (PD$\leq$)-problem can be found by performing $O(nD^+ + n\log n)$ operations. Theorem 6 is proved.

§ 4. Complexity of Problems with a Common Latest Start Time

In this section we show that the PD-, PDP-, and (PD$\leq$)-problems are NP-hard. We start with formulating the following known problem.

ZR-Problem. *Given an integer $a_i \geq 1$ ($i \in I$) and an integer $B \geq 1$, find a subset $J \subseteq I$ such that $\sum_{i\in J} a_i = B$.*

NP-completeness of the problem of the sum of sizes is shown in [4]. Note that the problem was mentioned as NP-hard in [5].

From the input of the ZR-problem we construct an input of the PD-problem. We set

$$a = \max_{i\in I} a_i, \quad A = \sum_{i\in I} a_i,$$

$$p_0 = 1, \quad v_0 = \mu^{-1},$$

$$p_i = \mu a_i, \; v_i = \varepsilon a_i, \quad i \in I,$$

$$D = \mu B,$$

where $\varepsilon = a^{-2}n^{-2}$ and $\mu = \varepsilon(28An)^{-1}$.

Denote by $\overline{t_i}$ the completion time of job k, i.e. $\overline{t_i} = s_i + y_i$.

Lemma 11. *Let $K = \{1, \dots, k\}$ be the list of the jobs that was not scheduled by time $T > D$. Suppose that since time T the jobs are processed successively, in the order of their positions in the list K. Then for any $l = 1, \dots, k$*

$$\overline{t_l} = D + (T-D)\prod_{i=1}^{l}(1+v_i) + \sum_{i=1}^{l} p_i \prod_{j=i+1}^{l}(1+v_j). \tag{15}$$

PROOF. By the definitions of $\overline{t_i}, y_i$ and the assumptions of the lemma for each $i = 1, \ldots, k$, we have

$$\overline{t_i} = \overline{t_{i-1}} + p_i + (\overline{t_{i-1}} - D)v_i. \tag{16}$$

We prove the lemma by induction on q $(1 \leq q \leq k)$. For $q = 1$ we have

$$\overline{t_1} = D + (T - D)(1 + v_1) + p_1 = T + p_1 + (T - D)v_1,$$

and (15) holds.

Assume that (15) holds for $q = l - 1$. Then from (16) for $q = l$ we obtain

$$\overline{t_l} = \overline{t_{l-1}} + p_l + (\overline{t_{l-1}} - D)v_l$$

$$= D + (T - D)\prod_{i=1}^{l-1}(1 + v_i) + \sum_{i=1}^{l-1} p_i \prod_{j=i+1}^{l-1}(1 + v_j) + p_l$$

$$+ \left[D + (T - D)\prod_{i=1}^{l-1}(1 + v_i) + \sum_{i=1}^{l-1} p_i \prod_{j=i+1}^{l-1}(1 + v_j) - D\right]v_l$$

$$= D + (T - D)\prod_{i=1}^{l-1}(1 + v_i) + v_l(T - D)\prod_{i=1}^{l-1}(1 + v_i)$$

$$+ \sum_{i=1}^{l-1} p_i \prod_{j=i+1}^{l-1}(1 + v_j) + v_l \sum_{i=1}^{l-1} p_i \prod_{j=i+1}^{l-1}(1 + v_j) + p_l$$

$$= D + (T - D)\left[\prod_{i=1}^{l-1}(1 + v_i)\right](1 + v_l) + \left[\sum_{i=1}^{l-1} p_i \prod_{j=i+1}^{l-1}(1 + v_j)\right](1 + v_l) + p_l$$

$$= D + (T - D)\prod_{i=1}^{l}(1 + v_i) + \sum_{i=1}^{l} p_i \prod_{j=i+1}^{l}(1 + v_j).$$

Lemma 11 is proved.

Theorem 7. *The* PD-*problem is* NP-*hard.*

PROOF. We reduce the ZR-problem to an instance of the PD-problem.

Let π be an optimal permutation in the PD-problem and Q be a set of the jobs, processed without violating the latest start times.

Consider two cases.

CASE 1: job 0 belongs to Q. Since $p_0 \geq p_i$ for any $i \in I$, we may assume that, among the jobs in Q, job 0 is last to be processed. If in the schedule defined by the permutation π job 0 is initialized at time μB, then the set $Q \setminus \{0\}$ of jobs is a solution to the ZR-problem. Suppose that 0 in the optimal permutation π is initialized at time T, $T < \mu B$. We show that there is no $J \subseteq I$ such that $\sum_{i \in J} a_i = B$. Assume to the contrary that such a subset exists. We construct another permutation π^* as follows. Suppose that the jobs in the set J are processed first, then job 0 and afterwards the remaining jobs in an arbitrary order. We suppose that the jobs are numerated in the permutation π in the order $j_0, j_1, \dots, j_n$; and in the permutation π^*, in the order $k_0, k_1, \dots, k_n$. All jobs in the set J are finished by time μB, since by assumption we have $\sum_{i \in J} p_i = \sum_{i \in J} \mu a_i = \mu B$. By Lemma 11 we obtain

$$C(\pi) = D + (T + 1 - D) \prod_{j_i \in I \setminus Q} (1 + v_{j_i}) + \sum_{j_i \in I \setminus Q} p_{j_i} \prod_{l=i}^{n} (1 + v_{j_l}),$$

$$C(\pi^*) = D + \prod_{k_i \in I \setminus J} (1 + v_{k_i}) + \sum_{k_i \in I \setminus J} p_{k_i} \prod_{l=i}^{n} (1 + v_{k_l}).$$

Hence,

$$\begin{aligned}
C(\pi) - C(\pi^*) &= (T + 1 - D) \prod_{j_i \in I \setminus Q} (1 + v_{j_i}) + \sum_{j_i \in I \setminus Q} p_{j_i} \prod_{i}^{n} (1 + v_{j_l}) \\
&\quad - \prod_{k_i \in I \setminus J} (1 + v_{k_i}) - \sum_{k_i \in I \setminus J} p_{k_i} \prod_{l=i}^{n} (1 + v_{k_l}) \\
&\geq (1 - (D - T)) \prod_{j_i \in I \setminus Q} (1 + v_{j_i}) - \prod_{k_i \in I \setminus J} (1 + v_{k_i}) - \sum_{k_i \in I \setminus J} p_{k_i} \prod_{l=i}^{n} (1 + v_{k_l}) \\
&\geq (1 - (D - T)) \prod_{j_i \in I \setminus Q} (1 + \varepsilon a_{j_i}) - \prod_{k_i \in I \setminus J} (1 + \varepsilon a_{k_i}) - \sum_{k_i \in I \setminus J} \mu a_{k_i} \prod_{l=i}^{n} (1 + \varepsilon a_{k_l}) \\
&\geq 1 - (D - T) + (1 - (D - T)) \sum_{j_i \in I \setminus Q} \varepsilon a_{j_i} - 1 - \sum_{k_i \in I \setminus J} \varepsilon a_{j_i}
\end{aligned}$$

$$-\sum_{i=2}^{n}\binom{n}{i}(\varepsilon a)^i - \sum_{k_i\in I\setminus J}\mu a_{k_i}\prod_{l=i}^{n}(1+\varepsilon a_{k_l}). \tag{17}$$

Next we have

$$\sum_{i=2}^{n}\binom{n}{i}(\varepsilon a_i)^i \le \varepsilon\sum_{i=2}^{n}\frac{n^i a^i \varepsilon^{i-1}}{i!} \le \varepsilon\sum_{i=2}^{n}\frac{n^{1-i}a^{1-i}}{i!} < 0.8\varepsilon. \tag{18}$$

Therefore,

$$\sum_{k_i\in I\setminus J}\mu a_{k_i}\prod_{l=i}^{n}(1+\varepsilon a_{k_l}) \le \varepsilon\sum_{k_i\in I\setminus J}\frac{1}{28n}\left(1+\sum_{i=1}^{n}\varepsilon a_i + 0.8\varepsilon\right)$$

$$\le \frac{1}{28n}\varepsilon\left(1+\frac{1}{an^2}+\frac{0.8}{a^2n^2}\right) \le 0.1\varepsilon. \tag{19}$$

By (18) and (19), from (17) it follows that

$$C(\pi)-C(\pi^*) \ge 1-(D-T)+(1-(D-T))\sum_{j_i\in I\setminus Q}\varepsilon a_{j_i} - 1 - \sum_{k_i\in I\setminus J}\varepsilon a_{j_i} - 0.9\varepsilon.$$

Since $D-T \le D = \mu B = \frac{\varepsilon}{28n}$ and $\sum_{j_i\in I\setminus Q} a_{j_i} - \sum_{k_i\in I\setminus J} a_{j_i} \ge 1$, we have

$$C(\pi)-C(\pi^*) \ge 0.1\varepsilon - \frac{\varepsilon}{28n} - \frac{\varepsilon}{28an^2} > 0,$$

which contradicts optimality of π.

CASE 2: job 0 does not belong to Q. Since a_i are integers, job 0 does not start by time $\mu B+\mu$ and $C(\pi) \ge \mu B + \mu + 1 + \mu v_0 > 2$. Let π^* be a permutation in which job 0 is the first to be processed, with the remaining jobs processed in arbitrary order. By Lemma 11 and (18), (19) we obtain

$$C(\pi^*) = D + (1-D)\prod_{i=1}^{n}(1+v_i) + \sum_{i=1}^{n}\left(p_i\prod_{j=i+1}^{n}(1+v_j)\right) \le D+1$$

$$+\sum_{i=1}^{n}\varepsilon a_i + 0.9\varepsilon < 2,$$

which contradicts optimality of π. Theorem 7 is proved.

Theorem 8. *The* (PD $\leq$)*-problem is* NP*-hard.*

PROOF follows from the fact that the parameters of the PD-problem constructed in the proof of Theorem 7 satisfy the following condition: if $p_i > p_j$, then $p_i/v_i \leq p_j/v_j$ for any $i, j \in I$.

Theorem 9. *The* PDP*-problem is* NP*-hard.*

PROOF. Take repairing in the PD-problem to be in the interval $[1 + \mu B, 1 + \mu B + \mu]$. Now it remains to repeat the proof of Theorem 7.

References

1. A. S. Kunnathur and S. K. Gupta (1990) Minimizing the makespan with late start penalties added to processing times in a single facility scheduling problem, *European J. Oper. Res.* **47**, No. 1, 56–64.
2. E. L. Lawler (1977) A "pseudopolynomial" algorithm for sequencing jobs to minimize total tardiness, *Ann. Discrete Math.* Vol. 1, North-Holland, Amsterdam, pp. 331–342.
3. J. Du and J. Y.-T. Leung (1990) Minimizing total tardiness on one machine is NP-hard, *Math. Oper. Res.* **15**, No. 3, 483–495.
4. R. M. Karp (1972) Reducibility among combinatorial problems, *Complexity of Computer Computations*, Plenum Press, New York, pp. 85–103.
5. M. R. Garey and D. S. Johnson (1979) *Computers and Intractability*, Freeman, San Francisco.

Sobolev Institute of Mathematics
Universitetskiĭ pr., 4
Novosibirsk 630090
RUSSIA

TRANSLATED BY A. V. RYAZANOV

PROBLEMS OF RECONSTRUCTING WORDS FROM THEIR FRAGMENTS AND APPLICATIONS[†)]

V. K. Leont'ev

Problems of reconstructing words from their fragments and applications to image recognition and information theory are considered. A number of new results characterizing the possibilities of recognition and decoding in terms of combinatorial and geometric constructs is presented.

Introduction

In the present paper, the statements of the research problems and results are discussed whose common concern is the problem of reconstructing words from their fragments. Interest in this subjects has arisen relatively recently and the concepts under investigation, problems, and methods for solving them are now at the stage of shaping. More generally, the problems of existence and of reconstructing an object of a certain class using some information about the set of its "parts" and the correlation of these "parts" may be considered as the problems of image recognition and information theory [1–3]. Such approach turns out to be useful and this paper discusses some new models that appear in connection with applications of the problems of reconstructing words from their fragments to speech recognition and to reliable message receiving.

In a theoretical aspect, the problems of reconstructing words lead to the necessity of investigating some new mathematical constructs that reflect correlations between the properties of sequences of symbols and the properties of their subsequences [4–10]. In the paper, a number of results are presented and also the combinatorial problems are formulated serving as a basis for the presented research. At the same time, the paper actually contains more than is pointed out in its title: many general concepts of "combinatorics of words" are elucidated.

Some of the results below are contained in the papers [5, 6, 11–13], while the others are new and appear here for the first time.

---

†) This research was supported by the Russian Foundation for Basic Research (Grant 93-01-00449).

A. D. Korshunov (ed.), Operations Research and Discrete Analysis, 123–144.

§ 1. The Covering Problem on a Set of Words

Let E^n be a set of binary words of length n, and let A^n be a set of binary words of length at most n. On the set A^n consider the partial order $\prec$ assuming, by definition, $a \prec b$ if the sequence of letters in a word a is a subsequence of the sequence of letters in a word b. In other words, a can be obtained from b by crossing out some letters in b. In this case, we say that the word a is a *fragment* of the word b [6, 10]. The structure of this partially ordered set is rather complicated and the results given below provide those trifles that we were able to establish as regards this set.

The Hasse diagram of a partially ordered finite set X is a *directed graph* whose vertices are the elements of X and whose two vertices x and y are connected by an arc from x to y if $x \succ y$ and there exists no $z \in X$ such that $x \succ z \succ y$.

The following statements may be considered as geometric facts concerning the Hasse diagram of A^n with the order $\prec$.

Lemma 1. *For every binary word b of length k, there are $\sum_{i=k}^{n} \binom{n}{i}$ binary words a of length n satisfying the condition $a \succ b$.*

The proof and application as well as various interesting interpretations of this statement can be found in [5, 14, 15].

In order to obtain a more detailed information on the Hasse diagram, we represent a binary word $a \in E^n$ as the series: $a = \gamma^{t_1}\bar{\gamma}^{t_2}\dots\gamma^{t_{r-1}}\bar{\gamma}^{t_r}$, where $\gamma \in \{0,1\}$ and $\bar{\gamma}$ is the logical negation of the letter γ. The word γ^t is called a *series* and the number t is called the *length of the series*. For example, if $a = 10011100$, then $a = 10^2 1^3 0^2$.

Lemma 2. *Let $a \in E^n$. Then the number of words b in E^{n-1} such that $b \prec a$ is equal to the number of series in the word a.*

It follows from Lemmas 1 and 2 that in the Hasse diagram of the partial order under study the outdegree of each vertex in the rth level is equal to $r+2$ and the indegree of the vertex corresponding to the word a is equal to the number of series in this word. The distribution of indegree is described by the following simple lemma.

Lemma 3. *In the rth level of the Hasse diagram, there are $2\binom{r-1}{k-1}$ vertices with indegree equal to k.*

It is easy to show that the "average" indegree of the vertices in the rth level is equal to $(r+1)/2$. Thus, the rth level of the Hasse diagram of the considered order looks approximately as follows: the outdegree of every vertex is equal to $r+2$ and the indegree is equal to $(r+1)/2$ on the average.

It should be note that the construction of Hasse diagrams for individual words is connected with computing the so-called Whitney parameters, i.e., the number of different fragments of a word a of the same length. Apparently, finding explicit Whitney parameters in a general case is a serious combinatorial problem [15, 16].

Let $t(a,k)$ be the number of different fragments of length k in a word a.

We give some possibly useful simple bounds for the value $t(a,k)$.

Put, by definition, $t(a,0) = 0$, and let ab be the concatenation of the words $a = \alpha_1 \dots \alpha_n$ and $b = \beta_1 \dots \beta_n$, i.e., $ab = \alpha_1\alpha_2 \dots \alpha_n\beta_1\beta_2 \dots \beta_n$.

Proposition. *For all a and b, the inequalities*

$$\max_p\{t(a,p)t(b,k-p)\} \le t(ab,k) \le \sum_{p=0}^{k} t(a,p)t(b,k-p)$$

hold. In particular, if the number of series in a is equal to m and $k \le |a| - 1$, then $t(a,k) \ge m$.

Note that it stands to reason to find the extrema of the function $t(a,k)$ of the following form.

Let word a consist of m series and the length of the ith series of a be equal to t_i, $1 \le i \le m$, i.e., $a = \gamma^{t_1}\bar{\gamma}^{t_2} \dots \gamma^{t_m}$, and let $t(k, t_1, \dots, t_m)$ denote the number of fragments of length k in a. Then we have $t(a,k) = t(k, t_1, \dots, t_m)$. Put

$$t(k,m,n) = \min(k, t_1, \dots, t_m), \tag{1}$$

where the minimum is taken over all the tuples $\{t_1, \dots, t_m\}$ of positive integers such that

$$\sum_{i=1}^{m} t_i = n, \quad t_i \ge 1, \ 1 \le i \le m.$$

It is clear that (1) is only one of the numerous examples of the Shannon-type functions associated with $t(a,k)$.

Often it is an interesting and difficult problem to calculate the Whitney parameters of special words.

EXAMPLE. Let $\xi_n = 1010\dots10$ be a word of length n, and let $t_{n,k}$ be the number of words of length k which are fragments of ξ_n.

Proposition. *For all n and k,*

$$t_{n,k} = \binom{k-1}{2k-n-1} + 2\sum_{r=2k-n+1}^{k} \binom{k-1}{r-1}. \tag{2}$$

This statement gives an answer to one of the questions in [15].

In (2), summation begins from zero if the lower bound for r is negative and $\binom{n}{m} = 0$ for $m > n$.

The word ξ_n has many interesting properties and, in particular, the universality property since, for $k \leq \lfloor n/2 \rfloor$, the word ξ_n contains as its fragments all the words of length k and, for $n = 2k$, the word ξ_n is the word of minimal length containing as its fragments all words of length k. It is easy to see that ξ_{2k} is not the unique universal word of minimal length, and therefore the problem arises of describing all such words of length $2k$. In connection with the "monotonicity" of the universality property, it is interesting to describe all the terminal universality words, i.e., the words losing the universality property after removing an arbitrary letter.

We consider the following problems.

COVERING PROBLEM. Given a set of words $\{a_1, a_2, \ldots, a_m\}$, generally of different lengths, in the alphabet $\{0, 1\}$, find a word a of minimal length such that this word contains as its fragments all the words a_i, $1 \leq i \leq m$. In [17], this problem is called the *minimal common supersequence problem.*

EXAMPLE. Let $a_1 = 1010$, $a_2 = 010$, and $a_3 = 111$. It is easy to check that the word $a = 10101$ is the minimal common supersequence of the words a_1, a_2, and a_3.

Another example of a minimal common supersequence is the word ξ_{2k} given above which is universal for the set of all binary words of length k.

For the set of binary words of length n such that each of these words contains exactly k units, a minimal common supersequence is given in [9]. Let us also mention the well-known unproven hypothesis that the length of the minimal common supersequence for the set of all permutations of an n-element set is equal to $n^2 - 2n + 4$ for each $n \geq 3$ (see, for example, [18]).

MULTIPLE COVERING PROBLEM. Given a set of words $a_1, a_2, \ldots, a_m$ with multiplicities $\nu_1, \nu_2, \ldots, \nu_m$, find a word a of minimal length such that each word a_i, $1 \leq i \leq m$, is contained as a fragment in this word at least ν_i times.

Note that for the covering problems formulated, it is interesting to find all the minimal terminal coverings, i.e., the coverings that stop to be universal after removing an arbitrary letter.

Another problem closely related to the covering problem is the *maximal common subsequence problem* [17].

MAXIMAL COMMON SUBSEQUENCE PROBLEM. Given a set of words $\{a_1, a_2, \ldots, a_m\}$ in the alphabet $\{0, 1\}$, find a word a of maximal length such that this word is a fragment of each word a_i, $1 \leq i \leq m$.

The minimal common supersequence and maximal common subsequence problems are closely related, and in [18], an algorithm of dynamic-programming type is given for solving the first problem. Generally, these problems are NP-complete [18].

However, for the problem formulated above unlike the covering problem of a set by a family of its subsets, there is a possibility of finding simple and acceptable universal bounds for the cardinality of a minimal covering.

Let $d(a_1, a_2, \ldots, a_m)$ be the length of a minimal common supersequence for a set of words $\{a_1, \ldots, a_m\}$ and

$$\gamma(a_1, a_2, \ldots, a_m) = \max\{|a_1|, |a_2|, \ldots, |a_m|\}.$$

Then the following obvious estimates are true:

$$\gamma(a_1, a_2, \ldots, a_m) \leq d(a_1, a_2, \ldots, a_m) \leq 2\gamma(a_1, a_2, \ldots, a_m).$$

The general multiple covering problem can be written as the following nonlinear problem of Boolean programming. Let

$$x^\sigma = \begin{cases} x & \text{for } \sigma = 1, \\ 1 - x & \text{for } \sigma = 0. \end{cases}$$

Then multiple covering problem can be formulated as follows. Given $a_r = (\sigma_1^r, \sigma_2^r, \ldots, \sigma_k^r)$, find the minimal positive integer r such that

$$\sum_{1 \leq i_1 < i_2 \cdots < i_k \leq z} x_{i_1}^{\sigma_1^{r_1}} x_{i_2}^{\sigma_2^{r_2}} \ldots x_{i_k}^{\sigma_k^{r_k}} \geq \nu_r, \quad 1 \leq r \leq m, \ x_i = \{0, 1\}.$$

EXAMPLE. Let $a_1 = 1010$, $a_2 = 0101$, and $a_3 = 111$. Then the optimization problem consists in finding the minimal value of z subject to

$$\sum_{1 \leq i_1 < i_2 < i_3 < i_4 \leq z} x_{i_1}(1 - x_{i_2})x_{i_3}(1 - x_{i_4}) \geq 1,$$

$$\sum_{1 \leq i_1 < i_2 < i_3 < i_4 \leq z} (1 - x_{i_1})x_{i_2}(1 - x_{i_3})x_{i_4} \geq 1,$$

$$\sum_{1 \leq i_1 < i_2 < i_3 \leq z} x_{i_1}x_{i_2}x_{i_3} \geq 1,$$

$$x_i = \{0, 1\}.$$

Without further theoretical and experimental research, it is difficult to predetermine whenever this formulation of the covering problem is useful. However, such formulation turns out to be useful when solving the problem of reconstructing words from their fragments (see below).

Returning to the covering problem, we consider the simplest case of $m = 2$. The length $d(a,b)$ of a minimal covering of words a, $b \in E^n$ satisfies the obvious inequalities

$$n+1 \leq d(a,b) \leq 2n. \tag{3}$$

Both bounds in (3) are attainable. In order to obtain the lower bound in (3), it is sufficient to take the word ξ_n as a and to take its logical negation as b, i.e. $b = \bar{\xi}_n$. The upper bound is attained at the pair $a = 1^n$, $b = 0^n$.

The minimal number of insertions and deletions of letters that transforms the word a into the word b determines some distance in E^n denoted by $\rho_L(a,b)$. This notion first introduced by V. I. Levenshtein in connection with research the codes capable of correcting deletions, insertions, and reversals.

EXAMPLE. Let $a = 0110$ and $b = 1001$. Then the word $c = 101010$ is a minimal covering for a and b, and $\rho_L(a,b) = 4$ (from a, the first and fourth symbols are removed and in the second and third places, the symbols zero and one are inserted, respectively).

There exists a simple connection between the functions $\rho_L(a,b)$ and $d(a,b)$.

Lemma 4. *For all a, $b \in E^n$, the equality*

$$d(a,b) - n = \frac{1}{2}\rho_L(a,b)$$

holds.

We call the distance $\rho(x,y) = d(x,y) - n$ the *covering metric.*

The following lemma is valid.

Lemma 5. *If (x,y) is the length of a minimal common supersequence for some words x and y in E^n, then*

$$d(x,y) + (x,y) = 2n. \tag{4}$$

From (4) it follows that the words $a = 1^n$ and $b = 0^n$ form the unique pair of words in E^n such that the distance between these words in the covering metric is equal to n (in the Hamming metric the number of such pairs is equal to 2^{n-1}). Note also that, unlike the Hamming metric, the covering metric is not invariant under automorphisms. Therefore, in particular, the number of points in a ball of a fixed radius in the covering metric depends on the center of the ball.

We call a word consisting of alternating 0s and 1s an *alternating* word.

EXAMPLES. 1. If some ball has radius 1 and is centered at 0^n, then this ball coincides with a ball in the Hamming metric and the number of its points is equal to $n+1$.

2. If a ball has radius 1 and is centered at an alternating word of length n, then the number of points in the ball is equal to $\binom{n+1}{2} + 1$.

In the general case, the number of points in a ball of radius 1 depends on the distribution of series in the word that is the center of this ball.

DEFINITION. An alternating subword b in a word a is called a *maximal* subword if there is no alternating subword c in a such that the word b is a proper subword of the word c.

EXAMPLE. The word $\gamma = \alpha_1\alpha_2\alpha_3\ldots\alpha_9 = 100110101$ has three maximal alternating subwords: $\alpha_1\alpha_2 = 10$, $\alpha_3\alpha_4 = 01$, and $\alpha_5\alpha_6\alpha_7\alpha_8\alpha_9 = 10101$.

Denote by k_i the number of maximal subwords of length i in a word γ.

Theorem 1[†)]. *Let γ be the center of a ball of radius 1 and the word γ consist of m series and k_i maximal alternating subwords of length i. Then the number of points in the ball of radius 1 centered at $\gamma \in E^n$ is equal to*

$$nm + 1 - \sum_{i=2}^{m} k_i \binom{i}{2}.$$

Corollary. *The maximal number of points in a ball of radius 1 in the covering metric is asymptotically equal to n^2.*

Thus, the inequalities

$$n + 1 \leq |S(\gamma)| < n^2(1 + o(1))$$

hold, where $|S(\gamma)|$ is the size of the unit ball centered at the point $\gamma \in E^n$.

It should be noted that finding the size of a ball of radius t in the covering metric is not a trivial problem due to the above-mentioned problem of computing the function $t(a, k)$. Below, we give bounds for the size of a ball of radius t which do not pretend to be universal and complete.

Let $v_t(n, m)$ denote the size of a ball of radius t centered at the point $a \in E^n$ consisting of m series.

Proposition. *For all n and m, the inequalities*

$$\sum_{i=0}^{t} \binom{n}{i} \leq v_t(n, m) < (n-1)^t \prod_{i=0}^{t-1} (m + i) \tag{5}$$

---

†) This result was obtained by the author in collaboration with J. Arocha (Cuba).

hold. If m and t are fixed and $n \to \infty$, then from (5) *it follows that*

$$\frac{n^t}{t!} \lesssim v_t(n,m) \lesssim m^t n^t.$$

Some inequalities for the function $v_t(n,m)$ can be extracted from the results of [11, 14] and also from other sources. However, there is no complete picture of the asymptotic behavior of the value $v_t(n,m)$ as a function of three parameters.

§ 2. Applications to Image Recognition and Information Theory

We consider a new decision-making model in information theory and image recognition.

Suppose that a given subset M in E^n is used as a source of messages. In addition, let the number of potential receivers of the messages in M be a fixed number q. Assume that each receiver has its own portion of information about an unknown message $a \in M$. Let this information be given by the vector $\{a_1, a_2, \ldots, a_q\}$. Using the information $\{a_1, a_2, \ldots, a_q\}$ and the data about the channel (way of receiving the information a_i) it is required a decision be made as to whether the message transmitted belongs to a fixed class, i.e., it is necessary to decode the message. It is clear that for $q = 1$ we have the classical Shannon model [1].

In the model formulated above, the "individuality" of perception is a "parameter" which is referred to as the "own" portion of information. We consider two examples in which the concept of individual information is given a precise sense.

1. Acoustic image recognition from fragments. We suppose that under the conditions of bad audibility, a message broadcast on the air is received by several stations. Furthermore, the noise of receiving the message is so strong that some letters are not read by receivers. So, in this case, each receiver has a fragment of the message (a fragment of a word) as input. The individuality of perception consists in the fact that the fragments received by different receivers are formed by deleting of different groups of symbols.

EXAMPLE. Let $q = 2$ and $a = 000111$, and assume it known that every message consists of five letters, i.e., each message is obtained from a by deleting one symbol. Then the restriction given above consists in the fact that the numbers of the symbols deleted must be different for the receivers, although the fragments themselves may coincide. For example, this will be the case if the first receiver reads the message a' that obtained from a by deleting the first symbol, and the second receiver reads the message a'' that obtained from a by deleting the third symbol.

Note that the model considered differs from the model of a channel with deleting and inserting symbols in the part related to the number of receivers [11]. The last model is well know in coding theory.

As one of the simplest restrictions on a communication channel, we assume to take the following: each of the messages received at an output is a fragment of length $n - t$ from the original word.

DEFINITION. A set of words $\mathcal{M} \subseteq E^n$ is called *q-decodable* if each word $a \in \mathcal{M}$ is uniquely reconstructed from q fragments under the above restrictions.

DEFINITION. The multiset of all fragments of length t of a word $a \in E^n$ is called the *t-neighborhood* of the word a. This neighborhood is denoted by $S_t(a)$.

REMARK. By the intersection of multisets we mean the set of their common elements; in addition, each element a is placed into the intersection with multiplicity equal to the minimal multiplicity of a in the intersected multisets.

According to the definitions and restrictions on the channel, it is possible to formulate the following condition for q-decodability of the set $\mathcal{M}$.

Proposition. *A set $\mathcal{M} \subseteq E^n$ is q-decodable if and only if there are at most $q - 1$ elements in the intersection of the $(n - t)$-neighborhoods of all elements $a, b \in M$.*

It is interesting to compare this condition with the usual condition of decodability for codes with deleting and inserting symbols [11].

The following statement shows that, for a sufficiently large q, the set E^n can be q-decodable.

Let $t = 1$, i.e., the information of the receiver is words of length $n - 1$.

Theorem 2 [19]. *The set E^n is q-decodable if and only if $q > \lfloor n/2 \rfloor + 1$.*

It is easy to give an example showing the necessity of the number of receivers in the theorem. Let $a = 0^{\lfloor n/2 \rfloor} 1 0^{n - \lfloor n/2 \rfloor - 1}$ and $b = 0^{\lfloor n/2 \rfloor - 1} 1 0^{n - \lfloor n/2 \rfloor}$. Then the fragments obtained by deleting one zero from the first series in the word a and one zero from the last series in the word b are the same.

According to Theorem 2, in the case of many receivers the results may principally differ from the classical coding theory, though, of course, the number of receivers does not evoke any optimism.

However, the examples below related to a binary symmetric communication channel show that this situation is not fatal. Following the usual procedure, we introduce the function $A(n, q, t)$ giving the size of a maximal q-decodable set with correcting at most t deletions of symbols in words of length n. From Theorem 2 it follows that $A(n, q, 1) = 2^n$ for $q > \lfloor n/2 \rfloor + 1$.

Note that in this variant it is possible to consider also the problem formulated in terms of probability when every symbol of the word $\alpha_1 \alpha_2 \ldots \alpha_n$ deletes with probability δ and preserves its place with probability $1-\delta$. In this case, the output is the vector $(\beta_1, \beta_2, \ldots, \beta_q)$ and we need to find a word in the code dictionary $\mathcal{M}$ "closest" to this vector which corresponds to the decoding procedure using the maximal likelihood principle. In this case, this procedure may be implemented as follows.

Maximal likelihood decoding.

1. For a set of words $\{a_1, a_2, \ldots, a_q\}$ in A^n, all the coverings of length n are found, i.e., the set $\mathcal{M}^0 = \{b_1, b_2, \ldots, b_N\}$ of all words in E^n such that each b_i contains all the words $a_1, a_2, \ldots, a_q$ as its fragments is constructed.

2. The distance between the sets $\mathcal{M}$ and $\mathcal{M}^0$ is found and every word a, for which this distance is "realized," is declared the desired message.

2. Coding with many receivers. Note that an analogous problem can be formulated also in the classical coding theory.

Consider a code set $\mathcal{M} \subseteq E^n$ as the source of messages and assume that, when transmitting via a communication channel, some letters can be transformed into their negations $(0 \to 1,\ 1 \to 0)$. Assume also that the message of the source is received at the output by two receivers and that there is at most one distortion in a word of length n.

The problem is standard: how large may the code set be when the source message is correctly reconstruction?

RESTRICTIONS. The messages obtained by different receivers result from the distortion of *different* positions in the source message.

So, the neighborhood of a code word $a \in \mathcal{M}$ in this case consists of the pairs (a', a''), where a', $a'' \in S_1(a)$ and $S_1(a)$ is the ball of radius 1 (in the Hamming metric) centered at the word a.

It is easily seen that the set $\mathcal{M}$ is decodable if and only if every two balls of radius 1 centered at the points of the code set intersect at most in a singleton.

Indeed, let a word $a \in \mathcal{M}$ was transmitted via the channel. Then the pair (a', a'') will appear at the output, where a', $a'' \in S_1(a)$. Further, the pair (a', a'') can belong to only *one* ball centered at in a point $\mathcal{M}$, since otherwise the condition of decodability would be broken (the cardinality of the intersection would be equal to two). Therefore, the ball $S_1(a)$ (and, therefore, its center) is uniquely determined from the pair (a', a'').

Note that, in the Hamming metric, balls of radius 1 either do not intersect at all or intersect in two points. Therefore, the previously formulated condition of decodability, in the case of one error and two receivers, transforms into the usual condition of nonintersection of balls of radius 1 centered at the points of the code set.

Observe that if the number of receivers is $q \geq 3$, then in the case of one error the set E^n is a decodable set.

Let the function $A(n, q, 1)$ denote the cardinality of the maximal decodable set in the case of one error and q receivers. Then from the above reasoning, we have the following relations for the function $A(n, q, 1)$:

$$A(n, 2, 1) = A(n, 1, 1) = A(n, 3), \quad A(n, q, 1) = 2^n \quad (q \geq 3).$$

In the general case, when there are t errors and q receivers, the condition of decodability looks as follows.

THE DECODABILITY CONDITION. A set $\mathcal{M}$ is decodable if and only if the intersection of two arbitrary balls of radius t centered at code points consists of at most $q - 1$ points.

Thus, in the case of many receivers, the classical condition that dense packing is replaced by a less burdensome condition for the intersection to be "small." The corresponding function $A(n, q, t)$ denotes, as above, the size of a maximal decodable set.

Note the following simple fact: for each $n \geq 4$ any two balls of radius 2 centered at points lying at distance $d = 4$ intersect in six points.

We now consider the code $\mathcal{M}$ with distance 4 in E^n (for example, the Hamming code with parity check). By virtue of the fact mentioned above, any two balls of radius 2 centered at points of the set $\mathcal{M}$ intersect in at most six points. Therefore, the set $\mathcal{M}$ is decodable and the following lower bound holds:

$$A(n, 7, 2) \geq \frac{2^n}{n}(1 + o(1)).$$

For comparison, we note that in the classical case of one receiver the relation

$$A(n, 1, 2) \asymp 2^n / n^2$$

holds for the function $A(n, 1, 2)$.

In the case of probabilistic formulation, it is possible to consider for simplicity a binary symmetric channel. Then at the output of the channel there appears a set of words $\{a_1, a_2, \ldots, a_q\}$ each of which is obtained from the same word a by replacing some of its letters by their negations with a given probability δ. The decoding based on the maximal likelihood principle may be done in the usual way. However, in the case of "ambiguity" of a nearest word in the variant with many receivers, there are still many ways for excluding this ambiguity.

REMARK. The lower bound given above can be easily generalized to find a general bound

$$A\left(n, \binom{2t+1}{t}, t\right) \succcurlyeq \frac{2^n}{n^{t-1}}$$

which is n times greater in order than the cardinality of the best t-error-correcting code. Undoubtedly, these are only the first results of such kind. Here, of course, it is interesting not only to obtain more precise bounds for cardinality of the code but also to establish a correlation of the problem with other problems, for example, in the case of the model with repetition as a kind of error correction. It is also important to consider different types of communication channels. We note that the results may differ substantially, as was shown in the two previous cases of the channel with deletion of letters and the binary symmetric channel.

§ 3. An Economical Form of a Set of Words and the Realizability Problem

In connection with many problems of computer mathematics, the problem arises concerning the economical way of storing words in the computer memory and their quick retrieval from the memory. This problem has many different aspects, and a number of combinatorial constructions concerning its solution are known. Among them we mention such well-known constructions as the Gray code [20], de Bruijn sequences [21] (in some papers they are called "circular codes"), cycles in the n-dimensional unit cube [22], and the Hamming codes. We consider one of the aspects of this problem.

The realizability problem. Let $\mathcal{M} = \{a_1, a_2, \ldots, a_m\} \subseteq E^k$. The question is whether there exist a number n and a word $a \in E^n$ such that the set of fragments of length k in the word a is the same of the set $\mathcal{M}$.

Every word a which is a solution of the above problem is called a *realization* of $\mathcal{M}$. The word a is like a package for the set $\mathcal{M}$ since it is an economical form of the set.

EXAMPLES. 1. Let $m = 1$. Then the set $\mathcal{M}$ is realizable only in the cases when $a_1 = 0^k$, $a_1 = 1^k$, and a_1 is an arbitrary word for $k = n$.

2. For $m = 2$, only the following sets are realizable:

$$\mathcal{M}_1 = \{1^k, 1^{k-1}0\},\ \mathcal{M}_2 = \{0^k, 0^{k-1}1\},\ \mathcal{M}_3 = \{10^{k-1}, 0^k\},\ \mathcal{M}_4 = \{01^{k-1}, 1^k\}.$$

This statement is easily justified since each realization of a two-element set cannot have more than two series.

It is natural to solve the problem of realizability with densest packing, i.e., to look for a word of minimal length which is a realization of the given set.

In the case when the fragments of a word are the subwords the realizability problem was investigated in a more general form in [23] in which the question was posed on the minimal number of words in common realizing the given set. Efficient solutions of the corresponding problems were also obtained for generalizations: the problems of covering graphs by paths [24].

In general, solution of the realizability problem can be represented as search of (0, 1)-solutions to a nonlinear equation.

Let $M = \{a_1, a_2, \ldots, a_m\} \subseteq E^k$ and $a_r = \alpha_1^r \alpha_2^r \ldots \alpha_k^r$, $1 \leq t \leq m$.

Proposition. *A word a is a realization of a set M if and only if a is a solution to the equation*

$$\sum_{r=1}^{m} \sum_{1 \leq i_1 < \cdots < i_k \leq n} x_{i_1}^{\alpha_1^r} x_{i_2}^{\alpha_2^r} \ldots x_{i_k}^{\alpha_k^r} = \binom{n}{k}. \tag{6}$$

EXAMPLE. Let $m = 1$ and $M = \{a_1 = \alpha_1 \alpha_2 \ldots \alpha_k\}$. Then (6) has the following form:

$$\sum_{1 \leq i_1 < \cdots < i_k \leq n} x_{i_1}^{\alpha_1} x_{i_2}^{\alpha_2} \ldots x_{i_k}^{\alpha_k} = \binom{n}{k}.$$

Since the number of terms in the sum is equal to $\binom{n}{k}$, all the terms are equal to one. Using this fact and the definition of the function x^σ, it is easy to see that $\alpha_1 = \alpha_2 = \cdots = \alpha_k$.

EXAMPLE. Let $M = \{(10), (01)\}$. Then (6) takes the following form:

$$\sum_{r=1}^{2} \sum_{1 \leq i_1 < i_2 \leq n} x_{i_1}^{\alpha_1^r} x_{i_2}^{\alpha_2^r} = \sum_{1 \leq i_1 < i_2 \leq n} x_{i_1}(1 - x_{i_2}) + \sum_{1 \leq i_1 < i_2 \leq n} (1 - x_{i_1}) x_{i_2} = \binom{n}{2}.$$

Using this fact and Lemma 4 in [6] or by straightforward calculation, we obtain

$$\sum_{1 \leq i_1 < i_2 \leq n} x_{i_1} = \sum_{i=1}^{n} (n - i) x_i, \qquad \sum_{1 \leq i_1 < i_2 \leq n} x_{i_2} = \sum_{i=1}^{n} i x_i.$$

After simple transformations, we arrive at the equation

$$2y^2 - 2y(n + 1) + n(n - 1) = 0,$$

where $y = \|x\| \leq n$, which has no solution for $n \geq 4$. From this it easily follows that the set M is not realizable. This illustrative example demonstrates the principle possibilities of using an algebraic approach.

Problem of reconstructing words from their fragments. This problem in the form presented here seems to have been first posed in [12]. Below we present a general formof it, some results, and a substantive interpretation. In nature the problem being considered is close to problem of Ulam [25] well know in graph theory.

Note that close to the problem under consideration we call the problem of reconstructing a word from the set of its subwords. This problem was studied in [26, 27] in which conditions were found for uniqueness reconstruction and a polynomial algorithm reconstructing a word from a set of overlapping subwords of fixed length. It is noted there that the problems was motivated by reconstructing the structure of polymer molecules from their overlapping fragments.

Let us return to the initial problem. Fix a set of subsequences of the sequence $1, 2, \dots, n$ and consider only those fragments of a word a that are generated by this set of subsequences. It is convenient to represent each subsequence by the characteristic tuples $(v_1, v_2, \dots, v_n)$ such that $v_i = 1$ if i belongs to the subsequence and $v_i = 0$ otherwise.

EXAMPLE. Let $n = 4$. Then the subsequences $(2, 4)$ and $(1, 3, 4)$ are defined by the binary tuples (0101) and (1011).

Denote by $V = \{v_1, v_2, \dots, v_N\}$ the given set of characteristic tuples and describe the procedure of generating a fragment e by a tuple v as the fragmentation operation $\langle a, v\rangle = e$.

EXAMPLE. Let $a = 1010$ and $v = 0011$. Then $\langle a, v\rangle = 10$.

We call the set V *characteristic* for the set of fragments.

Denote by $V^0(x)$ the set of fragments of a word x generated by all the possible tuples in V, and by $V^1(x)$, the corresponding multiset containing each fragment of the word x as many times as this fragment is obtained as a result of fragmentation of the word x by the set V.

DEFINITION. Words $a, b \in E^n$ are called *V^i-equivalent* if $V^i(a) = V^i(b)$, $i = 0, 1$.

The V^0-equivalence was defined in [5] and the V^1-equivalence was defined in [10]. V^i-equivalence of a and b is denoted as follows: $a \overset{V^i}{\sim} b$.

It is clear that the V^0-equivalence implies the V^1-equivalence but the converse fails.

The main problem of studying the V^i-equivalence consists in describing the equivalence classes and in finding efficient algorithms for establishment the V^i-equivalence.

For describing the equivalence classes we need few concepts.

Consider the equation

$$\langle x, v\rangle = e.$$

The set of solutions of this equation is an $(n - \|v\|)$-dimensional subcube in E^n.

Further, for an arbitrary system of sets $\{C_{ij}\}$, $1 \le i, j \le N$, we define a new set called the *set-theoretic permanent* of $\{C_{ij}\}$ as follows:

$$\operatorname{per} A = \operatorname{per} \|C_{ij}\| = \bigcup_{\{i_1,\dots,i_N\}} (C_{1i_1} \cap C_{2i_2} \cap \dots \cap C_{Ni_N}).$$

The set-theoretic permanent resembles the usual permanent of a matrix in which the operations of multiplication and addition are replaced with the operations of intersection and union.

Denote by V_a^0 the V^0-equivalence class containing a word a.

For the characteristic set $V = \{v_1, v_2, \dots, v_N\}$ and a word $a \in E^n$, we form a set of fragments $V^0(a) = \{e_i\}$. Further, consider the equation of the form

$$\langle x, v_i \rangle = e_j$$

and denote by A_{ij} the set of solutions of this equation.

Theorem 3 [5]. *For each* a,

$$V_a^0 = \operatorname{per} \|A_{ij}\|.$$

The procedure of describing equivalence classes in the second definition is a little more cumbersome although nothing changes substantially.

1. Using the characteristic set V, the set $V^1(a)$ is found. Let $V^1(a) = \{e_1, e_2, \dots, e_m\}$.

2. From the set $V^1(a)$ all the samples (multisets) of size N are formed each of which contains the set $V^1(a)$. Let the samples be $Q_1, Q_2, \dots, Q_t$, where $t = \binom{N-1}{m-1}$.

3. Associate with each sample Q_i a matrix A_i defined as follows. Put $Q_i = \{e_1^i, e_2^i, \dots, e_N^i\}$. Consider the equation

$$\langle x, v_r \rangle = e_s^i, \quad 1 \le s, r \le N.$$

Denote by A_{sr}^i the set of solutions of this equation and put $A_i = \|A_{sr}^i\|$.

Theorem 4 [10]. *For each* a,

$$V_a^1 = \bigcup_{i=1}^{t} \operatorname{per} A_i.$$

Theorems 3 and 4, although proving a description for V^i-equivalence classes, are not suitable enough for use in practice. However, in the case $i = 0$ there is another, more convenient, algebraic way for describing the equivalence classes to which we now turn.

Let $V = E_n^2$, i.e., all the pairs (α_i, α_j) are considered as fragments of the word $a = \alpha_1 \alpha_2 \dots \alpha_n$. In this case the following statement is true.

Proposition. *The words $a = \alpha_1\alpha_2\ldots\alpha_n$ and $b = \beta_1\beta_2\ldots\beta_n$ are E_n^2-equivalent if and only if the conditions*

$$\sum_{i=1}^{n}\alpha_i = \sum_{i=1}^{n}\beta_i, \quad \sum_{i=1}^{n}i\alpha_i = \sum_{i=1}^{n}i\beta_i \tag{7}$$

hold.

Thus, in the case $V = E_n^2$ each class V_a^0 is described by the equations

$$\sum_{i=1}^{n}x_i = p, \quad \sum_{i=1}^{n}ix_i = q,$$

where

$$p = \sum_{i=1}^{n}\alpha_i, \quad q = \sum_{i=1}^{n}i\alpha_i.$$

In general, the algebraic solution of the problem of reconstructing words from their fragments is as follows.

We return to the definition of the characteristic set as the collection of subsequences of the sequence $1, 2, \ldots, n$. We denote this set by V^*.

EXAMPLE. Let $V = E_n^2$. Then $V^* = \{(i,j) \mid i < j;\ 1 \le i,\ j \le n\}$.

Partition the set V^* into the subsets $V_1^*, V_2^*, \ldots, V_n^*$ including in the subset V_i^* all the subsequences in V^* of length i. Thus $V^* = \bigcup_{i=1}^{n} V_i^*$. Associate with each subset V_i^* the matrix A_i whose rows are fragments of the word $x = x_1x_2\ldots x_n$ corresponding to the characteristic set V_i^*.

EXAMPLE. Let $V^* = \{(1,2,3),(2,3,4),(3,4),(4,5)\}$. Then $V_1^* = \varnothing$, $V_2^* = \{(3,4),(4,5)\}$, $V_3^* = \{(1,2,3),(2,3,4)\}$, $V_4^* = \varnothing$, and

$$A_2 = \begin{Vmatrix} x_3 & x_4 \\ x_4 & x_5 \end{Vmatrix}, \quad A_3 = \begin{Vmatrix} x_1 & x_2 & x_3 \\ x_2 & x_3 & x_4 \end{Vmatrix}.$$

On the set of matrices $A = \|r_{ij}\|$ of size $m \times n$ define the function $\Theta(A)$:

$$\Theta(A) = \sum_{j=1}^{m} r_{j_1}r_{j_2}\ldots r_{j_n}.$$

Denote by $A_q(j_1, j_2, \ldots, j_s)$ the submatrix of the matrix A_q consisting of the columns with indices $j_1, j_2, \ldots, j_s$.

EXAMPLE. Let, as above, $A_3 = \begin{Vmatrix} x_1 & x_2 & x_3 \\ x_2 & x_3 & x_4 \end{Vmatrix}$. Then

$$A_3(1,2) = \begin{Vmatrix} x_1 & x_2 \\ x_2 & x_3 \end{Vmatrix}, \qquad \Theta(A_3(1,2)) = x_1x_2 + x_2x_3.$$

Denote by $\lambda^{j_1 j_2 \dots j_s}_{i_1 i_2 \dots i_s}$ the number of rows in the matrix $A_q(j_1, j_2, \dots, j_s)$ coinciding with the row $(x_{i_1} \dots x_{i_s})$.

EXAMPLE. Let $A = \begin{Vmatrix} x_1 & x_2 & x_3 \\ x_3 & x_2 & x_4 \\ x_1 & x_2 & x_4 \end{Vmatrix}$. Then $\lambda^{12}_{12} = 2$, $\lambda^{12}_{13} = 0$, $\lambda^{12}_{23} = 0$, and $\lambda^{12}_{32} = 1$.

In certain cases the parameters $\lambda^{j_1 j_2 \dots j_s}_{i_1 i_2 \dots i_s}$ which play a key role in the problem of V^0-equivalence can be given in explicit form.

Lemma 6 [6]. *If $V = E^k_n$, i.e., the characteristic set is the set of all subsequences of length k of the sequence $1, 2, \dots, n$, then*

$$\lambda^{j_1 j_2 \dots j_s}_{i_1 i_2 \dots i_s} = \binom{i_1 - 1}{j_1 - 1}\binom{i_2 - i_1 - 1}{j_2 - j_1 - 1} \cdots \binom{n - i_s}{k - j_s},$$

where $1 \le i_1 < i_2 < \dots < i_s \le n$.

We now turn to formulating an algebraic criterion of the equivalence.

1. Partition the characteristic set V^* into the subsets V_i^*, i.e. obtain $V^* = \bigcup_{i=1}^n V_i^*$. For each element in V_q^*, construct the matrix A_q. Let the cardinalities of nonempty subsets in $\{V_i^*\}$ be $q_1, q_2, \dots, q_m$.

2. For a fixed q, consider all the possible submatrices of the matrix A_q of the form $A_q(j_1, j_2, \dots, j_s)$ $(s = 1, 2, \dots, q)$ and for each such matrix construct the polynomial defined above

$$\Theta[A_q(j_1, j_2, \dots, j_s] = \sum_{1 \le i_1 < i_2 < \dots < i_s \le q} \lambda^{j_1 j_2 \dots j_s}_{i_1 i_2 \dots i_s} x_{i_1} x_{i_2} \dots x_{i_s}.$$

Theorem 5 [6]. *The words $a = \alpha_1\alpha_2 \dots \alpha_n$ and $b = \beta_1\beta_2 \dots \beta_n$ are V^0-equivalent if and only if*

$$\sum_{1 \le i_1 < i_2 < \dots < i_s \le q_i} \lambda^{j_1 j_2 \dots j_s}_{i_1 i_2 \dots i_s} \alpha_{i_1}\alpha_{i_2} \dots \alpha_{i_s} = \sum_{1 \le i_1 < i_2 < \dots < i_s \le q_i} \lambda^{j_1 j_2 \dots j_s}_{i_1 i_2 \dots i_s} \beta_{i_1}\beta_{i_2} \dots \beta_{i_s},$$

$$1 \le s \le q_i, \quad 1 \le i \le m.$$

In these equalities s ranges from 1 to q_i, and for all s and q_i, $\binom{q_i}{s}$ equations are written. Thus, the total number of equations in the system is equal to $\sum_{i=1}^m 2^{q_i}$.

EXAMPLE. Let $n = 4$ and $V = E_n^2$. Then

$$V^* = \{(1,2),(1,3),(1,4),(2,3),(3,4),(3,4)\}.$$

In this case, the length of all fragments is equal to two, $V^* = V_2^*$, and

$$A_2 = \begin{Vmatrix} x_1 & x_2 \\ x_1 & x_3 \\ x_1 & x_4 \\ x_2 & x_3 \\ x_2 & x_4 \\ x_3 & x_4 \end{Vmatrix}.$$

The matrix A_2 contains two submatrices $A_2(1)$ and $A_2(2)$, where

$$A_2(1) = \begin{Vmatrix} x_1 \\ x_1 \\ x_1 \\ x_2 \\ x_2 \\ x_3 \end{Vmatrix}, \qquad A_2(2) = \begin{Vmatrix} x_2 \\ x_3 \\ x_4 \\ x_3 \\ x_4 \\ x_4 \end{Vmatrix}.$$

Further, we have

$$\Theta[A_2(1)] = 3x_1 + 2x_2 + x_3, \;\; \Theta[A_2(2)] = x_2 + 2x_3 + 3x_4, \;\; \Theta[A_2] = \sigma_2(x_1x_2x_3x_4).$$

Now, let $\alpha = 1001$ and $\beta = 0110$. Then

$$\begin{aligned} \Theta_\alpha[A_2(1)] &= 3, & \Theta_\beta[A_2(1)] &= 3, \\ \Theta_\alpha[A_2(2)] &= 3, & \Theta_\beta[A_2(2)] &= 3, \\ \Theta_\alpha[A_2] &= 1, & \Theta_\beta[A_2] &= 1. \end{aligned}$$

From the last equalities it follows that $\alpha \overset{V_0}{\sim} \beta$. By similar calculations or using Lemma 6, it is possible to obtain general conditions of E_n^2-equivalence (7).

Completeness of fragmentary sets. The sets V whose V^i-equivalence classes are singletons represent of special interest. In this case each word $a \in E^n$ is uniquely reconstructed from its fragments.

DEFINITION. A characteristic set V is called *i-complete* if each V^i-equivalence class contains one word $(i = 0, 1)$.

EXAMPLE. In [12] it was shown that the set $V = E_n^k$ is 0-complete for $k > n/2$. Apparently, this example was the first nontrivial example of a 0-complete set. The same set is 1-complete for $k \geq \lfloor n/2 \rfloor + 1$ (see [10]).

From Theorem 5 we immediately infer

Theorem 6. *The characteristic set V^* is 0-complete if and only if the system of equations*

$$\sum_{1 \leq i_1 < i_2 < \cdots < i_s \leq q} \lambda_{i_1 i_2 \ldots i_s}^{j_1 j_2 \ldots j_s} x_{i_1} x_{i_2} \ldots x_{i_s} = \lambda(i_1, j_2, \ldots, j_s) \tag{8}$$

has at most one solution in $(0,1)$*-numbers for an arbitrary right-hand side.*

Note that for $V = E_n^k$ and $k > n/2$ the consideration of the linear part of the system of equations (8) only allows us to construct a simple algorithm of the majoritary type for reconstructing a word from its fragments of length k and thus allowing us to prove constructively the 0-completeness of E_n^k for $k > n/2$.

It is very likely that the available upper bound for the minimal number k such that the set E_n^k is 0-complete ($k < n/2 + 1$) overstated and its true value is equal in order to $\log_2 n$. The best recent lower bound is as follows.

Theorem 7. *Let $l(n)$ be the minimal value of k for which the set E_n^k is 0-complete. Then the inequality*

$$l(n) > c \log_2 n$$

holds, where $c = (\log_2(1 + \sqrt{5}\,) - 1)^{-1}$.

This bound was obtained in [28]. A most weak bound

$$l(n) \geq \log_2 n / \log_2 \log_2 n$$

appears in [5].

In [4] with a reference to [13], the concept of equivalence without restrictions on the length of words was introduced and investigated, and few profound results that may be considered as a solution of the equivalence problem was presented. There also the result solving the problem of reconstructing words in the case when the characteristic set is a ball of a fixed radius was obtained. In our terms, this result may be formulated as follows.

Theorem 8. *A set E_n^k is 1-complete if and only if the inequality $k \geq \lfloor n/2 \rfloor + 1$ holds.*

Some of the results mentioned above were also obtained in [8, 10].

Note that the property of completeness is not monotonous with respect to inclusion. In other words, there exist examples of complete characteristic sets such that adding to them some new characteristic tuples leads to the loss of the completeness property. This fact may be explained as follows. By adding a "new" characteristic tuple to a complete characteristic set, we on the one hand increase the amount of information about the unknown word and, on the other hand, also increase the uncertainty since it becomes less clear from the "actions" of which "elements" the fragments were obtained. The completeness of the characteristic set depends on the correlation between the "information" and "uncertainty" arising in this case.

With a set of fragments of an unknown word available, we have information about this word. How to evaluate the amount and the value of this information?

So, the input data is a set S of words, generally of different length, in the alphabet $\{0,1\}$. One is required to evaluate how completely the set S characterizes the unknown word. One way of characterization based on the concepts considered above is as follows.

1. Form an "information space" $V_n(S)$, i.e., find the set of all words of length n each of which contains all the words in S as its fragments.

2. Introduce a measure of uncertainty on the set S:

$$\mu(S) = \log_2 |V_n(S)|/n.$$

Given the characteristic set $V = \{v_1, v_2, \ldots, v_N\}$, the process of constructing the measure is as follows. Let $S = \{a_1, a_2, \ldots, a_N\}$ be a set of fragments and let $V = \{v_1, v_2, \ldots, v_N\}$ be the characteristic set. According to the procedure described above write the equations

$$\langle x, v_i \rangle = a_j, \quad 1 \le j \le N.$$

If A_{ij} is the set of solutions of the jth equation, then the information space for the set S is per $\|A_{ij}\|$ and, as above, define the measure

$$\mu(S) = \log_2 |\text{per}\, \|A_{ij}\|\,|/n.$$

With the help of the measure introduced it is also possible to measure the "value" of the information contained in the set S. The role of this measure related to the function $\pi(S) = 1 - \mu(S)$. In particular, if the set S uniquely determines the word, then the information contained in it has an "absolute" value and $\pi(S) = 1$.

REMARK. In the present paper we mentioned only some aspects and results that relate to the problem. In particular, we considered only the binary alphabet, although in some cases many of the results mentioned above may be directly generalized to an arbitrary alphabet. We believe that many problems and results deserve more attention, for example, the problem of economical representation of finite sets of words, the structural completeness problem, algorithmic problems of reconstruction, and so on.

Acknowledgments. The author expresses his gratitude to A. A. Evdokimov for his remarks on and additions to the paper which made more complete the exposition of the results on the topics under consideration and also for his pointing out the papers unknown to the author, in particular, [4, 13].

References

1. C. E. Shannon (1993) *Collected Papers*, IEEE Press, New York.
2. Yu. I. Zhuravlёv (1978) An algebraic approach to the solution of pattern recognition and identification problems (in Russian), in: *Problemy Kibernetiki.* Vol. 33, Nauka, Moscow, pp. 5–68
3. A. P. Gorelik, I. B. Gurevich, and V. A. Skripkin (1985) *A Modern State of a Pattern Problem* (in Russian), Radio i Svyaz', Moscow.
4. M. Lothaire *Combinatorics on Words*, Encyclopedia of Mathematics and its Applications. Vol.17 Addison-Wesley Publishing Co., Reading, Mass.
5. A. I. Zenkin and V. K. Leont'ev (1984) A nonclassical pattern recognition problem (in Russian), *Zh. Vychisl. Mat. i Mat. Fiz.* **24**, No. 6, 925–931.
6. V. K. Leont'ev and Yu. G. Smetanin (1988) Reconstruction of vectors from sets of their fragments (in Russian), *Dokl. Akad. Nauk SSSR* **302**, No. 6, 1319–1322.
7. V. I. Levenshtein (1991) Perfect codes in the metric of deletion and insertions (in Russian), *Diskret. Mat.* **3**, No. 1, 3–20.
8. B. Manvel, A. Meyerowitz, A. Schwenk, K. Smith, and P. Stockmeyer (1991) Reconstruction of sequences, *Discrete Math.* **94**, No. 3, 209–219.
9. A. A. Evdokimov and V. Nyu (1992) The length of a supersequence for the set of binary words with a given number of units (in Russian), *Metody Diskret. Anal.* **52**, 49–58.
10. V. K. Leont'ev (1993) Recognition of binary words from their fragments (in Russian), *Dokl. Akad. Nauk* **330**, No. 4, 434–436.
11. V. I. Levenshtein (1965) Binary codes capable of correcting deletions, insertions, and reversals (in Russian), *Dokl. Akad. Nauk SSSR* **163**, No. 4, 707–710.
12. V. V. Kalashnik (1973) Reconstruction of a word from their fragments (in Russian), *Vychisl. Mat. i Vychisl. Tekhnika.* Vol. 4, Khar'kov, pp. 56–57.
13. I. Simon (1975) Piecewise testable events. *Automata Theory and Formal Languages*, pp. 214–222. Lecture Notes in Comput. Sci. Vol. 33, Springer-Verlag, Berlin etc.
14. V. I. Levenshtein (1974) Elements of coding theory (in Russian), *Discrete Mathematics and Mathematical Questions of Cybernetics.* Vol. 1, Nauka, Moscow, pp. 207–305.
15. G. Burosch, U. Frankl, and S. Röhl (1990) Über Ordnungen von Binäworten, *Rostock. Math. Kolloq.* No. 39, 53–64.
16. K. Engel and H.-D. Gronau (1985) *Sperner Theory in Partially Ordered Sets*, BSB B. G. Teubner Verlagsgesellschaft, Leipzig.
17. M. R. Garey and D. S. Johnson (1979) *Computers and Intractability*, Freeman, San Francisco.

18. D. Maier (1978) The complexity of some problems on subsequences and supersequences, *J. Assoc. Comput. Mach.* **25**, No. 2, 322–336.
19. Sh. M. Aliev (1993) Algorithmic questions for the construction of optimal codes for many receivers (in Russian), *Dis. Kand. Fiz.-Mat. Nauk*, Moscow.
20. E. M. Reingold, J. Nievergeld, and N. Deo (1977) *Combinatorial Algorithms: Theory and Practice*, Prentice-Hall, Inc., Englewood Cliffs, New Jork.
21. M. Hall, Jr. (1967) *Combinatorial Theory*, Blaisdell Publishing Company, Waltham, Toronto, and London.
22. A. A. Evdokimov (1969) The maximal length of a chain in the n-dimensional unit cube (in Russian), *Mat. Zametki* **6**, No. 3, 309–319.
23. V. Nyu (1990) Covering of a set of words by chains (in Russian), *Dis. Kand. Fiz.-Mat. Nauk*, Institute of Mathematics, Novosibirsk.
24. V. Nyu and A. A. Evdokimov (1983) Covering of graphs by paths (in Russian), *Metody Diskret. Anal.* **40**, 72–86.
25. W. T. Tutte (1984) *Graph Theory*, Addison-Wesley Publishing Company, Reading, Massachusetts.
26. Ya. S. Smetanich (1971) The reconstruction of words (in Russian), *Dokl. Akad. Nauk SSSR* **201**, No. 4, 798–800.
27. Ya. S. Smetanich (1973) On the reconstruction of words (in Russian), Mathematical Logic, Theory of Algorithms, and Theory of Sets. *Trudy Mat. Inst. Steklov.* Vol. 133, Nauka, Moscow, pp. 183–202.
28. Yu. G. Smetanin (1986) On some problems for the reconstruction of words from their fragments (in Russian), *Dis. Kand. Fiz.-Mat. Nauk*, Vychisl. Tsentr Akad. Nauk SSSR, Moscow.

Computing Center
ul. Vavilova, 40
Moscow, GSP-1, 117967
RUSSIA

TRANSLATED BY E. BODIN

A VERTEX VARIANT OF THE KLEITMAN–WEST PROBLEM†)

B. List

We study the problem of finding an m-element set A of vertices in the nth layer of the d-dimensional unit cube with the minimum number of vertices in the nth layer lying at distance 2 from A (i.e., in the frontier of A). Our main results are lower and upper bounds on the ratio of the size of the frontier to the size of the set and exact formulas for the size of the boundaries of ideals.

The isoperimetric problems (IP) are the problems of finding a geometric figure of maximum area given a fixed perimeter. In the case of discrete spaces, the problems of finding a set of maximum size given a frontier or of finding a set of a given size with minimum frontier (discrete IP) are studied. For example, J. B. Kruskal [1], G. Katona [2], L. H. Harper [3], and others investigated these problems.

In certain cases [1, 2], a discrete IP admits lexicographical intervals to be its solutions. On the other hand, there exist discrete IP for which this is not so. One of them is the well-known Kleitman–West problem [3] consisting in estimating the number of pairs of vertices lying at distance 2 with one vertex in and the other out of a given subset of vertices in a unit cube layer.

In this paper, we study a vertex variant of the Kleitman–West problem on minimization of the vertex frontier. Bounds for the size of this frontier are used, for example, for obtaining asymptotic formulas for the number of monotone Boolean functions in different classes [4, 5]. Moreover, this problem is of independent interest. The aim of the present work is to test the hypothesis that the minimum of the ratio of the size of the vertex frontier to the size of the set subject to certain restrictions on its size is equal in order to $1/\sqrt{d}$, where d is the dimension of the cube. We obtain upper and lower bounds for the minimum of the ratio of the size of the frontier to the size of the set on condition that the size of the set is not greater than half of the size of the layer B_n^d. It is proved that the upper bound is achieved for the balls whose size is close to half of the size of the layer and that the balls are locally optimal with respect to the size of the frontier of the set. We failed however to prove or disprove the above hypothesis in the general case.

†) This research was supported by the Russian Foundation for Basic Research (Grant 95–01–01595).

A. D. Korshunov (ed.), Operations Research and Discrete Analysis, 145–170.

1. Statement of the Problem

We first introduce the notions that are used in formulating the problem. The set

$$B^d = \{\tilde{x}^d = (x_1, \ldots, x_d) \mid x_i \in \{0,1\},\ 1 \le i \le d\}$$

is called the *d-dimensional unit cube* and the set

$$B_n^d = \left\{\tilde{x}^d \in B^d \mid \sum_{i=1}^{d} x_i = n\right\}$$

is called the nth *layer of the cube* B^d. We put $m = d - n$. Without loss of generality, we may and shall assume that $m \ge n$. Let A be an arbitrary subset of vertices in B_n^d. The set

$$\delta(A) = \{\tilde{\beta} \in B_n^d \setminus A \mid (\exists \tilde{\alpha} \in A)\, \rho(\tilde{\alpha}, \tilde{\beta}) = 2, \text{ where } \rho \text{ is the Hamming distance}\}$$

is referred to as the *vertex frontier* of A.

Put

$$\Phi(A) = |\delta(A)|/|A|,$$
$$\Phi(p,n) = \min_{A \in B_n^d : |A| = p} \Phi(A),$$
$$\varphi(d,n) = \min_{p \le \frac{1}{2}|B_n^d|} \Phi(p,n).$$

The problem is to estimate the functionals $\Phi(A)$, $\Phi(p,n)$, and $\varphi(d,n)$ for each $n \le d/2$.

2. Structure of Optimal Sets

In this section, it is proved that the optimal sets may be sought among the sets representable as the unions of the so-called initial intervals and certain properties of these intervals are established.

It follows from Theorem 1 that the method of stabilization which is described in [3] for the edge variant of the Kleitman–West problem applies also to the vertex variant. It is possible to indicate a finite number of mappings in the family of all the subsets of B_n^d into itself under which the size of the frontier $\delta(A)$ of an arbitrary set A does not increase. As a result of consecutive cyclical application of these mappings, each set $A \subseteq B_n^d$ is transformed in a finite number of steps into a set of the same cardinality which is stable under these mappings (i.e., a set which is a fixed point of all these mappings). Thus, the optimal sets may be sought from among stable sets.

Consider the following sets $T_{i,j}(A)$ (see [6]) generated by the mappings T_{ij}, $i < j$:

$$T_{i,j}(A) = (A \setminus A_{(i,j)}) \cup \pi_{i,j}(A_{(i,j)}),$$

where $\pi_{i,j}$ is the mapping transposing the ith and jth coordinates and

$$A_{(i,j)} = \{\tilde{\alpha} = (\alpha_1, \dots, \alpha_d) \in A \mid \pi_{i,j}(\tilde{\alpha}) \notin A,\ \alpha_i = 0,\ \alpha_j = 1\}.$$

A set A is called *stable* if $T_{i,j}(A) = A$ for all i, j $(1 \le i, j \le d, i \ne j)$.

The following theorem is similar to the Harper theorem [3] (the proof appears in [7]).

Theorem 1. *Let A be an arbitrary subset of B_n^d, and let $\{T_{i(k),j(k)}\}_{k=0}^{\infty}$ be a sequence of mappings such that $i(k) = \lfloor k/d \rfloor \pmod{d+1}$ and $j(k) = k \pmod{d+1}$. Put $A^{(0)} = A$ and $A^{(k+1)} = T_{i(k),j(k)}(A^{(k)})$. Then there exists a k_0 such that $A^{(k+1)} = A^{(k)}$ for each $k \ge k_0$, where $|A^{k_0}| = |A|$ and $|\delta(A^{k_0})| \le |\delta(A)|$.*

A set A is said to be an *ideal* (an *upper set*) with respect to an arbitrary partial order $\le$ if, for each $\tilde{\alpha}$ and $\tilde{\beta}$ such that $\tilde{\alpha} \in A$ and $\tilde{\beta} \le \tilde{\alpha}$ $(\tilde{\beta} \ge \tilde{\alpha})$, it follows that $\tilde{\beta} \in A$.

On the set of vertices in B_n^d, introduce a partial order $\le_1$. We say that a vector $\tilde{v}_1$ *precedes* a vector $\tilde{v}_2$ (denoted by: $\tilde{v}_1 \le_1 \tilde{v}_2$) if there exists a sequence π of permutations $\pi_{i,j}$ of coordinates of vertices in B_n^d such that $\tilde{v}_1 = \pi(\tilde{v}_2)$, only those permutations $\pi_{i,j}$ being allowed which move a unit to the left and a zero to the right. In other words, a vector $\tilde{v}_1$ precedes a vector $\tilde{v}_2$ if, for each $i = 1, 2, \dots, d$, the number of ones among the first i coordinates of $\tilde{v}_2$ does not exceed the number of ones among the first i coordinates of $\tilde{v}_1$.

Theorem 2. *Each stable set is an ideal with respect to the order $\le_1$ and each ideal is stable.*

For the proof, see [3].

Define a one-to-one correspondence f of the set B_n^d onto the space

$$L_{m,n} = \{\tilde{i} = (i_1, i_2, \dots, i_n) \mid 0 \le i_1 \le i_2 \le \cdots \le i_n \le m = d - n\}$$

as follows: if $\tilde{v} \in B_n^d$, then $f(\tilde{v}) = i(\tilde{v}) = (i_1(\tilde{v}), i_2(\tilde{v}), \dots, i_n(\tilde{v})) \in L_{m,n}$, where $i_k(\tilde{v})$ is the number of zeros preceding the kth one in the vector $\tilde{v} = (v_1, v_2, \dots, v_d)$ [3].

On the set $L_{m,n}$, introduce the partial order $\le_2$. We say that a vector $i(\tilde{v}) = (i_1(\tilde{v}), \dots, i_n(\tilde{v}))$ precedes a vector $i(\tilde{w}) = (i_1(\tilde{w}), \dots, i_n(\tilde{w}))$ if $i(\tilde{v}) \ne i(\tilde{w})$ and $i_j(\tilde{v}) \le i_j(\tilde{w})$ for each $j, 1 \le j \le n$.

It is easy to see that $\tilde{v} <_1 \tilde{w}$ if and only if $i(\tilde{v}) <_2 i(\tilde{w})$. Hence, the image of a stable subset of B_n^d is an ideal in $L_{d-n,n}$.

A point $\tilde{\alpha}^*$ is called an *upper point* of an ideal A if A contains no points preceding $\tilde{\alpha}^*$. The *initial interval* with the greatest point $\tilde{\alpha}^*$ is the ideal $I(\tilde{\alpha}^*) = \{\tilde{\alpha} \in B_n^d \mid \tilde{\alpha} \leq_1 \tilde{\alpha}^*\}$.

The definitions of ideal and initial interval imply

REMARK 1. Each ideal A can be represented as the union of certain initial intervals

$$A = \bigcup_{\tilde{\alpha} \in A} I(\tilde{\alpha}) = \bigcup_{\tilde{\alpha}^* \in A^*} I(\tilde{\alpha}^*),$$

where A^* is the set of all upper points in A.

Subsequently, we make use of the following evident

REMARK 2. The least vertex $\tilde{\beta}^0$ in $\delta(\tilde{\alpha})$ (i.e., $\tilde{\beta}^0 \leq_1 \tilde{\beta}$ for each vertex $\tilde{\beta} \in \delta(\{\tilde{\alpha}\})$ other than $\tilde{\beta}^0$) is obtained from $\tilde{\alpha}$ by transposing the first (from the left) zero and the last one. Analogously, the greatest vertex $\tilde{\beta}^1$ in $\delta(\{\tilde{\alpha}\})$ is obtained from $\tilde{\alpha}$ by transposing the first one and the last zero.

The *closure* of a set A is the set $\Delta(A) = \delta(A) \cup A$.

Clearly, the problems of minimizing the cardinality of the frontier and the cardinality of the closure and the functional $\Phi(A)$, given a fixed cardinality of the set, are equivalent. Note that $\Phi(A) = |\delta(A)|/|A| = |\Delta(A)|/|A| - 1$.

The next theorem states that the closure of an initial interval is an initial interval.

Theorem 3. *For each point $\tilde{\alpha}^* \in B_n^d$, there exists a point $\tilde{\beta}^* \in B_n^d$ such that $\tilde{\beta}^* \geq_1 \tilde{\alpha}^*$ and $\Delta(I(\tilde{\alpha}^*)) = I(\tilde{\beta}^*)$.*

PROOF. Define $\tilde{\beta}^*$ as the greatest point in $\delta(\tilde{\alpha}^*)$. Note that $\tilde{\beta}^*$ is obtained from $\tilde{\alpha}^*$ by transposing the first one and the last zero.

We first prove that $I(\tilde{\alpha}^*) \cup \delta(I(\tilde{\alpha}^*)) \subseteq I(\tilde{\beta}^*)$. To this end, it suffices to show that $\tilde{\beta} \in I(\tilde{\alpha}^*) \cup \delta(I(\tilde{\alpha}^*))$ implies $\tilde{\beta} \in I(\tilde{\beta}^*)$, i.e., $\tilde{\beta} \leq_1 \tilde{\beta}^*$. Since $I(\tilde{\alpha}^*) \cup \delta(I(\tilde{\alpha}^*)) = I(\tilde{\alpha}^*) \cup \bigcup_{\tilde{\alpha} \in I(\tilde{\alpha})} \delta(\{\tilde{\alpha}\})$, it suffices to show that $\tilde{\beta} \leq_1 \tilde{\beta}^*$ for each vertex $\tilde{\alpha} \in I(\tilde{\alpha}^*)$ and the greatest point $\tilde{\beta}$ in $\delta(\tilde{\alpha})$. By the definition of initial interval, $\tilde{\alpha} \leq_1 \tilde{\alpha}^*$ for any point $\tilde{\alpha} \in I(\tilde{\alpha}^*)$. Therefore, by the definition of the order $\leq_1$ on B_n^d, it follows that there exists a sequence π of permutations $\pi_{i,j}$ which transforms $\tilde{\alpha}^*$ into $\tilde{\alpha}$ (i.e., $\pi(\tilde{\alpha}^*) = \tilde{\alpha}$). Accordingly, we have a sequence of *neighboring vertices* (different in precisely two coordinates): $\tilde{\alpha}^* = \tilde{\alpha}^0, \tilde{\alpha}^1, \dots, \tilde{\alpha}^s = \tilde{\alpha}$. Define the sequence $\{\tilde{\beta}^k\}$ such that $\tilde{\beta}^k$ is the greatest point in $\delta(\{\tilde{\alpha}^k\})$ (i.e., $\tilde{\beta}^k$ is obtained from $\tilde{\alpha}^k$ by moving the first one to the right and the last zero to the left). Let us now see which $\tilde{\alpha}^{k+1}$ can be obtained from $\tilde{\alpha}^k$ and show that if $\tilde{\alpha}^k, \tilde{\alpha}^{k+1}$ are neighboring vertices and $\tilde{\alpha}^k >_1 \tilde{\alpha}^{k+1}$, then $\tilde{\beta}^k \geq_1 \tilde{\beta}^{k+1}$.

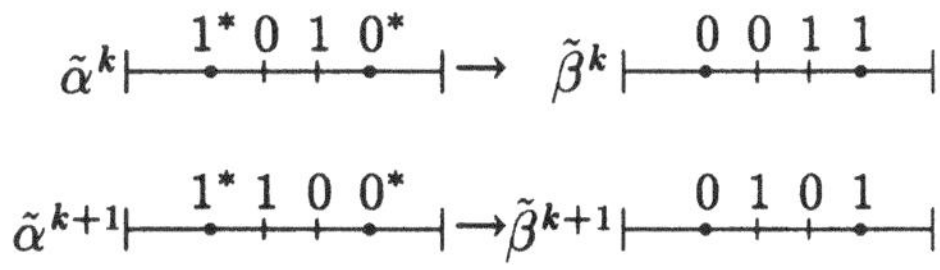

$\tilde{\alpha}^{k+1}$ results from $\tilde{\alpha}^k$ by transposing two "interior" coordinates. Obviously, $\rho(\tilde{\beta}^k, \tilde{\beta}^{k+1}) = 2$ and $\tilde{\beta}^k \geq_1 \tilde{\beta}^{k+1}$.

(First ones and last zeros in the vectors $\tilde{\alpha}^k$ and $\tilde{\alpha}^{k+1}$ are marked with a star.)

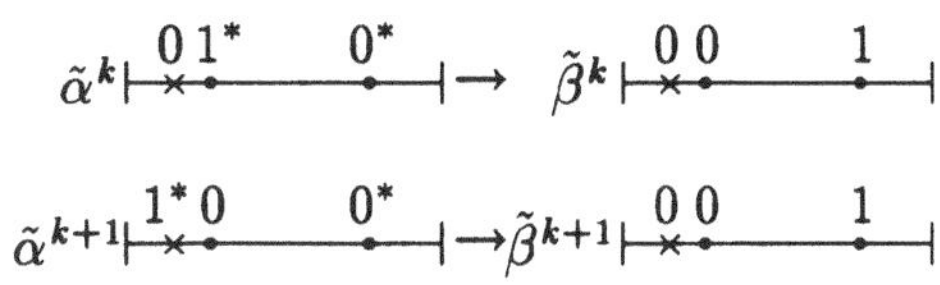

$\tilde{\alpha}^{k+1}$ results from $\tilde{\alpha}^k$ by transposing the first one and a zero before the first one; $\tilde{\beta}^k = \tilde{\beta}^{k+1}$.

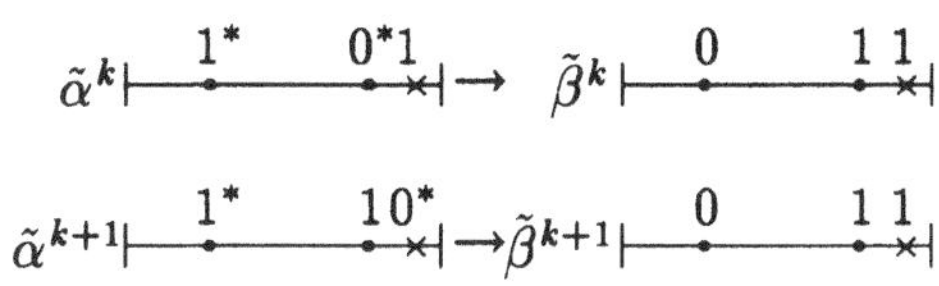

$\tilde{\alpha}^{k+1}$ results from $\tilde{\alpha}^k$ by transposing the last zero and a one after the last zero; $\tilde{\beta}^k = \tilde{\beta}^{k+1}$.

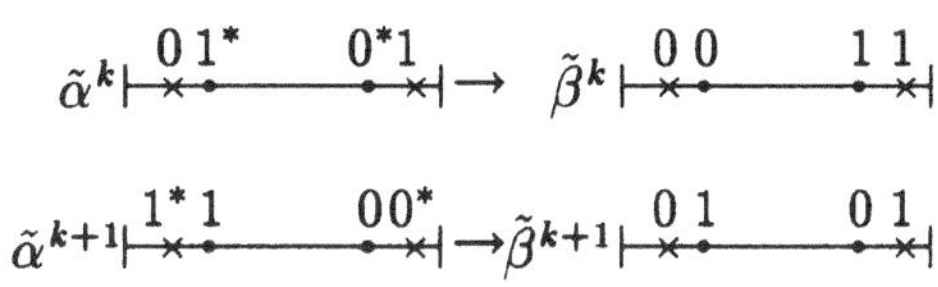

$\tilde{\alpha}^{k+1}$ results from $\tilde{\alpha}^k$ by transposing a zero before the first one and a one after the last zero; $\rho(\tilde{\beta}^k, \tilde{\beta}^{k+1}) = 2$, $\tilde{\beta}^k \geq_1 \tilde{\beta}^{k+1}$.

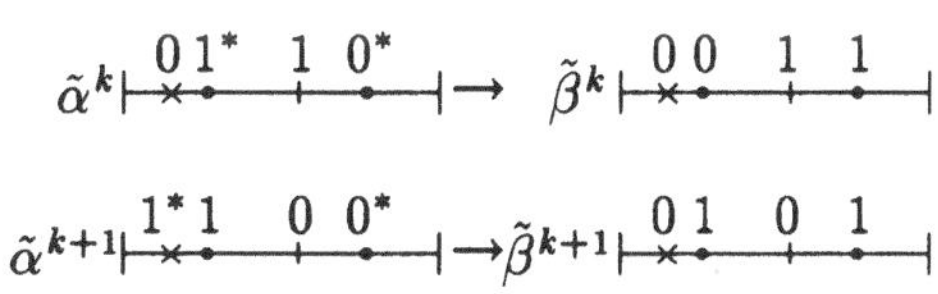

$\tilde{\alpha}^{k+1}$ results from $\tilde{\alpha}^k$ by transposing a zero before the first one and an "interior" one; $\rho(\tilde{\beta}^k, \tilde{\beta}^{k+1}) = 2$, $\tilde{\beta}^k \geq_1 \tilde{\beta}^{k+1}$.

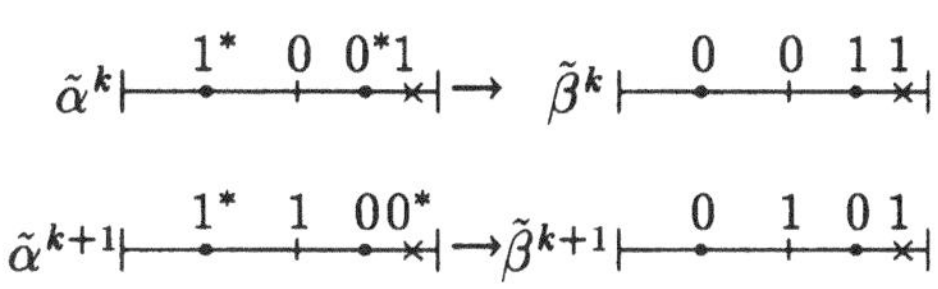

$\tilde{\alpha}^{k+1}$ results from $\tilde{\alpha}^k$ by transposing an "interior" zero and a one after the last zero; $\rho(\tilde{\beta}^k, \tilde{\beta}^{k+1}) = 2$, $\tilde{\beta}^k \geq_1 \tilde{\beta}^{k+1}$.

So, in all cases, $\tilde{\beta}^k \geq_1 \tilde{\beta}^{k+1}$ and thus, $\tilde{\beta}^* \geq_1 \tilde{\beta}$.

We next verify that $I(\tilde{\alpha}^*) \cup \delta(I(\tilde{\alpha}^*)) \supseteq I(\tilde{\beta}^*)$. To this end, it suffices to show that each point $\tilde{\beta} \in I(\tilde{\beta}^*)$ belongs to the set $I(\tilde{\alpha}^*) \cup \delta(I(\tilde{\alpha}^*))$. By definition, $\tilde{\beta} \leq_1 \tilde{\beta}^*$ for any $\tilde{\beta} \in I(\tilde{\beta}^*)$, i.e., there exists a sequence π of permutations $\pi_{i,j}$ which realizes the chain of mappings: $\tilde{\beta}^* = \tilde{\beta}^0 \to \tilde{\beta}^1 \to \cdots \to \tilde{\beta}^k \to \cdots \to \tilde{\beta}^t =$

$\tilde{\beta}$. Define the sequence $\{\tilde{\alpha}^k\}$ so that $\tilde{\alpha}^k$ is the least point in $\delta(\{\tilde{\beta}^k\})$ (i.e., $\tilde{\alpha}^k$ is obtained from $\tilde{\beta}^k$ by transposing the first zero and the last one). It is evident that $\tilde{\alpha}^* \geq_1 \tilde{\alpha}^0$. We claim that for each vertex $\tilde{\beta} \leq_1 \tilde{\beta}^*$, there exists an $\tilde{\alpha}$ such that $\tilde{\alpha} \leq_1 \tilde{\alpha}^*$ (i.e., $\tilde{\alpha} \in I(\tilde{\alpha}^*)$, $\rho(\tilde{\alpha}, \tilde{\beta}) = 2$, and $\tilde{\beta} \in \delta(\{\tilde{\alpha}\})$). Put $\tilde{\alpha} = \tilde{\alpha}^t$. Since $\rho(\tilde{\alpha}^k, \tilde{\beta}^k) = 2$ for each k, we have $\rho(\tilde{\alpha}, \tilde{\beta}) = 2$. Inspecting all the cases, we show that $\tilde{\beta}^k \geq_1 \tilde{\beta}^{k+1}$ implies $\tilde{\alpha}^k >_1 \tilde{\alpha}^{k+1}$.

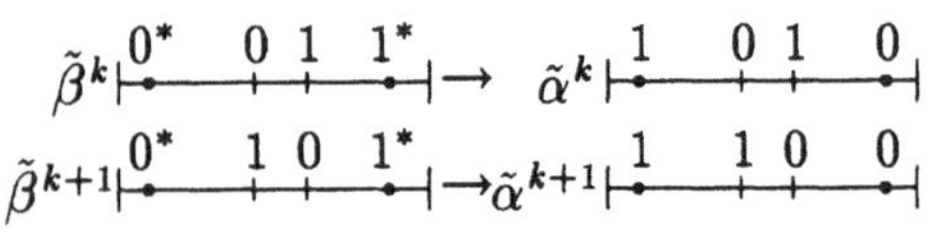

$\tilde{\beta}^{k+1}$ results from $\tilde{\beta}^k$ by transposing two "interior" coordinates; $\tilde{\alpha}^k >_1 \tilde{\alpha}^{k+1}$.

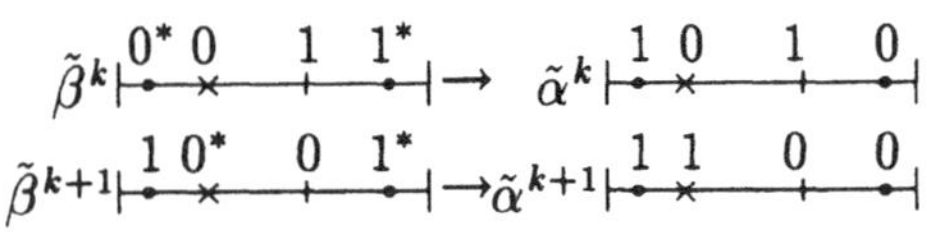

$\tilde{\beta}^{k+1}$ results from $\tilde{\beta}^k$ by transposing the first zero and an "interior" one; $\tilde{\alpha}^k >_1 \tilde{\alpha}^{k+1}$.

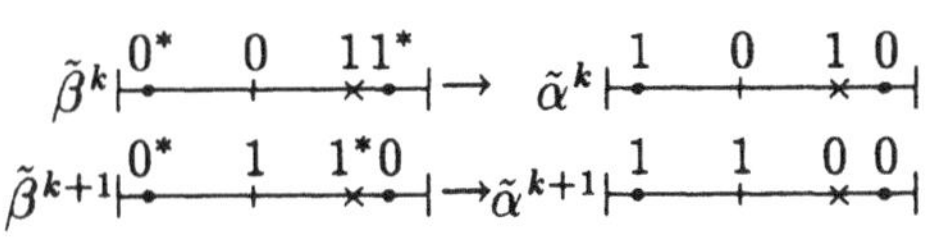

$\tilde{\beta}^{k+1}$ results from $\tilde{\beta}^k$ by transposing the last one and an "interior" zero; $\tilde{\alpha}^k >_1 \tilde{\alpha}^{k+1}$.

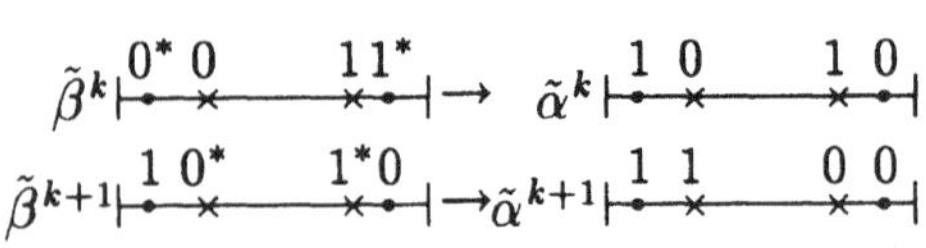

$\tilde{\beta}^{k+1}$ results from $\tilde{\beta}^k$ by transposing the first zero and the last one; $\tilde{\alpha}^k >_1 \tilde{\alpha}^{k+1}$.

So, $\tilde{\alpha}^* \geq_1 \tilde{\alpha}^0 >_1 \tilde{\alpha}^1 >_1 \cdots >_1 \tilde{\alpha}^k >_1 \tilde{\alpha}^{k+1} >_1 \cdots >_1 \tilde{\alpha}^t = \tilde{\alpha}$ and $\rho(\tilde{\alpha}, \tilde{\beta}) = 2$. Hence, $\tilde{\beta} \in I(\tilde{\alpha}^*) \cup \delta(I(\tilde{\alpha}^*))$. Theorem 3 is proved.

The next two lemmas assert that the closure of a union (an intersection) of ideals coincides with the union (the intersection) of the closures of these ideals.

Lemma 1. *Let $I(\tilde{\alpha})$ and $I(\tilde{\beta})$ be arbitrary initial intervals in the layer B_n^d. Then*

$$I(I(\tilde{\alpha}) \cap I(\tilde{\beta})) = I(I(\tilde{\alpha})) \cap I(I(\tilde{\beta})). \tag{1}$$

PROOF. It is clear that $I(\tilde{\alpha}) \cap I(\tilde{\beta}) = I(\tilde{\gamma})$ and also $\tilde{\gamma}$ is obtained from the vectors $\tilde{\alpha}$ and $\tilde{\beta}$ as follows:

$$i_j(\tilde{\gamma}) = \min\{i_j(\tilde{\alpha}), i_j(\tilde{\beta})\}, \quad 1 \leq j \leq n. \tag{2}$$

Consider the left-hand side of (1). By (2), we have $I(I(\tilde{\alpha}) \cap I(\tilde{\beta})) = I(I(\tilde{\gamma}))$, where $\tilde{\gamma}$ is such that $\tilde{i}(\tilde{\gamma}) = (i_1(\tilde{\gamma}), \ldots, i_n(\tilde{\gamma}))$ and $i_j(\tilde{\gamma}) = \min\{i_j(\tilde{\alpha}), i_j(\tilde{\beta})\}$. It follows from Theorem 3 that $\Delta(I(\tilde{\gamma})) = I(\tilde{\varepsilon})$, where $\tilde{\varepsilon}$ is such that

$$\tilde{i}(\tilde{\varepsilon}) = ((i_2(\tilde{\gamma}) + 1), (i_3(\tilde{\gamma}) + 1), \ldots, (i_n(\tilde{\gamma}) + 1), m),$$

i.e.,

$$\begin{aligned}\tilde{i}(\tilde{\varepsilon}) = ((\min\{i_2(\tilde{\alpha}), i_2(\tilde{\beta})\} + 1), (\min\{i_3(\tilde{\alpha}), i_3(\tilde{\beta})\} + 1),\\ \ldots, (\min\{i_n(\tilde{\alpha}), i_n(\tilde{\beta})\} + 1), m).\end{aligned}$$

Consider the right-hand side of (1). By Theorem 3, $\Delta(I(\tilde{\alpha})) \cap \Delta(I(\tilde{\alpha})) = I(\tilde{\lambda}) \cap I(\tilde{\mu})$, and (2) implies that $I(\tilde{\lambda}) \cap I(\tilde{\mu}) = I(\tilde{\nu})$. In addition, λ, μ, and ν are obtained from the relations

$$\begin{aligned}\tilde{i}(\tilde{\lambda}) &= ((i_2(\tilde{\alpha}) + 1), (i_3(\tilde{\alpha}) + 1), \ldots, (i_n(\tilde{\alpha}) + 1), m),\\ \tilde{i}(\tilde{\mu}) &= ((i_2(\tilde{\beta}) + 1), (i_3(\tilde{\beta}) + 1), \ldots, (i_n(\tilde{\beta}) + 1), m),\\ \tilde{i}(\tilde{\nu}) &= (i_1(\tilde{\nu}), \ldots, i_n(\tilde{\nu})), i_j(\tilde{\nu}) = \min\{i_j(\tilde{\lambda}), i_j(\mu)\}.\end{aligned}$$

Hence,

$$\begin{aligned}\tilde{i}(\nu) = ((\min\{i_2(\tilde{\alpha}), i_2(\tilde{\beta})\} + 1), (\min\{i_3(\tilde{\alpha}), i_3(\tilde{\beta})\} + 1),\\ \ldots, (\min\{i_3(\tilde{\alpha}), i_3(\tilde{\beta})\} + 1), m).\end{aligned}$$

Therefore, $\tilde{i}(\tilde{\varepsilon}) = \tilde{i}(\tilde{\nu})$ and thus, $I(\tilde{\varepsilon}) = I(\tilde{\nu})$. Lemma 1 is proved.

Lemma 2. *Let A and B be arbitrary ideals in the layer B_n^d. Then*

$$\Delta(A \cup B) = \Delta(A) \cup \Delta(B), \tag{3}$$
$$\Delta(A \cap B) = \Delta(A) \cap \Delta(B). \tag{4}$$

The assertion (3) is directly verified whereas (4) follows from the representation of an ideal as a union of initial intervals and Lemma 1.

The *ball of radius* r with center the point $\tilde{\beta}$ is the set $S_r(\tilde{\beta}) = \{\tilde{\alpha} \in B_n^d \mid \rho(\tilde{\alpha}, \tilde{\beta}) \leq r\}$.

Clearly, balls can be only of even radius.

For a number of isoperimetric problems, balls are optimal sets. We prove that the balls centered at the least point in a layer are ideals.

Proposition 1. *A ball with center* $\tilde{\alpha}^0 = (\overbrace{1,\dots,1}^{n},\overbrace{0,\dots,0}^{m=d-n})$ *is an initial interval. If an ideal S is a ball, then* $\tilde{\alpha}^0 = (\underbrace{1,\dots,1}_{n},\underbrace{0,\dots,0}_{m})$ *is the center of the ball S.*

PROOF. We first verify that if $\tilde{\alpha} \in S_{2r}(\tilde{\alpha}^0)$, then every point $\tilde{\beta}$ preceding $\tilde{\alpha}$ is also in the ball $S_{2r}(\tilde{\alpha}^0)$.

Consider an arbitrary point $\tilde{\alpha} \in S_{2r}(\tilde{\alpha}^0)$. It is clear that among the first n coordinates of $\tilde{\alpha}$ there is at least $n-r$ ones and at most r zeros. But then at most r zeros occur before the $(n-r)$th one of $\tilde{\alpha}$, thus $i_{n-r}(\tilde{\alpha}) \le r$. For each point $\tilde{\beta}$ which precedes $\tilde{\alpha}$, by the definition of the order $\le_2$, the inequality $i_j(\tilde{\beta}) \le i_j(\tilde{\alpha})$ holds for each j, $1 \le j \le n$, and there exists a j^* such that $i_{j^*}(\tilde{\beta}) \le i_{j^*}(\tilde{\alpha})$. Hence, $i_{n-r}(\tilde{\beta}) \le i_{n-r}(\tilde{\alpha})$ and so, $i_{n-r}(\tilde{\beta}) \le r$. The above shows that $\tilde{\beta} \in S_{2r}(\tilde{\alpha}^0)$ and thus, $S_{2r}(\tilde{\alpha}^0)$ is an ideal. It is easy to verify that the set

$$S_{2r}(\tilde{\alpha}^0) = \{\tilde{\alpha} \in B_n^d \mid \rho(\tilde{\alpha},\tilde{\alpha}^0) \le 2r\} = \{\tilde{\alpha} \in B_n^d \mid i_{n-r}(\tilde{\alpha}) \le r\}$$
$$= \{\tilde{\alpha} \in B_n^d \mid \tilde{i}(\tilde{\alpha}) \le (\underbrace{r,\dots,r}_{n-r},\underbrace{m,\dots,m}_{r})\} = I(\tilde{\alpha}^0)$$

is the initial interval with the extreme point

$$\tilde{\alpha}^* = (\underbrace{0,\dots,0}_{r},\underbrace{1,\dots,1}_{n-r},\underbrace{0,\dots,0}_{m-r},\underbrace{1,\dots,1}_{r}).$$

The first assertion is proved by way of contradiction. We prove the second assertion. Let a set $S = S_{2r}(\tilde{\alpha}^0)$ be a ball and an ideal. Assume that there exists a ball $S' = S_{2r}(\tilde{\alpha}^1)$ with center a point $\tilde{\alpha}^1$ other than the point $\tilde{\alpha}^0$, and that this ball is an ideal. Then there exist two points $\tilde{\beta}$ and $\tilde{\gamma}$ such that $\tilde{\beta} \in S' \setminus S$ and $\tilde{\gamma} \in S \setminus S'$. We see that $\tilde{i}(\tilde{\beta}) >_2 \tilde{i}(\tilde{\gamma})$, and hence $\tilde{\gamma} \in S'$, which contradicts the assumption.

So, let $\tilde{\beta}$ be an arbitrary point lying at distance $2(r+k)$ from $\tilde{\alpha}^0$ and at distance $2r$ from $\tilde{\alpha}^1$. Suppose that $\tilde{\alpha}^1$ differs from $\tilde{\alpha}^0$ in coordinates with the indices $j_1,\dots,j_k,j_{k+1},\dots,j_{2k}$ $(1 \le j_1 < \dots < j_{2k} \le d)$. Clearly, $\tilde{\alpha}^1$ has zeros at the positions $j_1,\dots,j_k$ and ones at the positions $j_{k+1},\dots,j_{2k}$. Choose $\tilde{\gamma}$ to be a point that differs from $\tilde{\beta}$ only in the coordinates with the indices $j_1,\dots,j_k,j_{k+1},\dots,j_{2k}$. It is clear that the vector $\tilde{\gamma}$ must have ones in the positions $j_1,\dots,j_k$ and zeros in the positions $j_{k+1},\dots,j_{2k}$. This point lies at distance $2r$ from $\tilde{\alpha}^0$ and at distance $2(r+k)$ from $\tilde{\alpha}^1$. Obviously, $\tilde{\beta} \in S' \setminus S$ and $\tilde{\gamma} \in S \setminus S'$. Since $\tilde{\gamma}$ is obtained from $\tilde{\beta}$ by k permutations of the form $\pi_{i,j}$, the point $\tilde{\gamma}$ precedes the point $\tilde{\beta}$ (in order $\leq_1$). As S' is an ideal, $\tilde{\gamma}$ must belong to S'. But this contradicts $\tilde{\gamma} \in S \setminus S'$. Hence, the assumption is false that there exists a ball with center other than $\tilde{\alpha}^0$. Proposition 1 is proved.

3. Exact Formulas for the Size of the Frontiers of the Ideals

In this section, we establish exact formulas for the cardinality of an initial interval and its closure depending on the upper point of the initial interval and also a formula for the cardinality of the closure of an arbitrary ideal.

REMARK 3. Recall that $i_j(\tilde{\alpha})$ is the number of zeros preceding the jth one in $\tilde{\alpha} \in B_n^d$. Let $\tilde{\beta}^*$ be the greatest point in the frontier of a point $\tilde{\alpha}^* \in B_n^d$. Then with regard to Remark 2, we obtain the following formula:

$$i_j(\tilde{\beta}^*) = \begin{cases} i_{j+1}(\tilde{\alpha}^*) + 1 & \text{if } (j < n) \& (i_{j+1}(\tilde{\alpha}^*) < d - n); \\ d - n & \text{if } ((j < n) \& (i_{j+1}(\tilde{\alpha}^*) = d - n)) \vee (j = n). \end{cases} \tag{5}$$

Theorem 4. *For all n and d and each $\tilde{\alpha}^*$,*

$$|I(\tilde{\alpha}^*)| = \sum_{i_1=0}^{i_1(\tilde{\alpha}^*)} \sum_{i_2=i_1}^{i_2(\tilde{\alpha}^*)} \cdots \sum_{i_n=i_{n-1}}^{i_n(\tilde{\alpha}^*)} 1, \tag{6}$$

$$|\Delta(I(\tilde{\alpha}^*))| = \sum_{i_1=0}^{i_2(\tilde{\alpha}^*)+1} \sum_{i_2=i_1}^{i_3(\tilde{\alpha}^*)+1} \cdots \sum_{i_{j^*-1}=i_{j^*-2}}^{i_{j^*}(\tilde{\alpha}^*)+1} \sum_{i_{j^*}=i_{j^*-1}}^{d-n} \cdots \sum_{i_n=i_{n-1}}^{d-n} 1, \tag{7}$$

where j^* is the greatest number j such that $i_j(\tilde{\alpha}^*) < d - n$.

PROOF. By definition, we have $I(\tilde{\alpha}^*) = \{\tilde{\alpha} \in B_n^d \mid i_j(\tilde{\alpha}) \leq i_j(\tilde{\alpha}^*), 1 \leq j \leq n\}$. In addition, the coordinates of each point $\tilde{i} = (i_1, \ldots, i_n) \in L_{m,n}$ do not decrease. Count all the points $\tilde{i} \in L_{m,n}$ such that $i_j \leq i_j(\tilde{\alpha}^*)$ for all j, $1 \leq j \leq n$, and $0 \leq i_1 \leq i_2 \leq \cdots \leq i_n \leq m = d - n$; here the index i_1 varies from 0 to $i_1(\tilde{\alpha}^*)$, for each fixed i_1 the index i_2 varies from i_1 to $i_2(\tilde{\alpha}^*)$, etc. The equality follows from this.

Next, it follows from Theorem 3 that $\Delta(I(\tilde{\alpha}^*)) = \delta(I(\tilde{\alpha}^*)) \cup I(\tilde{\alpha}^*)$ is an initial interval with an upper point $\tilde{\beta}^*$, with $\tilde{\beta}^*$ being the greatest point in $\delta(\tilde{\alpha}^*)$. With regard to (5), we now obtain (7) from the formula (6) for $I(\tilde{\beta}^*)$. Theorem 4 is proved.

Now we obtain a formula for the cardinality of the closure of an arbitrary ideal.

If $\tilde{i} \in L_{m,n}$, $A' = I(\tilde{i}) \setminus \{\tilde{i}\}$, and $A'' = I(\tilde{i})$, then we write $\eta(\tilde{i}) = |\Delta(A'') \setminus \Delta(A')|$ and $\eta(\tilde{\alpha}) = \eta(\tilde{i})$, where $\tilde{i} = \tilde{i}(\tilde{\alpha})$.

Proposition 2. *Let $\tilde{i}^0$ be the least point in $L_{m,n}$ (recall that, to the vector $\tilde{i}^0 \in L_{m,n}$, there corresponds the point $\tilde{\alpha}^0 = (\underbrace{1,\dots,1}_{n},\underbrace{0,\dots,0}_{m}) \in B_n^d$), and let the point $\tilde{i} \in L_{m,n}$ be of the form $\tilde{i} = (\underbrace{0,\dots,0}_{s}, i_{s+1},\dots,i_n)$. Then*

$$\eta(\tilde{i}) = \begin{cases} nm+1 & \text{if } \tilde{i} = \tilde{i}^0; \\ s(m-i_n) & \text{if } \tilde{i} \neq \tilde{i}^0. \end{cases}$$

PROOF. The frontier of the point $\tilde{\alpha}^0$ consists of the points among whose first n coordinates there is one 0 and among whose last m coordinates there is one 1. Hence, $\eta(\tilde{\alpha}_0^0) = nm+1$.

Suppose $\tilde{\alpha} \neq \tilde{\alpha}^0$ and $\tilde{\beta}$ is the greatest point in the frontier $\delta(\{\tilde{\alpha}\})$. Then it follows from (5) that

$$\tilde{i}(\tilde{\beta}) = (\underbrace{1,\dots,1}_{s-1}, \min\{(i_{s+1}+1),m\}, \min\{(i_{s+2}+1),m\}, \\ \dots, \min\{(i_n+1),m\}, m).$$

Since $\eta(\tilde{\alpha}) = |\Delta(I(\tilde{\alpha})) \setminus \Delta(I(\tilde{\alpha}) \setminus \{\tilde{\alpha}\})|$, we have

$$\eta(\tilde{\alpha}) = |\{\tilde{\gamma} | \tilde{\gamma} \le \tilde{\beta}, \tilde{\gamma} \notin \delta(I(\tilde{\alpha}) \setminus \{\tilde{\alpha}\})\}| = |\{\tilde{\gamma} | \tilde{i}(\tilde{\gamma}) \\ = (\underbrace{0,\dots,0}_{p\ge 0}, \underbrace{1,\dots,1}_{q \ge 0}, \min\{(i_{s+1}+1),m\},\dots,\min\{(i_n+1),m\}, j)\}|,$$

where $p+q = s-1$ and $j \in \{\min\{(i_n+1),m\}, \min\{(i_n+2),m\},\dots,m\}$. Therefore, $\eta(\tilde{\alpha}) = s(m-i_n)$. Proposition 2 is proved.

Theorem 5. $|\Delta(A)| = \sum_{\tilde{\alpha}\in A} \eta(\tilde{\alpha})$.

PROOF. The theorem is proved by induction on the cardinality p of the set A. For $p=1$, the assertion is obvious. Assume that it is proved for all sets of cardinality $p-1$. Consider an arbitrary p-element ideal. It can be represented as follows

$$A = I(\tilde{\alpha}') \cup \bigcup I(\tilde{\alpha}^j), \tag{8}$$

where $\tilde{\alpha}' \notin I(\tilde{\alpha}^j)$ for all j (i.e., $A \setminus \{\tilde{\alpha}'\}$ is an ideal). Then it follows from Lemma 1 that

$$\Delta(A) = \Delta(I(\tilde{\alpha}')) \cup \bigcup \Delta(I(\tilde{\alpha}^j)). \tag{9}$$

Since $A \setminus \{\tilde{\alpha}'\} = (I(\tilde{\alpha}') \setminus \{\tilde{\alpha}'\}) \cup \bigcup I(\tilde{\alpha}^j)$, we have

$$\Delta(A \setminus \{\tilde{\alpha}'\}) = \Delta(I(\tilde{\alpha}') \setminus \{\tilde{\alpha}'\}) \cup \bigcup \Delta(I(\tilde{\alpha}^j)). \tag{10}$$

So, $\tilde{\alpha}' \notin I(\tilde{\alpha}^j)$ for each j, we have $I(\tilde{\alpha}') \cap I(\tilde{\alpha}^j) \subseteq I(\tilde{\alpha}') \setminus \{\tilde{\alpha}'\}$. From this and Lemma 2, we obtain

$$\Delta(I(\tilde{\alpha}')) \cap \Delta(I(\tilde{\alpha}^j)) = \Delta(I(\tilde{\alpha}') \cap I(\tilde{\alpha}^j)) \subseteq \Delta(I(\tilde{\alpha}') \setminus \{\tilde{\alpha}'\}).$$

Note that if $B \cap C \subseteq D$, then

$$(B \cup C) \setminus (D \cup C) = (B \cup C) \setminus D \cap (B \cup C) \setminus C = B \setminus D \cap B \setminus C = B \setminus D.$$

Then it follows from (9) and (10) that

$$\Delta(A) \setminus \Delta(A \setminus \{\tilde{\alpha}'\}) = \Delta(I(\tilde{\alpha}')) \setminus \Delta(I(\tilde{\alpha}') \setminus \{\tilde{\alpha}'\}).$$

Therefore, $|\Delta(A) \setminus \Delta(A \setminus \{\tilde{\alpha}'\})| = \eta(\tilde{\alpha}')$ which implies

$$|\Delta(A)| = \eta(\tilde{\alpha}') + |\Delta(A \setminus \{\tilde{\alpha}'\})| = \eta(\tilde{\alpha}') + \sum_{\tilde{\alpha} \in A \setminus \{\tilde{\alpha}'\}} \eta(\tilde{\alpha}) = \sum_{\tilde{\alpha} \in A} \eta(\tilde{\alpha}).$$

Theorem 5 is proved.

For an arbitrary set A, introduce the following special subsets:

$$[A]_t^s = \{\tilde{i} = (i_1, \dots, i_n) \in A \mid i_1 = \cdots = i_s = 0,\ i_n \leq m - t\}, \tag{11}$$

$$\begin{aligned} A_t^s &= \{\tilde{i} \in A \mid i_1 = \cdots = i_s = 0,\ i_{s+1} > 0,\ i_n = m - t\} \\ &= \{\tilde{\alpha} \in A \mid \tilde{\alpha} = (\underbrace{1, \dots, 1}_{s}, 0, \alpha_{s+2}, \dots, \alpha_{d-t-1}, 1, \underbrace{0, \dots, 0}_{t})\}. \end{aligned} \tag{12}$$

The above definitions of the sets A_t^s and $[A]_t^s$ directly imply

Proposition 3. (a) $A = [A]_0^0$ *and for all* s *and* t, $0 \leq s \leq n$, $0 \leq t \leq m$, *the following inclusions hold* $[A]_t^{s+1} \subset [A]_t^s$, $[A]_{t+1}^s \subset [A]_t^s$, $[A]_{t+1}^{s+1} \subset [A]_t^s$.

(b) *The sets* A_t^s *are pairwise disjoint and* $A = \bigcup_{s=0}^n \bigcup_{t=0}^m A_t^s$.

(c) $\eta(\tilde{i}) = st$ *for each point* $\tilde{i} \in A_t^s$, $\tilde{i} \neq \tilde{i}^0$, *where* $\tilde{i}^0$ *is the least point in* $L_{m,n}$.

(d) $\eta(\tilde{i}) = 0$ *if and only if* $\tilde{i} \in A_0^0 \cup A_1^0 \cup A_0^1$; $\eta(\tilde{i}) \neq 0$ *if and only if* $\tilde{i} \in [A]_1^1$. (*Note that* $[A]_1^1 = A \setminus (A_0^0 \cup A_1^0 \cup A_0^1)$.)

(e) $|A_t^s| = |[A]_t^s| - |[A]_t^{s+1}| - |[A]_{t+1}^s| + |[A]_{t+1}^{s+1}|$.

Theorem 6. $|\Delta(A)| = \left(\sum_{s=1}^{n}\sum_{t=1}^{m}|[A]_t^s|\right) + 1.$

PROOF. By Theorem 5 and statements (b) and (c) in Proposition 3, we have

$$|\Delta(A)| = \sum_{\tilde{i}\in A}\eta(\tilde{i}) = \sum_{s=0}^{n}\sum_{t=0}^{m}\sum_{\tilde{i}\in A_t^s}\eta(\tilde{i}) = \left(\sum_{s=0}^{n}\sum_{t=0}^{m}|A_t^s|st\right) + 1.$$

Using this fact and statement (b) in Proposition 3, by standard transformations of sums, we obtain

$$\begin{aligned}
|\Delta(A)| &= \left(\sum_{s=0}^{n}\sum_{t=0}^{m}(|[A]_t^s| - |[A]_t^{s+1}| - |[A]_{t+1}^s| + |[A]_{t+1}^{s+1}|)st\right) + 1 \\
&= \left(\sum_{s=0}^{n}\sum_{t=0}^{m}|[A]_t^s|st\right) - \left(\sum_{s=1}^{n}\sum_{t=0}^{m}|[A]_t^s|(s-1)t\right) \\
&\quad - \left(\sum_{s=0}^{n}\sum_{t=1}^{m}|[A]_t^s|s(t-1)\right) + \left(\sum_{s=1}^{n}\sum_{t=1}^{m}|[A]_t^s|(s-1)(t-1)\right) + 1 \\
&= \left(\sum_{s=1}^{n}\sum_{t=1}^{m}|[A]_t^s|(st - s(t-1) - (s-1)t + (s-1)(t-1))\right) + 1 \\
&= \left(\sum_{s=1}^{n}\sum_{t=1}^{m}|[A]_t^s|\right) + 1.
\end{aligned}$$

Theorem 6 is proved.

4. An Upper Bound for the Functional $\varphi(d, n)$

In this section, we prove that $\varphi(2n, n) \leq c/\sqrt{n}$, where c is an appropriate constant and present an example of a set which shows that this bound cannot be improved in order. We consider initial intervals $I(\tilde{\alpha}^*) = \{\tilde{\alpha} \in B_n^d \mid \tilde{\alpha} \leq \tilde{\alpha}^*\}$ in the middle layer ($d = 2n$) whose greatest point $\tilde{\alpha}^* \in B_n^d$ has the form:

$$\tilde{\alpha}^* = (\underbrace{0,\ldots,0}_{n-k},\underbrace{1,\ldots,1}_{n-k+1},\underbrace{0,\ldots,0}_{k},\underbrace{1,\ldots,1}_{k-1}).$$

By Theorem 3, we have $\delta(I(\tilde{\alpha}^*)) \cup I(\tilde{\alpha}^*) = I(\tilde{\beta}^*)$, where $\tilde{\beta}^*$ is the greatest point in the frontier of the point $\tilde{\alpha}^*$ ($\tilde{\alpha}^*, \tilde{\beta}^* \in B_n^d$). In our case, $\tilde{\beta}^*$ has the form

$$\tilde{\beta}^* = (\underbrace{0,\ldots,0}_{n-k+1},\underbrace{1,\ldots,1}_{n-k},\underbrace{0,\ldots,0}_{k-1},\underbrace{1,\ldots,1}_{k}).$$

It follows from this that

$$|I(\tilde{\alpha}*)| = \sum_{i=0}^{k-1} \binom{2k-1}{i} \binom{2n-2k+1}{n-i} \tag{13}$$

and

$$|\delta(I(\tilde{\alpha}^*))| = \binom{2k-1}{k} \binom{2n-2k+1}{n-k}. \tag{14}$$

By definition, $\Phi(I(\tilde{\alpha}^*)) = |\delta(I(\tilde{\alpha}^*))|/|I(\tilde{\alpha}^*)|$. Using this relation and also (13) and (14), we obtain

$$\begin{aligned}\Phi(I(\tilde{\alpha}^*)) &= \frac{\binom{2k-1}{k}\binom{2n-2k+1}{n-k}}{\sum_{i=0}^{k-1}\binom{2k-1}{i}\binom{2n-2k+1}{n-i}} \\ &= \frac{\binom{2k-1}{k}\binom{2n-2k+1}{n-k}}{\sum_{i=0}^{k-1}\binom{2k-1}{i}\binom{2n-2k+1}{n-k}\prod_{j=1}^{k-i}\frac{2n-2k+1-(n-k+j-1)}{n-k+j}} \\ &= \frac{\binom{2k-1}{k}}{\sum_{i=0}^{k-1}\binom{2k-1}{i}\prod_{j=1}^{k-i}\frac{n-k-j+2}{n-k+j}} = \binom{2k-1}{k}/S,\end{aligned} \tag{15}$$

where

$$S = \sum_{i=0}^{k-1} \binom{2k-1}{i} \prod_{j=1}^{k-1} \frac{n-k-j+2}{n-k+j}.$$

Theorem 7. 1. *If k is a constant ($0 \le k \le n$), then*

$$\lim_{n\to\infty} \Phi(I(\tilde{\alpha}^*)) = \frac{2}{\sqrt{\pi k}}.$$

2. *If $k = k(n) \le \lambda n$, where λ is an arbitrary fixed constant $0 < \lambda < 1$, then*

$$\Phi(I(\tilde{\alpha}^*)) = \Omega(1/\sqrt{k}).$$

PROOF. If k is a constant, then

$$\lim_{n\to\infty} \prod_{j=1}^{k-1} \frac{n-k-j+2}{n-k+j} = 1.$$

This and (15) imply

$$\lim_{n\to\infty} \Phi(I(\tilde{\alpha}^*)) = \binom{2k-1}{k} / \sum_{i=0}^{k-1} \binom{2k-1}{i}$$

$$= 2^{-2k+2}\binom{2k-1}{k} = 2^{-2k+1}\binom{2k}{k}.$$

Using Stirling's formula, we obtain

$$\binom{2k}{k} \sim \frac{2^{2k}}{\sqrt{\pi k}}.$$

Therefore,

$$\lim_{n\to\infty} \Phi(I(\tilde{\alpha}^*)) = \frac{2}{\sqrt{\pi k}}. \tag{16}$$

Now consider the second case. We first establish a lower bound for S in (15). We have

$$S \geq \sum_{i=\lfloor k-\sqrt{k}\rfloor}^{k-1} \binom{2k-1}{i} \prod_{j=1}^{k-i}\left(1 - \frac{2j}{n-k}\right)$$

$$> \sum_{i=\lfloor k-\sqrt{k}\rfloor}^{k-1} \binom{2k-1}{i} \left(1 - \frac{2(k-i)}{n-k}\right)^{k-i}$$

$$> \sum_{i=\lfloor k-\sqrt{k}\rfloor}^{k-1} \binom{2k-1}{i} \left(1 - \frac{2\sqrt{k}}{n-k}\right)^{\sqrt{k}}.$$

Since $n - 1 \geq n(1-\lambda)$, where $\lambda < 1$,

$$\left(1 - \frac{2\sqrt{k}}{n-k}\right)^{\sqrt{k}} = \Omega(1).$$

Therefore,

$$S > \Omega(1) \sum_{i=\lfloor k-\sqrt{k}\rfloor}^{k-1} \binom{2k-1}{i} = \Omega(2^{2k}).$$

Since

$$\prod_{j=1}^{k-i} \frac{n-k-j+2}{n-k+j} < 1,$$

we have

$$S = \sum_{i=0}^{k-1} \binom{2k-1}{i} = \Omega(2^{2k}).$$

Hence,

$$S = \Omega(2^{2k}). \tag{17}$$

Substituting (17) into (15), we obtain

$$\Phi(I(\tilde{\alpha}^*)) = \frac{1}{S}\binom{2k-1}{k} = \Omega(2^{-2k})\binom{2k-1}{k} = \Omega\left(\frac{1}{\sqrt{k}}\right). \tag{18}$$

The theorem now follows from (16) and (18).

So, we have proved that, for each constant $\lambda < 1$, there exists a constant c such that $\Phi(I(\tilde{\alpha}^*))\,|_{k(n)=\lambda n} \leq c/\sqrt{n}$. With this and the fact that the cardinality of the intervals considered does not exceed $\frac{1}{2}|B^d_n|$, we obtain the following assertion.

Theorem 8. *There exists a constant c such that*

$$\varphi(2n, n) \leq c/\sqrt{n}. \tag{19}$$

We next prove that in the case of the middle layer the upper bound in (19) is achieved in order for the balls of radius $n - O(\sqrt{n})$ centered at the point $\tilde{\alpha}^0$. The volume of such balls is close to half of the cardinality of the whole layer B^d_n. Note that the largest ball whose volume does not exceed $\frac{1}{2}|B^d_n|$ has radius $r = 2\lfloor n/2 \rfloor$.

Proposition 4. *If $n = \lfloor d/2 \rfloor$ and $r = \lfloor n/2 \rfloor - O(\sqrt{n})$, then*

$$\Phi(S_{2r}(\tilde{\alpha}^*)) = \Omega(1/\sqrt{d}).$$

Proof. If $n = \lfloor d/2 \rfloor$ and $r = \lfloor n/2 \rfloor - O(\sqrt{n})$, then

$$|S_{2r}(\tilde{\alpha}^0)| = \sum_{k=0}^{r} \binom{n}{n-k}\binom{d-n}{k} = \Omega(4n). \tag{20}$$

We now show that a ball is a "locally optimal" set with respect to the cardinality of the frontier.

Proposition 5. *For a fixed* $p \le \frac{1}{2}|B_n^d|$ *such that there exists a ball* $S = S_{2r}(\{\tilde\alpha^0\})$ *of volume* p *and for an arbitrary* p*-set* $A \subseteq B_n^d$ *such that* $|(S \setminus A) \cup (A \setminus S)| = 2$, *it is true that* $\Phi(A) \ge \Phi(S)$.

PROOF. Suppose A is a set such that $A \setminus S = \{\tilde\beta\}$ and $S \setminus A = \{\tilde\alpha\}$. We claim that $|\delta(A)| \ge |\delta(S)|$. Four cases are possible.

Case 1. Let $\rho(\tilde\alpha, \tilde\alpha^0) < 2r$ and $\rho(\tilde\beta, \tilde\alpha^0) < 2(r+1)$. Then, since $\delta(A) = (\delta(S) \setminus \{\tilde\beta\}) \cup \{\tilde\alpha\} \cup (\delta(\{\tilde\beta\}) \setminus A)$ and $\tilde\alpha$ and $\tilde\beta$ are distinct points in $\delta(S)$, we have $|\delta(A)| = |\delta(S)| + |\delta(\tilde\beta) \setminus A|$. Therefore, $|\delta(A)| \ge |\delta(S)|$.

Case 2. Let $\rho(\tilde\alpha, \tilde\alpha^0) < 2r$ and $\rho(\tilde\beta, \tilde\alpha^0) = 2(r+1)$. Obviously, $\delta(A) = \delta(S) \cup \{\tilde\alpha\} \cup (\delta(\{\tilde\beta\}) \setminus A)$ and $\delta(S) \cap \{\tilde\alpha\} \cap (\delta(\{\tilde\beta\}) \setminus A) = \varnothing$. But then $|\delta(A)| = |\delta(S)| + |\delta(\{\tilde\beta\}) \setminus A| + 1$. Hence, $|\delta(A)| > |\delta(S)|$.

Case 3. Let $\rho(\tilde\alpha, \tilde\alpha^0) = 2r$ and $\rho(\tilde\beta, \tilde\alpha^0) < 2(r+1)$. Since $\delta(A) \supseteq (\delta(S) \cup \{\tilde\alpha\}) \setminus \{\tilde\beta\}$, we have $|\delta(A)| \ge |\delta(S)|$.

Case 4. Let $\rho(\tilde\alpha, \tilde\alpha^0) = 2r$ and $\rho(\tilde\beta, \tilde\alpha^0) = 2(r+1)$. Since $\delta(A) = \delta(S) \cup \{\tilde\alpha\} \cup (\delta(\{\tilde\beta\}) \setminus A)$, we have $|\delta(A)| > |\delta(S)|$.

Thus, in each case $|\delta(A)| \ge |\delta(S)|$. Hence, $|\Phi(A)| \ge |\Phi(S)|$. Proposition 5 is proved.

5. Lower Bounds for the Functional $\varphi(d, n)$

The bound appearing in this section is based on the well-known Kruskal–Katona theorem [1, 2]. The *lower shade* of a set $A \subseteq B_n^d$ is understood to be the set $\partial(A)$ of all the vertices lying at distance 1 from this set and belonging to the $(n-1)$th layer, i.e.,

$$\partial(A) = \{\tilde\beta \in B_{n-1}^d | (\exists \tilde\alpha \in A) |\tilde\alpha \oplus \tilde\beta| = 1\}.$$

The *upper shade* is defined similarly:

$$d(A) = \{\tilde\beta \in B_{n+1}^d | (\exists \tilde\alpha \in A) |\tilde\alpha \oplus \tilde\beta| = 1\}. \tag{21}$$

Note that $\partial(d(A)) = \delta(A) \cup A$. The set $L(p)$ of the first p vertices in the nth layer (with respect to the lexicographical order) is called the *left lexicographical interval.*

Theorem 9 (Kruskal–Katona theorem [1, 2]). *For an arbitrary set* $A \subseteq B_n^d$, $n \le d - n$, *with* $|A| = p$, $|\partial(A)| \ge |\partial(L_n(p))|$.

The Kruskal–Katona theorem implies the following lemma (for the proof, see, for example, [8]).

Lemma 3. *If* $p \leq \binom{d-k}{n}$, *then* $|\partial(L_n(p))| \geq \frac{np}{d-n-k+1}$.

The Kruskal–Katona theorem and Lemma 3 imply the following

Proposition 6. *If* $|A| \leq \binom{d-k}{d-n}$, $0 \leq k \leq n$, *then*

$$\Phi(A) \geq \frac{(d-n)(n+1)}{(d-n-k+2)(n-k+1)} - 1. \tag{22}$$

If $k \leq \frac{3-\sqrt{5}}{2}n - (1 + \frac{1}{\sqrt{5}})$, *then*

$$\Phi(A) \geq \frac{(d-n)(n+1)}{(d-n-k+1)(n-k+1)} - 1. \tag{23}$$

PROOF. Let $A \subseteq B_n^d$ and $|A| = p$. Denote by A^* the set consisting of the points opposite to the points in A. It is clear that $A^* \subseteq B_{d-n}^d$ and $|A^*| = |A| = p$. By the Kruskal–Katona theorem, $|\partial(A^*)| \geq |\partial(L_{d-n}(p))| =: q$. Obviously, $\partial(L_{d-n}(p)) \in B_{d-n-1}^d$. Since $|\partial(L_{d-n}\binom{d-k}{d-n})| = \binom{d-k}{d-n-1}$, we have $q \leq \binom{d-k}{d-n-1}$. In addition, $|d(A)| = |\partial(A^*)| \geq |\partial(L_{d-n}(p))| = q$. It follows from the inclusion $\partial(L_{n+1}(q)) \subseteq \partial(L_{n+1}(|d(A)|))$ and the Kruskal–Katona theorem that

$$|\partial(d(A))| \geq |\partial(L_{n+1}(|d(A)|))| \geq |\partial(L_{n+1}(q))|. \tag{24}$$

Fix a constant $c = c(k)$ such that $\binom{d-k}{d-n-1} \leq \binom{d-c(k)}{n+1}$. Then $q \leq \binom{d-k}{d-n-1}$ implies $q \leq \binom{d-c(k)}{n+1}$. Without loss of generality, we may and shall assume that $n \leq d/2$. Then

$$\frac{\binom{d-k}{d-n-1}}{\binom{d-c(k)}{n+1}} = \frac{(n+1)n(n-1)\dots(n-k+2)}{(d-n-1)\dots(d-n-c(k))} \frac{(d-k)!}{(d-c(k))!}. \tag{25}$$

We now show that (25) does not exceed 1 for $c(k) = k-2$ and $k \leq n$:

$$\begin{aligned} \frac{\binom{d-k}{d-n-1}}{\binom{d-k+2}{n+1}} &= \frac{(n+1)n(n-1)\dots(n-k+2)}{(d-n-1)\dots(d-n-k+2)} \frac{1}{(d-k+1)(d-k+2)} \\ &\leq \frac{(n-1)}{(n-1)} \cdots \frac{(n-k+2)}{(n-k+2)} \frac{(n+1)n}{(d-k+1)(d-k+2)} \\ &= \frac{(n+1)n}{(d-k+1)(d-k+2)} \leq 1. \end{aligned} \tag{26}$$

It can be proved analogously that if $k \leq \frac{3-\sqrt{5}}{2}n$, then (25) is not greater than 1 for $c(k) = k-1$. Since $|A| \leq \binom{d-k}{d-n}$, it follows from Lemma 3 and (24) that

$$\begin{aligned}\Phi(A) &= \frac{|\partial(d(A))|}{|A|} - 1 \geq \frac{|\partial(L_{n+1}(q))|}{|A|} - 1 \\ &= \frac{|\partial(L_{n+1}(q))|}{q}\,\frac{|\partial(L_{d-n}(p))|}{p} - 1 \geq \frac{n+1}{d-n-c}\,\frac{d-n}{n-k+1} - 1.\end{aligned} \tag{27}$$

Substituting $c = k-2$ ($c = k-1$) into (27), we obtain (22) ((23)). Statement 6 is proved.

Corollary. *Let $n = \lfloor d/2 \rfloor$ and $|A| \leq \binom{d-k}{d-n}$, $1 \leq k \leq n$. Then, for large k ($k = n - \Delta$ and $\Delta = o(n)$),*

$$\Phi(A) \geq \frac{n(n+1)}{(\Delta+1)(\Delta+2)} \tag{28}$$

and, for small k ($1 \leq k << n$),

$$\Phi(A) \geq \frac{2k-1}{n+1}. \tag{29}$$

PROOF. 1. If k is large ($k = n - \Delta$ and $\Delta = o(n)$), then the cardinality of the set A is small as compared to the cardinality of the layer ($|A| \leq \binom{n+\Delta}{n}$). In this case, by (22), for each A in the middle layer,

$$\Phi(A) \geq \frac{k}{n-k+1} + \frac{(k-2)(n+1)}{(n-k+2)(n-k+1)} \geq \frac{n(n+1)}{(\Delta+1)(\Delta+2)}.$$

2. If k is small and A is contained in the middle layer, then it follows from (23) that

$$\begin{aligned}\Phi(A) &\geq \frac{(d-n)k}{(d-n-k+1)(n-k+1)} + \frac{k-1}{(d-n-k+1)} \\ &\geq \frac{k}{n-k+1} + \frac{k-1}{d-n} \geq \frac{2k-1}{d-n}.\end{aligned}$$

The proof is complete.

(23) implies the following

Theorem 10. *If* $k = 1$ *(hence,* $|A| \leq \frac{n}{d}|B_n^d|$*), then* $\Phi(A) \geq 1/n$*, i.e.,* $\varphi(d,n) \geq 1/n$.

Theorem 6 establishes a relation between $\Delta(A)$ and certain special sets $[A]_t^s$. Thus, estimating the cardinalities of the sets $[A]_t^s$, one may obtain a bound for the cardinality of the closure $\Delta(A)$. The next two lemmas present optimal sets for the problem of minimizing the sets $[A]_0^1$ and $[A]_1^0$. Consider the following linear order on the vertices in the layer B_n^d. A vector $\tilde{\alpha}$ *precedes* a vector $\tilde{\beta}$ ($\tilde{\alpha} <_3 \tilde{\beta}$) if $(i_n(\tilde{\alpha}), \ldots, i_1(\tilde{\alpha})) < (i_n(\tilde{\beta}), \ldots, i_1(\tilde{\beta}))$ in the usual lexicographical order. The *left lexicographical interval of cardinality* p in a layer with respect to some order is the set of the first (in this order) p vertices in the layer.

Lemma 4. *The minimum of* $|[A]_0^1|$ *over all the sets* $A \subseteq L_{m,n}$ *of a fixed cardinality is attained at a left lexicographical interval in order* $<_3$.

PROOF. We proceed by induction on n. For $n = 1$, there exists only one ideal. It is an initial interval in $<_3$. Assume that the lemma is true for $n = k$ and for every m and show that it is also true for $L_{m,k+1}$. We say that $A^* \subseteq L_{m,k+1}$ is an optimal set for a fixed cardinality p if

$$|[A^*]_0^1| = \min_{A \subseteq L_{m,k+1},\, |A|=p} |[A]_0^1| = r. \tag{30}$$

Select a set A' so that

$$|A'| = \min_{A \subseteq L_{m,k+1},\, |[A]_0^1|=r} |A|. \tag{31}$$

Then all the p-element subsets of A' are optimal. It follows from (31) that if $(0, x_2, \ldots, x_{k+1})$ is an arbitrary point in the set $[A']_0^1$, then all the points of the form $(j, x_2, \ldots, x_{k+1})$, where $1 \leq j \leq x_2$, belong to A'. Otherwise, by adding them to the set A', we obtain an ideal. Thus, $|A'|$ increases while $|[A']_0^1|$ remains the same, which contradicts (31).

For the sets $B \subseteq [L_{m,k+1}]_0^1$, introduce the functional

$$f(B) = \sum_{\tilde{i} \in B} f(\tilde{i}), \quad \text{where } \tilde{i} = (i_1, i_2, \ldots, i_n),\ f(\tilde{i}) = i_2.$$

It follows from Lemma 7 and the definition of $f(B)$ that

$$f([A']_0^1) = |A'| - |[A']_0^1| = |A'| - r = \max_{B \subseteq [L_{m,k+1}]_0^1,\, |B|=r} f(B). \tag{32}$$

The space $[L_{m,k+1}]_0^1 = \{(0, x_2, \ldots, x_{k+1}) \mid 0 \leq x_2 \leq \cdots \leq x_{k+1 \leq m}\}$ is equivalent to the space $L_{m,k} = \{(x_1, \ldots, x_k) \mid 0 \leq x_1 \leq \cdots \leq x_k \leq m\}$. Therefore, $[A']_0^1$

can be considered as a set in $L_{m,k}$. On $L_{m,k}$, the functional f is defined as follows: $f(\tilde{i}) = i_1$. We show that an initial interval in $L_{m,k}$ in order $<_3$ can be taken as the set $[A']_0^1$. To this end, it suffices to prove that it satisfies (32). Indeed, then the set A' which is obtained from $[A']_0^1$ by adding the points of the form $(i, x_2, \ldots, x_{k+1})$ is an initial interval in $L_{m,k+1}$ in the same order $<_3$. Deleting now the $|A'| - p$ greatest points from A', we obtain an optimal initial interval.

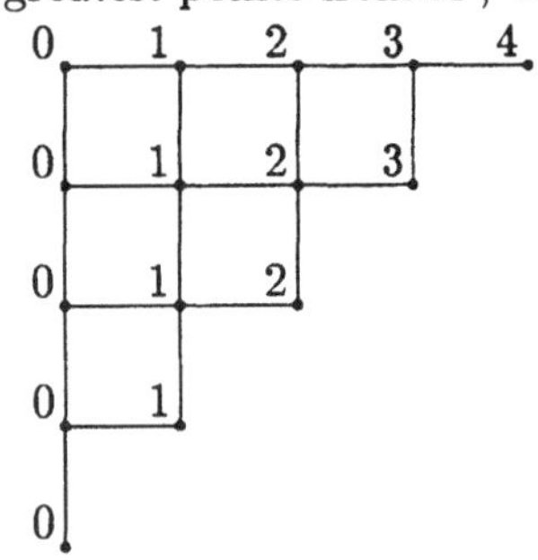

Each point in $L_{m,k}$ is assigned a value of the functional f (see the picture for $k = 2$, $m = 4$). For each set $B \in L_{m,k}$, consider the "truncated" sets

$$B_c = \{\tilde{i} \in B \mid f(\tilde{i}) \geq c\}.$$

Put $b_0 = |B_0| = r$ and $b_c = |B_c|$. Since for an initial interval D in $L_{m,k}$ (in order $<_3$), the "truncated" sets D_c are initial intervals in $(L_{m,k})_c$ $(L_{m-c,k})$, we have the following: Given a fixed cardinality $|B_c|$, the cardinality $|B_{c+1}|$ is maximum if B is an initial interval (this follows from the induction assumption for $L_{m-c,k}$). Considering an initial interval D and an arbitrary set B such that $|D| = |B| = r$, we successively conclude that $|D_1| \geq |B_1|$, $|D_2| \geq |B_2|, \ldots, |D_m| \geq |B_m|$. Hence,

$$f(D) = |D_1| + |D_2| + \cdots + |D_m| \geq |B_1| + |B_2| + \cdots + |B_m| = f(B),$$

i.e., the initial interval D can be taken as $[A']_0^1$. Lemma 4 is proved.

Lemma 5. *The minimum of $|[A]_1^0|$ over all the sets $A \in L_{m,n}$ of a fixed cardinality is attained at the left lexicographical interval in the usual lexicographical order $\leq_4$ on $L_{m,n}$.*

PROOF. We show that the optimality of the order $\leq_4$ in the problem of minimizing $|[A]_1^0|$ with $|A| = |L_{m,n}| - p$ is implied by the optimality of the order $<_3$ in the problem of minimizing $|[A]_0^1|$ with $|A| = p$. For each set $A \in L_{m,n}$, consider the sets $\bar{A} = L_{m,n} \setminus A$ and $A^* = \{\tilde{j} \in L_{m,n} \mid \exists \tilde{i} = (i_1, \ldots, i_n) \in A$ such that $\tilde{j} = (m - i_n, \ldots, m - i_1)\}$. It is clear that $\overline{A^*} = \bar{A}^*$.

To begin with we show that a set A solves the problem of minimizing $|[B]_0^1|$ with $|B| = p$ if and only if the set $\bar{A}^*$ solves the problem of minimizing $|[B]_1^0|$ with $|B| = |L_{m,n}| - p$.

Indeed, let A minimize $|B \cap [L_{m,n}]_0^1|$ over all the ideals B of cardinality p. Then $\bar{A}$ maximizes $|B \cap [L_{m,n}]_0^1|$ over all the upper sets B of cardinality $|L_{m,n} - p|$. Hence, $\bar{A}$ minimizes $|B \cap (L_{m,n} \setminus [L_{m,n}]_0^1)|$ over all the upper sets B of cardinality $|L_{m,n} - p|$. This, in turn, implies that $\bar{A}^*$ minimizes $|B \cap (L_{m,n} \setminus [L_{m,n}]_0^1)^*| = |B \cap [L_{m,n}]_1^0| = |[B]_1^0|$ over all the ideals B of cardinality $|L_{m,n} - p|$.

We next verify that A is an initial interval with respect to $\leq_3$ if and only if $\bar{A}^*$ is an initial interval in $\leq_4$. Indeed, $\tilde{\alpha}^*(\tilde{\beta}^*)$ is obtained from $\tilde{\alpha}^* = (m - \tilde{\alpha}_n, \ldots, m - \tilde{\alpha})(\tilde{\beta}^* = (m - \tilde{\beta}_n, \ldots, m - \tilde{\beta}_1)$. Further, $\tilde{\alpha} <_3 \tilde{\beta}$ if and only if $\tilde{\alpha}^* >_4 \tilde{\beta}^*$. Therefore, the following statements are equivalent:

(a) A is an initial interval in order $<_3$;
(b) $A = \{\tilde{\alpha} \mid \tilde{\alpha} \leq_3 \tilde{\beta}\}$ for some point $\tilde{\beta}$;
(c) $\bar{A} = \{\tilde{\alpha} \mid \tilde{\alpha} >_3 \tilde{\beta}\}$ for some point $\tilde{\beta}$;
(d) $\bar{A}^* = \{\tilde{\alpha}^* \mid \tilde{\alpha}^* <_4 \tilde{\beta}^*\}$ for some point $\tilde{\beta}^*$;
(e) $\bar{A}^*$ is an initial interval in order $<_4$.

Thus, the claims are proved and the lemma now follows from Lemma 4 since p is arbitrary.

Introduce the notation:

$$a_{s,t} = \min_{A:|A|\leq(1/2)|L_{m,n}|} \frac{|[A]_t^s|}{|A|} \quad \text{and} \quad l_{s,t} = \frac{|[L_{m,n}]_t^s|}{|L_{m,n}|} = \frac{\binom{d-s-t}{n-s}}{\binom{d}{n}}.$$

Proposition 7. *Given arbitrary m and n, for each set $A \subseteq L_{m,n}$,*

$$|[A]_t^s| / |A| \geq |[L_{m,n}]_t^s| / |L_{m,n}|, \quad \text{i.e.} \quad a_{s,t} \geq l_{s,t}.$$

For the proof we need the following two lemmas.

Lemma 6. *Given arbitrary m and n, for each set $A \subseteq L_{m,n}$,*

$$|[A]_0^1| / |A| \geq |[L_{m,n}]_0^1| / |L_{m,n}| = n/(m+n), \quad \text{i.e.} \quad a_{1,0} \geq l_{1,0}.$$

PROOF. The space $L_{m,n}$ decomposes into the layers $L_{m,n}^k$ with respect to the last coordinate i_n:

$$L_{m,n}^k = \{\tilde{i} \in L_{m,n} \mid i_n = k\} = \{\tilde{\gamma} \mid \tilde{\gamma} = (\gamma_1, \ldots, \gamma_{n+k-1}, 1, \underbrace{0, \ldots, 0}_{m-k})\}. \tag{33}$$

Put

$$I_r = \bigcup_{k=0}^{r} L_{m,n}^k = I(\tilde{\alpha})\Big|_{i(\tilde{\alpha})=(r,\ldots,r)}. \tag{34}$$

The above definitions imply the following equalities:

$$|L_{m,n}| = \binom{m+n}{n}, \quad |[L_{m,n}]_0^1| = \binom{m+n-1}{n-1}, \tag{35}$$

$$|L^k_{m,n}| = \binom{n+k-1}{k}, \quad |[L^k_{m,n}]^1_0| = \binom{n+k-2}{k}, \tag{36}$$

$$|I_r| = \sum_{k=0}^{r} \binom{n+k-1}{k} = \binom{n+r}{n}, \tag{37}$$

$$|[I_r]^1_0| = \sum_{k=0}^{r} \binom{n+k-2}{k} = \binom{n+r-1}{n-1}. \tag{38}$$

By Lemma 4, the minimum $a_{1,0}$ is attained at an initial interval A in order $\leq_3$. It can be represented in the form $A = I_{r-1} \cup (A \cap L^r_{m,n})$. Put $C = A \cap I_{r-1} = I_{r-1}$ and $D = A \cap L^r_{m,n}$. Then $A = C \cup D$, where $A \subseteq L_{m,n}$, $D \subset L^r_{m,n}$, and $L^r_{m,n}$ is isomorphic to $L_{r,n-1}$.

The assertion is proved by induction on dimension n. For $n = 1$ and an arbitrary m, for each nonempty lower set A we have $|[A]^1_0| = 1$. Therefore, $|[A]^1_0| / |A| = 1/|A| \geq 1/|L_{m,n}| = |[L_{m,n}]^1_0| / |L_{m,n}|$.

Assume that the assertion is proved for $n-1$. Consider D as an ideal in $L^r_{m,n} = L_{r,n-1}$. By the induction assumption,

$$|[D]^1_0| / |D| \geq |[L_{r,n-1}]^1_0| / |L_{r,n-1}|. \tag{39}$$

We have to show that $|[A]^1_0| / |A| \geq n/(m+n)$. We have

$$\frac{|[A]^1_0|}{|A|} = \frac{|[C]^1_0| + |[D]^1_0|}{|C| + |D|} = \frac{|[C]^1_0|}{|C|} \frac{1 + |[D]^1_0| / |[C]^1_0|}{1 + |D| / |C|} = \frac{|[C]^1_0|}{|C|} \frac{1+xy}{1+x},$$

where $x = \frac{|D|}{|C|}$ and $y = \frac{|[D]^1_0|/|D|}{|[C]^1_0|/|C|}$. Since $D \subset L^r_{m,n}$ and $L^r_{m,n}$ is isomorphic to $L_{r,n-1}$, we have $x \leq x_0 = \frac{|L_{r,n-1}|}{|C|}$. This and (39) imply

$$y \geq y_0 = \frac{|[L_{r,n-1}]^1_0|}{|L_{r,n-1}|} \frac{|C|}{|[C]^1_0|} = \frac{|[L_{r,n-2}]^1_0|}{|L_{r,n-1}|} \frac{|C|}{|[C]^1_0|}.$$

If $y \geq 1$, then

$$|[D]^1_0| / |D| \geq |[C]^1_0|| / |C| = n/(n+r-1) \geq n/(n+m).$$

It follows from this that

$$(|[C]^1_0| + |[D]^1_0|) / (|C| + |D|) \geq n/(n+m).$$

If $y < 1$, then the function $f(x,y) = \frac{1+xy}{1+x} = 1 - (1-y)(1+\frac{1}{1+x})$ decreases in x and increases in y. Therefore, $f(x,y) \geq f(x_0, y_0)$. Hence,

$$\frac{|[C]_0^1| + |[D]_0^1|}{|C| + |D|} \geq \frac{|[C]_0^1|}{|C|} \frac{1 + \frac{|L_{r,n-1}|}{|C|} \frac{|L_{r,n-2}|}{|L_{r,n-1}|} \frac{|C|}{|[C]_0^1|}}{1 + \frac{|L_{r,n-1}|}{|C|}} = \frac{|[C]_0^1| + |L_{r,n-2}|}{|C| + |L_{r,n-1}|}.$$

From this, by (35), (37), and (38), we obtain

$$\frac{|[C]_0^1| + |[D]_0^1|}{|C| + |D|} \geq \frac{\binom{n+r-2}{n-1} + \binom{n+r-2}{n-2}}{\binom{n+r-1}{n} + \binom{n+r-1}{n-1}} = \frac{\binom{n+r-1}{n-1}}{\binom{n+r}{n}} = \frac{n}{n+r} \geq \frac{n}{n+m}.$$

Lemma 6 is proved.

Lemma 7. *Given arbitrary m and n, for each set $A \subseteq L_{m,n}$,*

$$|[A]_1^0| / |A| \geq |[L_{m,n}]_1^0| / |L_{m,n}| = n/(m+n), \quad \text{i.e.} \quad a_{0,1} \geq l_{0,1}.$$

PROOF. The proof is similar to that of Lemma 12, but now we consider the decomposition of the space $L_{m,n}$ into the layers $\widehat{L}_{m,n}^k$ with respect to the first coordinate i_1:

$$\widehat{L}_{m,n}^k = \{\tilde{i} \in L_{m,n} \mid i_1 = k\} = \{\tilde{\gamma} \mid \tilde{\gamma} = (\underbrace{0, \ldots, 0}_{k}, 1, \gamma_{k+2}, \ldots, \gamma_{m+n})\}. \tag{40}$$

Put

$$\widehat{I}_r = \bigcup_{k=0}^{r} \widehat{L}_{m,n}^k = I(\tilde{\alpha}) \big|_{i(\tilde{\alpha})=(r,m,\ldots,m)} . \tag{41}$$

We make use of Lemma 5 and the following equalities which are implied by the above definitions (40) and (41):

$$|\widehat{L}_{m,n}^k| = \binom{m+n-k-1}{n-1}, \quad |[\widehat{L}_{m,n}^k]_1^0| = \binom{m+n-k-2}{n-1}, \tag{42}$$

$$|\widehat{I}_r| = \sum_{k=0}^{r} \binom{m+n-k-1}{m-k} = \binom{m+n}{n} - \binom{m+n-r-1}{n}, \tag{43}$$

$$|[\widehat{I}_r]_0^1| = \sum_{k=0}^{r} \binom{m+n-k-2}{m-k-1} = \binom{m+n-1}{n} - \binom{m+n-r-2}{n}. \tag{44}$$

Note that $[L]_t^s$ is isomorphic to $L_{m-t,n-s}$ and $[A]_t^s = [\ldots[[[\ldots[[A]_0^1]_0^1 \ldots]_0^1]_1^0]_1^0 \ldots]_1^0$. The rest of the argument is similar to that of Lemma 6.

Proposition 7 is proved by induction on s and t. By Lemmas 6 and 7, $a_{1,0} \geq l_{1,0}$ and $a_{0,1} \geq l_{0,1}$. Since $[A]_1^1 = [[A]_0^1]_1^0$, where $[A]_0^1 \subseteq L_{m,n-1}$, we obtain

$$a_{1,1} = \min \frac{|[A]_1^1|}{|A|} \geq \min \frac{|[A]_0^1|}{|A|} \frac{|[L_{m,n-1}]_0^1|}{|L_{m,n-1}|} \geq \frac{|[L_{m,n}]_0^1|}{|L_{m,n}|} \frac{|[L_{m,n-1}]_0^1|}{|L_{m,n-1}|}$$
$$= \frac{|L_{m,n-1}|}{|L_{m,n}|} \frac{|[[L_{m,n}]_0^1]_1^0|}{|L_{m,n-1}|} = \frac{|[L_{m,n}]_1^1|}{|L_{m,n}|} = l_{1,1}.$$

Assume that the inequality $a_{s,t} \geq l_{s,t}$ is already proved. We have to show that $a_{s+1,t} \geq l_{s+1,t}$ and $a_{s,t+1} \geq l_{s,t+1}$. Since $[A]_t^{s+1} = [[A]_t^s]_0^1$, where $[A]_t^s \subseteq L_{m-t,n-s}$, we have

$$a_{s+1,t} = \min \frac{|[A]_t^{s+1}|}{|A|} \geq \min \frac{|[[A]_t^s]_0^1|}{|A|} \geq \min \frac{|[A]_t^s|}{|A|} \frac{|[L_{m-t,n-s}]_0^1|}{|L_{m-t,n-s}|}$$
$$\geq \frac{|[L_{m,n}]_t^s|}{|L_{m,n}|} \frac{|[L_{m,n}]_t^{s+1}|}{|L_{m-t,n-s}|} = l_{s+1,t}.$$

The inequality $a_{s,t+1} \geq l_{s,t+1}$, is proved in a similar way, this time, taking account of $[A]_{t+1}^s = [[A]_t^s]_1^0$. The proof of Proposition 7 is now complete.

The next statement is a strengthening of Lemma 11 for $a_{1,1}$.

Lemma 8. *For the middle layer, $a_{1,1}/l_{1,1} \geq 1 + \Omega(1/n)$.*

PROOF. We have

$$a_{1,1} = \min_{A \subseteq L_{m,n}} \frac{|[A]_1^1|}{|A|} = \min_{A \subseteq L_{m,n}} \frac{|[[A]_0^1]_1^0|}{|A|} = \min_{A \subseteq L_{m,n}} \frac{|[[A]_0^1]_1^0|}{|[A]_0^1|} \frac{|[A]_0^1|}{|A|}. \tag{45}$$

By Lemma 4, the minimum of $|[A]_0^1|/|A|$ over all the sets $A \subseteq L_{m,}n$ is attained at an initial lexicographical interval in order $\leq_3$. Denote this interval by A'. Then we have $|A'| \leq \frac{1}{2}|L_{m,n}|$ and $|L_{m,n}^m| \leq \frac{1}{2}|L_{m,n}|$. Therefore, $A' \subseteq L_{m-1,n}$ and hence $[A']_0^1 \subseteq L_{m-1,n-1}$. By Lemma 4, this implies

$$\frac{|[A]_0^1|}{|A|} \geq \frac{|[A']_0^1|}{|A'|} \geq \frac{|[L_{m-1,n}]_0^1|}{|L_{m-1,n}|} = \frac{|L_{m-1,n-1}|}{|L_{m-1,n}|} = \frac{n}{m+n-1}. \tag{46}$$

Denote $B = [A]_1^0 \subseteq L_{m,n-1}$. Then, by Lemma 4,

$$\frac{|[B]_1^0|}{|B|} \geq \frac{|[L_{m,n-1}]_1^0|}{|L_{m,n-1}|} = \frac{|L_{m-1,n-1}|}{|L_{m,n-1}|} = \frac{m}{m+n-1}. \tag{47}$$

Substituting (46) and (47) into (45), we obtain $a_{1,1} \geq \frac{mn}{(m+n-1)^2}$. Finally, taking into account that $l_{1,1} \geq \frac{mn}{(m+n)(m+n-1)}$ and $m \leq n+1$, we obtain

$$\frac{a_{1,1}}{l_{1,1}} \geq \frac{m+n}{m+n-1} = 1 + \frac{1}{m+n-1} = 1 + \Omega\left(\frac{1}{n}\right).$$

Lemma 8 is proved.

The following lemma establishes a relationship between the problem of minimizing $|[A]_1^1|$ and the original problem of minimizing the frontier functional.

Lemma 9. 1. *If* $n = \lfloor n/2 \rfloor$ *and* $a_{1,1}/l_{1,1} \geq 1 + \Omega(1/n)$, *then* $\varphi(d,n) \geq \Omega(1/n)$.

2. *If* $n = \lfloor n/2 \rfloor$ *and* $a_{1,1}/l_{1,1} \geq 1 + \Omega(1/\sqrt{n})$, *then* $\varphi(d,n) \geq \Omega(1/\sqrt{n})$.

PROOF. Both statements are proved analogously. We prove the second. By Theorem 7,

$$\varphi(d,n) = \min_{A:|A| \leq (1/2)|L_{m,n}|} \left(\frac{|\Delta(A)|}{|A|} - 1\right)$$

$$= \min_{A:|A| \leq (1/2)|L_{m,n}|} \left(\sum_{s=1}^{n}\sum_{t=1}^{m} \frac{|[A]_t^s|}{|A|} + \frac{1}{|A|} - 1\right) \geq \sum_{s=1}^{n}\sum_{t=1}^{m} a_{s,t} + \frac{1}{|L_{m,n}|} - 1.$$

Since

$$1 = \frac{\Delta(L_{m,n})}{|L_{m,n}|} = \frac{1 + \sum_{s,t} |[A]_t^s|}{|L_{m,n}|} = \frac{1}{|L_{m,n}|} + \sum_{s,t} l_{s,t},$$

we have

$$\sum_{s,t} l_{s,t} = 1 - \frac{1}{|L_{m,n}|}.$$

For the middle layer,

$$l_{1,1} = \frac{|[L_{m,n}]_1^1|}{|L_{m,n}|} = \binom{d-2}{n-1} \Big/ \left(\binom{d}{n} \frac{n(d-n)}{d(d-1)}\right) \geq \frac{1}{4}.$$

Then

$$\sum_{s=1}^{n}\sum_{t=1}^{m} a_{s,t} \geq l_{1,1}\Omega\left(\frac{1}{\sqrt{n}}\right) + \sum_{s=1}^{n}\sum_{t=1}^{m} l_{s,t} \geq \frac{1}{4}\Omega\left(\frac{1}{\sqrt{n}}\right) + 1 - \frac{1}{|L_{m,n}|},$$

from which we obtain that $\varphi(d,n) \geq \Omega(1/\sqrt{n})$ for $n = \lfloor d/2 \rfloor$. Lemma 9 is proved.

Lemmas 8 and 9 directly imply

Theorem 11. *If* $n = \lfloor d/2 \rfloor$, *then* $\varphi(d,n) \geq \Omega(1/n)$.

References

1. J. B. Kruskal (1963) The number of simplices in a complex, *Mathematical Optimization Techniques*, University of California Press, Berkeley, pp. 251–278.
2. G. Katona (1968) A theorem of finite sets, *Theory of Graphs*, Akad. Kiado, Budapest, pp. 187–207.
3. L. H. Harper (1991) On a problem of Kleitman and West, *Discrete Math.* **93**, No. 2–3, 169–182.
4. A. D. Korshunov (1981) The number of monotone Boolean functions (in Russian), in: *Problemy Kibernetiki.* Vol. 38, Nauka, Moscow, pp. 5–108.
5. A. A. Sapozhenko (1992) The boundary functional method for counting isoperimetric problems (in Russian), Ph. D. Thesis, Moscow University, Moscow.
6. L. H. Harper (1977) Stabilization and the edgesum problem, *Ars Combinatorica* **4**, pp. 225–270.
7. B. List (1994) A vertex variant of the Kleitman–Vest problem (in Russian), *Diploma Paper*, Moscow University, Moscow.
8. A. A. Sapozhenko (1989) On the number of antichains in multilayered ranked sets (in Russian), *Diskret. Mat.* **1**, No. 2, 110–128.

Faculty of Informatics
Ulm University
Einstein Alley
Ulm 89081
GERMANY

TRANSLATED BY S. I. SUSLOV

REGULAR GRAPHS WHOSE EVERY EDGE BELONGS TO MANY TRIANGLES†)

A. A. Makhnëv

An undirected graph is said to be edge-regular with parameters (v, k, λ) if it consists of v vertices, the degree of each vertex is k, and each edge belongs to precisely λ triangles. It is proved that every edge-regular graph with parameters (v, k, λ) in which $3\lambda \geq 2k - 5$ either has the diameter 2, or is a polygon or else the icosahedron, unless $k = 4$ and $\lambda = 1$.

Introduction

We consider undirected graphs without loops and multiple edges. If a and b are vertices of a graph Γ, then the distance between a and b is denoted by $d(a, b)$, and the subgraph of Γ induced by the set of vertices lying at the distance i from a is denoted by $\Gamma_i(a)$. The subgraph $\Gamma_1(a)$ is said to be the *neighborhood* of the vertex a and is denoted by $[a]$. The subgraph that is the ball of the radius 1 centered at a is denoted by $a^{\perp}$.

A graph Γ is called a *regular graph of degree* k if, for every vertex a in Γ, the neighborhood $[a]$ has precisely k vertices. A graph Γ is called an *edge-regular* graph with parameters (v, k, λ) if Γ consists of v vertices, Γ is regular of degree k, and each edge of Γ belongs to λ triangles. A graph Γ is called a *completely regular* graph with parameters (v, k, λ, μ) if Γ is edge-regular and, for all vertices a and b at distance two, $[a] \cap [b]$ consists of μ vertices. A completely regular graph with the diameter 2 is called a *strongly regular* graph. For all vertices a and b if $d(a, b) = 1$ $(d(a, b) = 2)$, then we denote the number of vertices in $[a] \cap [b]$ by $\lambda(a, b)$ $(\mu(a, b))$ and call this subgraph a λ-(μ-)subgraph.

A *Terwilliger graph* is a noncomplete graph Γ such that for all vertices a and b at distance two, the subgraph $[a] \cap [b]$ is a μ-clique (i.e. a clique of size μ) for some μ.

Denote by $(m_1, \ldots, m_n)$ the complete n-partite graph whose parts have cardinalities $m_1, \ldots, m_n$. If $m_1 = \cdots = m_n = m$ then the graph $(m_1, \ldots, m_n)$ is denoted by $K_{n\times m}$. A *triangular* graph $T(m)$ is a graph whose vertices are unordered pairs of elements in X, $|X| = m$, where vertices $\{a, b\}$ and $\{c, d\}$ are

†) This research was supported by the Russian Foundation for Basic Research (Grant 94–01–00802-a) and the Foundation of the State Committee for Higher Education of the Russian Federation.

A. D. Korshunov (ed.), *Operations Research and Discrete Analysis*, 171–182.

adjacent if and only if they have a common element. A *Taylor* graph is a completely regular graph of diameter 3 such that, for every two vertices a and b with $d(a,b) = 3$, each vertex belongs to $a^\perp \cup b^\perp$. The 12-vertex Taylor graph is called the *graph of the icosahedron* if the neighborhood of each vertex is a pentagon.

The α-*extension* of a graph Γ is the graph Γ' constructed from Γ by replacing each vertex a in Γ by the complete subgraph (a) with α vertices (a subgraph of such kind is called α-*clique*), where every two vertices (a) and (b) are adjacent in Γ' if and only if a and b are adjacent in Γ. The *kernel* $K(a)$ of a vertex a in a regular graph Γ is the vertex set $\{x \in \Gamma \mid x^\perp = a^\perp\}$. A vertex a is called *reduced* if $K(a) = \{a\}$. A graph Γ is called *reduced* if all vertices of Γ are reduced. For a subgraph Δ, the number of vertices in Δ is denoted by $|\Delta|$.

If the distance between vertices u and w in Γ is i, then we denote by $b_i(u,w)$ ($c_i(u,w)$) the number of vertices in the intersection of $\Gamma_{i+1}(u)$ ($\Gamma_{i-1}(u)$) with $[w]$. Put $b_1(\Gamma) = \max\{b_1(u,w) \mid uw$ is an edge in $\Gamma\}$. Note that, in an edge-regular graph with parameters (v,k,λ), the value of $b_1(u,w)$ does not depend on a choice of an edge uw and equals $k - \lambda - 1$.

It is proved in [1] that the diameter of each noncomplete connected edge-regular graph of degree k, $k \geq 3b_1$, is at most 2, and this graph consists of at most $2k - 2$ vertices. In the present paper, a bound for the diameter of reduced graphs is obtained for $k \geq 3b_1 - 2$.

Theorem. *Let Γ be a connected regular reduced graph of degree k, and let $k \geq 3b_1 - 2$. Then either the diameter of this graph is less then 4, or Γ is a n-polygon for $n \geq 8$, or else $k = 4$, $b_1 = 2$, and $\lambda(a,b) = 1$ or 2 for each edge ab.*

Corollary. *Let Γ be a connected edge-regular graph with parameters (v,k,λ) in which $k \geq 3b_1 - 2$ (or equivalently: $3\lambda \geq 2k - 5$). Then either Γ is a polygon or the icosahedron, or $k = 4$ and each edge belongs to a unique triangle, or else Γ has diameter 2 and consists of at most $2k + 4$ vertices.*

This corollary reduces the problem of describing edge regular graphs with $\lambda \geq k - \lfloor k/3 \rfloor - 2$ to the case $k \equiv 0 \pmod 3$. Observe that if the neighborhood of each vertex in Γ is hexagons, then $b_1(\Gamma) = 3$ and $k = 3b_1 - 3$. V. D. Paduchikh has shown (which is unpublished) that such graphs can be of an arbitrary diameter. Indeed, consider an $m \times n$ rectangular table such that the vertex with coordinates (i,j) is adjacent to the vertices $(i-1,j)$, $(i+1,j)$, $(i,j-1)$, $(i,j+1)$, $(i-1,j-1)$, $(i+1,j+1)$, and then identify the bottom row with the top one, as well as the first column with the last one. (The Shrikhande graph is constructed from 5×5 table in the same way.)

§ 1. Auxiliary Statements

The following statement is used in the proof of the Theorem.

Proposition 1. *Let Γ be a connected regular graph of diameter greater than 2 such that $\mu(u,w) \geq b_1$ for all two vertices u and w lying on a shortest path of length 3. Then either Γ is a b_1-extension of a polygon or each vertex of Γ belongs to $u^\perp \cup w^\perp$ for every u and every w such that $d(u,w) = 3$.*

This statement generalizes the Neumaier Theorem [1, Theorem 1.5.5]: a completely regular graph of diameter at least 3 with $\mu = k - \lambda - 1$ is a Taylor graph.

Some auxiliary claims (Lemmas 1.1–1.4) are used in the proof of Proposition 1. Let Γ be a graph meeting the conditions of Proposition 1, and let $eacw$ be a shortest 3-path between the vertices e and w, $\lambda = k - b_1 - 1$ and $\Delta(x) = \Gamma_3(w) \cap x^\perp$ for $x \in \Gamma_2(w)$, $\Delta = \Delta(a)$.

Lemma 1.1. *For a graph Γ meeting the conditions of Proposition 1, the following statements are true:*

(a) $[a] \subset e^\perp \cup w^\perp$, $\mu(e,c) = \mu(a,w) = b_1$, *and* $\lambda(e,a) = \lambda(c,w) = \lambda$;

(b) $[a] - w^\perp \subset x^\perp$ *for* $x \in \Delta$. *In particular,* Δ *is a clique and* $\Delta(y) = \Delta$ *for each* $y \in [e] \cap [c]$;

(c) $\Delta(y) = \Delta$ *for* $y \in [a] \cap \Gamma_2(w)$ *and* $\Delta = \Gamma_2(c) \cap \Gamma_3(w)$.

PROOF. On the one hand, the number of vertices in $[a] - e^\perp$ is at most b_1. On the other hand, $\mu(a,w) \geq b_1$. Therefore, $[a] - e^\perp = [a] \cap w^\perp$, $\mu(u,w) = b_1$, $|[a] \cap [e]| = \lambda$, and (a) is proved.

Let $x \in \Delta$. By (a), we have $[a] - w^\perp \subset x^\perp$. So $[a] \cap e^\perp = [a] \cap x^\perp$ for $x \in \Delta$. In particular, Δ is a clique. If $y \in [e] \cap [c]$, then $[c] - w^\perp \subset [z]$ for $z \in \Delta(y)$. Hence, $z \in [a]$ and $\Delta(y) = \Delta$, which implies (b).

Let $y \in [a] \cap \Gamma_2(w)$. By (a), we have $[a] - [w] \subset x^\perp$ for each $x \in \Delta$, in particular, $y \in [x]$ and $\Delta(y) = \Delta$.

Let $x \in \Gamma_2(c) \cap \Gamma_3(w)$. Then $[c] - [w] \subset x^\perp$, in particular, $a \in [x]$, and (c) is proved.

Lemma 1.2. *If* $[e] \cap \Gamma_3(w) \neq \Delta$, *then* Γ *is a* b_1*-extension of the heptagon.*

PROOF. Assume that $z \in [e] \cap \Gamma_3(w) - \Delta$. By (c) of Lemma 1.1, we have $d(c,z) = 3$. Take some path $zuyw$ in Γ. Then $\Lambda = eacwyuz$ is a path of length 7. By Lemma 1.1, $[a] \cap [w] \subset \Gamma_3(z)$ and similarly $[a] \cap [z] \subset \Gamma_3(w)$. Hence $[a] \subset [w] \cup e^\perp$ and $[a] \subset [z] \cup c^\perp$ for any $e \in [a] \cap [z]$ and $c \in [a] \cap [w]$. In particular, this implies that $[a] \subset c^\perp \cup e^\perp$ and that the μ-subgraphs $[a] \cap [w]$ and $[a] \cap [z]$ are cliques, $k \leq 3b_1 - 1$, and $\lambda \leq 2b_1 - 2$.

Prove that the distance between each two vertices in the heptagon Λ is the same as in Γ. Suppose that $d(u,a) = 2$ and choose an arbitrary $x \in [a] \cap [u]$. Due to (c) of Lemma 1.1, we have $x \notin [w]$. If $x \in \Gamma_2(w)$, then $x \in [e] \cap [z]$

which once again contradicts (c) of Lemma 1.1. Hence, $x \in \Gamma_3(w)$. Observe that $d(c,u) = 2$, for otherwise $e \in u^{\perp} \cup c^{\perp}$ due to (a) of Lemma 1.1. By applying (a) of Lemma 1.1 to a 3-path through c, u, z, we obtain $y \in c^{\perp}$. But in this case $([a] \cap [w]) \cup ([u] \cap [w])$ is a clique on $2b_1$ vertices, and $\lambda(c,y) \geq 2b_1 - 1$. This contradicts the choice of $cyuz$ taken as the shortest 3-path between the vertices c and z. Hence, $d(u,a) = 3$.

Suppose that $d(a,y) = 2$ and $x \in [a] \cap [y]$. For a 3-path $axyu$, the vertex $w \in [y]$ must belong to $[a] \cup u^{\perp}$. A contradiction. Hence, $d(y,a) = 3$. Now it is obvious that Λ is a heptagon, all μ-subgraphs of Λ are cliques, $k = 3b_1 - 1$, and $\lambda = 2b_1 - 2$. Observe that $[z] \cap [a] = \Delta$, $[u] \cap [e] = \Delta(u)$, and $e^{\perp} \cup z^{\perp}$ is a $(\lambda + 2)$-clique. Therefore, $\Delta = K(e)$, $\Delta(u) = K(z)$, and Γ is a b_1-extension of a heptagon. Lemma 1.2 is proved.

Lemma 1.3. *If the diameter of a graph is greater than* 3, *then* Γ *is a* b_1-*extension of a polygon.*

PROOF. Let $zeacw$ be a shortest path of length 4 between vertices z and w. Then $[a]\cap[z] \subset \Gamma_3(w)$. By Lemma 1.1, $\Delta \subset [e]-[c] = [e]\cap z^{\perp}$, $[a]-c^{\perp} = [a]\cap[z]$, where $[a] \cap z^{\perp}$ lies in Δ. So $[a] \cap z^{\perp} = \Delta$. By Lemma 1.2, $\Delta = e^{\perp} \cap \Gamma_3(w)$ and $x^{\perp} = e^{\perp}$ for each vertex x in Δ. Thus, $[a] \cap [z] = K(e)$ and $b_2(w,a) = b_1$. By symmetry, $[a] \cap [w] = K(c)$. Hence, $[a] \subset K(c) \cup K(e) \cup ([c] \cap [e])$ and $[c] \cap [e] = K(a)$. Thus, $k = 3b_1 - 1$ and a vertex in $[w] - c^{\perp}$ belongs to $\Gamma_4(e)$. Since Γ is a connected graph for arbitrary vertex x, we have $|K(x)| = b_1$, and Γ is a b_1-extension of a polygon. Lemma 1.3 is proved.

Lemma 1.4. *If* Γ *is not a* b_1-*extension of a polygon, then* $\Gamma_3(w) = K(e)$.

PROOF. Assume that Γ is not a b_1-extension of a polygon. As follows from Lemma 1.3 the diameter of Γ is equal to 3 and in accordance with Lemma 1.2 we have $\Gamma_3(w) \cap [e] = \Delta$. Let $u \in \Gamma_2(w) \cap [e]$. Since $\Gamma_3(w) \cap [e] = \Delta(u)$ by Lemma 1.2, we have $\Delta = K(e)$. Assume that $\Gamma_3(w)$ contains a vertex $y \notin \Delta$ and choose a path $yuzw$. Due to (c) of Lemma 1.1, we have $d(e,z) = 3$. On the other hand, $z \in \Gamma_3(e) \cap [w]$, and hence $z \in K(w)$. A contradiction with $d(y,w) = 3$.

Now we prove Proposition 1. By Lemma 1.3, we can assume that the diameter of Γ is 3. As follows from Lemmas 1.1 and 1.4, $e^{\perp} \cup w^{\perp}$ induces a connected component of Γ. Hence, for all vertices u and w such that $d(u,w) = 3$, each vertex of Γ belongs to $e^{\perp} \cup w^{\perp}$.

§ 2. Reduced Graphs of Large Diameter

We now turn to proving the Theorem. Let Γ be a connected regular reduced graph of degree k, where $k \geq 3b_1 - 2$, and let $eacw$ be a shortest path between e and w, where $\lambda = k - b_1 - 1$. Recall that a graph is reduced if and only if for each vertex a, the kernel $K(a)$ consists of a single vertex.

Lemma 2.1. *Let y and z be nonadjacent vertices in $[a]$. If $\mu(y,z) < b_1$ then $[y] \cap [z]$ is a clique on $b_1 - 1$ vertices, $k = 3b_1 - 2$, $\lambda = 2b_1 - 3$, and for each $x \in [y] \cap [z]$ the inclusion $[x] \subset y^\perp \cup z^\perp$ holds.*

PROOF. Note that both $[x] - y^\perp$ and $[x] - z^\perp$ consist of at most b_1 vertices. Hence, $[y] \cap [z]$ contains at least $k - 2b_1$ vertices in $[x]$. If $\mu(y,z) < b_1$, then $[y] \cap [z]$ is a clique on $b_1 - 1$ vertices, $k = 3b_1 - 2$, $\lambda = 2b_1 - 3$ and, for each $x \in [y] \cap [z]$ the inclusion $[x] \subset y^\perp \cup z^\perp$ holds.

Lemma 2.2. *If $\mu(y,z) = b_1$ holds whenever $d(y,z) = 2$, then Γ is a polygon or the icosahedron.*

PROOF. It is sufficient to prove that for each x, the subgraph $\Gamma_3(x)$ is not empty. Otherwise $\Gamma_2(x)$ would consist of $k+1$ vertices by Proposition 1. The number of edges between $[a]$ and $\Gamma_2(a)$ may vary from kb_1 to $(k+1)b_1$. Since Γ is reduced, Γ is a polygon or a Taylor graph due to Proposition 1. If Γ is a Taylor graph, then $\mu = b_1$, $k = \lambda + \mu + 1$, and $\lambda \geq 2\mu - 3$. By Theorem 1.5.3 in [1], $k(\Delta) = 2\mu(\Delta)$ holds for the neighborhood Δ of each vertex in Γ. Therefore, each μ-subgraph of a Taylor graph is a regular graph of degree $(1/2)\lambda(\Gamma) \geq \mu - 3/2$. Hence, Γ is a Terwilliger graph of degree $k = 2\lambda + 3 - \mu$. By Theorem 1.2.3 in [1], Γ is either the icosahedron or has $\mu = 1$. In the second case $k = 2$, and Γ is a hexagon. Lemma 2.2 is proved.

If $b_1 = 1$ then Γ is either a polygon or $K_{n\times 2}$ by Corollary 1.1.6 in [1]. If $b_1 = 2$ then $k = 4$ and $\lambda(x,y) \in \{1,2\}$ for each edge xy. It follows from Lemmas 2.1 and 2.2 that $k = 3b_1 - 2$, $\lambda = 2b_1 - 3$, and $b_1 > 2$.

Lemma 2.3. *The degree of every vertex of a μ-graph is at least $b_1 - 2$.*

PROOF. Let $d(y,z) = 2$, and let $x \in [y] \cap [z]$. Then $[x]$ contains at least $2(2b_1 - 2) - \alpha$ vertices in $y^\perp \cup z^\perp$, where α is the degree of x in the graph $y^\perp \cap z^\perp$. Due to Lemma 2.1, we have $2(2b_1 - 2) - \alpha \geq 3b_1 - 2$. Hence, $\alpha \geq b_1 - 2$.

Lemma 2.4. *The subgraph $e^\perp \cap c^\perp$ is a clique.*

PROOF. Let a and x be nonadjacent vertices in $e^\perp \cap c^\perp$. From Lemma 2.1, we have $|e^\perp \cap c^\perp| = b_1$. Hence, $|[c] \cap [w]| = \lambda$. Due to Lemma 2.3, the degree of a and x in the μ-graph $e^\perp \cap c^\perp$ is $b_(1) - 2$. Therefore, $|[c] \cap [y]| = \lambda = |[y] \cap [e]|$, and $[y] \subset e^\perp \cup c^\perp$ for $y \in \{a, x\}$. Hence, $[c]$ contains a and at least $2(2b_1 - 2) - \alpha$ vertices in $x^\perp \cup w^\perp$, where α is the degree of c in the graph $x^\perp \cap w^\perp$. Then $\alpha = b_1 - 1$, $|[c] \cap [x]| = \lambda = |[c] \cap [w]|$, and $[x] \cap [w]$ is a subgraph of $c^\perp$. Using symmetry, we conclude that $|[c] \cap [a]| = \lambda$, $[a] \cap [w]$ consists of b_1 vertices and is a subgraph of $c^\perp$. Finally, the graph $[c]$ contains the vertex w and at least $2(2b_1 - 2) - \beta$ vertices in $x^\perp \cup a^\perp$, where β is the degree of c in the graph $x^\perp \cap a^\perp$. Hence, $\beta \geq b_1 - 1$, and $x^\perp \cap a^\perp$ contains some vertex $d \in [c] \cap [w]$.

The neighborhood of the vertex d contains nonadjacent vertices a and x in $[e]$. Therefore, this vertex can be chosen as c. Let $[d]$ contain γ vertices in $[w]\cap[c]-a^{\perp}$. Then $[d]$ contains $b_1 - 1 - \gamma$ vertices in $[w] - c^{\perp}$. Hence, $[x] \cap [w]$ contains at most γ vertices in $a^{\perp}$, and $[x]$ contains at least $b_1 - \gamma$ vertices in $[a] \cap [w]$. Thus $\Gamma_2(a)$ contains x, as well as $b_1 - 1$ vertices in each of the $[e]$ and $[c]$ which are distinct from x, and also $b_1 - 1 - \gamma$ vertices in $[d]$. Hence, $|\Gamma_2(a)| \geq 3b_1 - 2 - \gamma$. The number of edges between $\Gamma_2(a)$ and $[a]$ is at most $(2b_1 - 3)b_1$ and at least $2b_1 - 1 - \gamma + b_1 + (3b_1 - \gamma - 4)(b_1 - 1)$. This implies that $(\gamma + 4)(b_1 - 1) \geq b_1^2 + 3b_1 - 1 - \gamma$, which contradicts $\gamma \leq b_1 - 1$. Lemma 2.4 is proved.

In the following Lemmas 2.5–2.7 it is assumed that the diameter of the graph Γ is not less then 4. Let $ueacw$ be a shortest path of length 4 between some vertices u and w, and let $\Delta = a^{\perp} - (u^{\perp} \cup w^{\perp})$.

Lemma 2.5. *At least two among three μ-subgraphs for the path $ueacw$ contain $b_1 - 1$ vertices.*

PROOF. Firstly, assume that $\mu(u,a) = \mu(w,a) = b_1$. Then the graph Δ consists of $b_1 - 1$ vertices, and $e^{\perp} \cap c^{\perp} = \Delta$ holds for any vertices $e \in a^{\perp} \cap u^{\perp}$ and $c \in a^{\perp} \cap w^{\perp}$. Therefore, $\Delta = K(a)$, which contradicts the assumption that the graph is reduced.

Let $\mu(u,a) = \mu(e,c) = b_1$. Then $\mu(w,a) = b_1 - 1$ and $|\Delta| = b_1$. If $d \in a^{\perp} \cap w^{\perp}$, then $[a] \cap [d]$ contains $b_1 - 2$ vertices in $[w]$ and $b_1 - 1$ vertices in Δ. So $\Delta = e^{\perp} \cap d^{\perp}$ for each vertex $d \in a^{\perp} \cap w^{\perp}$.

If f is a vertex in $a^{\perp}\cap u^{\perp}$ distinct from e, then $[a]\cap[f]$ contains $b_1 - 1$ vertices in $[u] \cap [a]$ and at least $b_1 - 2$ vertices in $\Delta - \{a\}$. Hence, each vertex in $a^{\perp} \cap u^{\perp}$ distinct from e is not adjacent to at most one vertex in Δ. Conversely, since the graph is reduced, each vertex in Δ distinct from a is not adjacent to at most one vertex in $a^{\perp} \cap u^{\perp}$.

Let $x \in \Delta - \{a\}$ and $y \in [x] - a^{\perp}$. If $[y]$ contains the vertices $e \in [u] \cap [a]$ and $c \in [w] \cap [a]$, then the μ-graph $[e] \cap [c]$ is not a clique. Since $[x] \cap [y]$ consists of at most $b_1 - 2$ vertices in Δ, then $[x] \cap [y]$ contains exactly $b_1 - 2$ vertices in Δ and either $b_1 - 1$ vertices in $[u] \cap [a] - \{e\}$ or all $b_1 - 1$ vertices in $[w] \cap [a]$. In the first case we come to a contradiction with the fact that x is not adjacent to a vertex in $[u] \cap [a]$. In the second case, $|[x] \cap [d]| = \lambda + 1$ for $d \in a^{\perp} \cap w^{\perp}$. Therefore, $\mu(x,w) = b_1$, and by the argument of the previous paragraph every vertex in $[u] \cap [x]$ must be adjacent to y. A contradiction. Lemma 2.5 is proved.

Lemma 2.6. $\mu(u,a) = \mu(a,w) = b_1 - 1$.

PROOF. Assume that $\mu(u,a) = b_1$. Then $|\Delta| = b_1$. Similarly as above, $\Delta \subset [d]$ for $d \in [a] \cap [w]$ and $\mu(x,d) = b_1 - 1$ for $x \in \Delta$ and $d \in [w] \cap [a]$. If $f \in [u] \cap [a]$ then $[f] \cap [a]$ contains $b_1 - 1$ vertices in $[u]$ and $b_1 - 2$ vertices in Δ. Hence, f is not adjacent to only one vertex of Δ.

Assume that a vertex z in Δ nonadjacent to two vertices in $[u] \cap [a]$, and let each vertex in Δ distinct from z be not adjacent with a sole vertex in $[u] \cap [a]$. Let $x \in \Delta - \{a, z\}$ and $y \in [x] - a^{\perp}$. As above, $[x] \cap [y]$ can not contain vertices both in $[u]$ and $[w]$. Hence, $[x] \cap [y]$ contains $b_1 - 1$ vertices in $[u] \cap [a]$ and $b_1 - 2$ vertices in Δ. So $\Delta - \{a\}$ is a subgraph of a graph $[y]$ and y is not adjacent to only one vertex f in $[u] \cap [a]$. Hence, $b_1 = 3, k = 7$, and $[x] \cap [a]$ includes some vertices c and d in $[w]$, the vertex z in Δ and the vertices e and g in $[u]$ nonadjacent to z. Thus, $[z]$ contains vertices a, c, d, x, and y, a vertex f in $[u] \cap [a]$, and a vertex h in $[u]$ not adjacent to a, c, d, and x. The contradiction with $\lambda(z, h) \geq \lambda$. Lemma 2.6 is proved.

Lemma 2.7. *If the diameter of* Γ *is greater than* 3, *then* $b_1 \leq 2$.

PROOF. Assume that $\mu(u, a) = \mu(a, w) = b_1 - 1$. Then $|\Delta| = b_1 + 1$, and each vertex in $[w] \cap [a]$ is nonadjacent to only one vertex in Δ. Therefore, some vertex x in Δ distinct from a is adjacent to all vertices in $[w] \cap [a]$. It follows from Lemma 2.4 that the graph $[x]$ contains all vertices of Δ which are adjacent to a vertex in $[w] \cap [a]$. If Δ is not a subgraph of the graph $x^{\perp}$, then $\Delta - x^{\perp}$ consists of a unique vertex y, and y is not adjacent to every vertices in $[a] \cap [w]$. In this case $a^{\perp} \cap y^{\perp}$ contains $b_1 - 1$ vertices in $[u]$ and all b_1 vertices in $\Delta - \{x\}$. Therefore, x is adjacent to no vertices in $[u] \cap [a]$, and $\Delta - \{x, y\} = K(a)$. The graph being reduced implies $b_1 \leq 2$.

So, $\Delta \subset x^{\perp} \cap y^{\perp}$ for a $y \in \Delta$ adjacent to all vertices in $[u] \cap [a]$. If z is a vertex in Δ and z is not adjacent to $[a] \cap [w]$, then the graphs $[z]$ and $[a] \cap [w]$ are disjoint. Otherwise, a vertex d in $[a] \cap [w] \cap [z]$ is adjacent to x, and $|[x] \cap [w]| = b_1$. A contradiction with Lemma 2.6. Thus, $z = y$ and $\Delta - \{x, y\} = K(a)$. Lemma 2.7 and the Theorem are proved.

§ 3. Edge-Regular Graphs with Large λ

Let a graph Γ be a counterexample for the Corollary, and let $eacw$ be a shortest 3-path between vertices e and w. Then $b_1 \geq 3$, and due to the Theorem, the diameter of this graph is 3. It follows from Lemmas 2.1, 2.2 that $k = 3b_1 - 2$ and $\lambda = 2b_1 - 3$. Observe that b_1 is even, for otherwise k and λ would be odd and the number of edges in the neighborhood of a vertex would be $(1/2)k\lambda$. For $x \in \Gamma_2(w)$, we let $\Delta(x) = [x] \cap \Gamma_3(w)$.

Lemma 3.1. *If* $\mu(e, c) = \mu(a, w) = b_1 - 1$ *then the graphs* $[c] \cap [a]$ *and* $\Gamma_2(e) \cap \Gamma_2(w)$ *have a single common vertex* x. *Moreover, either* $[e] \cup [w]$ *contains* *a vertex in $\Gamma_3(x)$ or $[x]$ is contained in $[e] \cup [w]$, $\mu(x,y) = b_1 - 1$ for each vertex y in $\Gamma_2(x)$ such that y differs from e and w, and $\Gamma_2(x) \subset e^\perp \cup w^\perp$.*

PROOF. Let $\mu(e,c) = \mu(a,w) = b_1 - 1$. Then $[c] \subset a^\perp \cup w^\perp$ and $[a] \subset e^\perp \cup c^\perp$. Hence, $[a] - e^\perp$ consists of b_1 vertices and only one of them does not belong to $[w]$. It is clear that the vertex $x \in [a] - e^\perp$ nonadjacent to w belongs to $[c]$.

Assume that $[e] \cup [w]$ does not contain vertices in $\Gamma_3(x)$. Put $\mu(e,x) = \mu_1$ and $\mu(x,w) = \mu_2$. Then $\Gamma_2(x)$ contains $k + 1 - \mu_1$ vertices in $e^\perp$ and $k + 1 - \mu_2$ vertices in $w^\perp$. The number of edges between $[x]$ and $\Gamma_2(x)$ is $kb_1 \geq \mu_1 + \mu_2 + (2k - \mu_1 - \mu_2)(b_1 - 1)$. From this it follows that $(\mu_1 + \mu_2)(b_1 - 2) \geq k(b_1 - 2)$. Hence, $\mu_1 + \mu_2 = k$, $|\Gamma_2(x)| = k+2$, and $\mu(x,y) = b_1 - 1$ for each vertex $y \in \Gamma_2(x)$ distinct from b and e. Lemma 3.1 is proved.

Lemma 3.2. *Let $\mu(e,c) = b_1$. Then there exists a unique vertex f in $[a]$ such that $f \in \Gamma_2(w) - ([e] \cup [w])$.*

PROOF. Assume that a vertex $f \in [a] - ([e] \cup [w])$ belongs to $\Gamma_3(w)$. Then $[a] \subset \{f\} \cup [w] \cup e^\perp$. Therefore, $[e] \cap [a] = [e] \cap [f]$ and for each vertex x in $[e] \cap [c]$ the inclusion $[x] - [w] \subset \{e,f\} \cup ([e] \cap [f])$ holds.

If a vertex y in $[w] \cap \Gamma_2(e)$ is not adjacent to $[e] \cap [f]$, then $|[e] \cap [y]| \geq b_1 - 1$. Hence, $\Delta = [e] - [y]$ contains at most $2b_1 - 1$ vertices, and $|\Delta - x^\perp| \leq 1$ for each vertex $x \in [e] \cap [c]$. Therefore, there are vertices x, z of $[e] \cap [c]$ such that $\Delta - x^\perp = \Delta - z^\perp$ and $|[x] \cap [z]| > \lambda$.

So, each vertex y in $[w] \cap \Gamma_2(e)$ is adjacent to a vertex in $[e] \cap [f]$, $[e] \cap \Gamma_2(w) = [f] \cap \Gamma_2(w)$, and $\mu(e,y) = b_1$. Hence, $\mu(u,w) = b_1 - 1$ for every vertex $u \in [e] \cap \Gamma_2(w)$. Therefore, the number of edges between $[e] \cap \Gamma_2(w)$ and $[w] \cap \Gamma_2(e)$ is $c_3(e,w)b_1 = c_3(w,e)(b_1 - 1)$. If $c_3(e,w) = b_1 - 1$ then $c_3(w,e) = b_1$ and, for the distinct vertices a and x in $[e] \cap [c]$, the graph $[a] \cap [x]$ contains $b_1 - 1$ vertices in $[w]$, $b_1 - 2$ vertices in $[e] \cap [c]$, and also the vertices e and f. A contradiction. Hence, $c_3(e,w) = 2b_1 - 2$ and $c_3(w,e) = 2b_1$. If a and x are nonadjacent vertices in $[e] \cap \Gamma_2(w)$, then due to Lemma 2.4 each vertex in $[w] \cap \Gamma_2(e)$ belongs to $[a] \cup [x]$.

Thus, a is not adjacent to at least two vertices x and z in $[e] \cap \Gamma_2(w)$, and $[x] \cap [w] = [z] \cap [w]$. Therefore, $[x] \cap [z]$ contains $b_1 - 1$ vertices in $[w]$, at least $b_1 - 2$ vertices in $\Gamma_2(w) \cap [e]$, and also the vertices e and f. A contradiction. Lemma 3.2 is proved.

Lemma 3.3. *Let $\mu(e,c) = b_1$. Then there exists a unique vertex f in $[a]$ that does not belong to $[e] \cup [w]$ and is adjacent to a vertex in $[x] \cap [w]$ for each $x \in [e] \cap [c] \cap [f]$.*

PROOF. Assume that the graphs $[f]$ and $[a]\cap[w]$ are disjoint. Then $[a]-[w] \subset \{e\} \cup (f^{\perp} \cap [e])$ and $[a] \cap [e]$ is a subgraph of $[f]$. Hence, as c we can take any vertex in $[a]\cap[w]$. Assume that a vertex $d \in [f]\cap[w]$ is not adjacent to a vertex c. Then $[d]$ contain no vertices in $[e]\cap[c]$. Therefore, $[f]-[c] = [f]\cap d^{\perp}$. If the vertex $d \in [f] \cap [w]$ is not adjacent to the vertices c and g in $[a] \cap [w]$, then $[f] - d^{\perp}$ is a subgraph of $[c] \cap [g]$, the graph $[c] \cap [g]$ contains b_1 vertices in $[e]$, $b_1 - 3$ vertices in $[a] \cap [w]$, and also a vertex w. A contradiction.

If $\mu(f, w) = b_1$, then there are vertices $d, z \in [w] \cap [f]$ such that subgraph $[d] \cap [z]$ contains $b_1 - 2$ vertices in each of $[w] \cap [a]$ and $[w] \cap [f]$, and also vertices w and f. A contradiction. Thus, $\mu(f, w) = b_1 - 1$.

On the other hand, if c is not adjacent to vertices d and g in $[w] \cap [f]$, then $[f] - [c] = [f] \cap d^{\perp} = [f] \cap g^{\perp}$, and a graph $[d] \cap [g]$ contains w and λ vertices from $f^{\perp}$. So each vertex in $[f] \cap [w]$ can fail to be adjacent to a unique vertex in $[a]\cap[w]$, and, conversely, each vertex in $[a]\cap[w]$ can fail to be adjacent to a unique vertex in $[f] \cap [w]$.

Each vertex of $\Delta = ([a] \cap [w]) \cup ([f] \cap [w])$ is adjacent to a vertex in $[w] - \Delta$, and some vertex y in $[w] - \Delta$ is adjacent to at most one vertex in Δ. Therefore, $[y] \cap [w]$ contains at most b_1 vertices. Hence, $b_1 \leq 3$. A contradiction. Lemma 3.3 is proved.

Lemma 3.4. *Let $[w]$ contain a vertex z in $\Gamma_3(e) \cap \Gamma_3(a)$. Then no vertex in $[e] \cap [c]$ is adjacent to $[c] \cap [z]$.*

PROOF. Put $\Delta = c^{\perp} - ([e] \cup [z])$. A vertex in $[c] - \Delta$ is said to be *bad* if it is not adjacent to vertices in $\Delta - \{c\}$; otherwise the vertex is said to be *good.*

Assume that a vertex $x \in [e] \cap [c]$ is adjacent to a vertex $y \in [z] \cap [c]$. Then $[x] \cap [y]$ contains $b_1 - 2$ vertices from each of $[e]$ and $[z]$, and the unique vertex $c \in \Delta$. By Lemma 3.1 applied to the path $exyz$, the graphs $[x]$ and $[y]$ include the unique vertex c in $e^{\perp} \cup z^{\perp}$. Therefore, $\Delta \cap [x] = \Delta \cap [y] = \{c\}$, and x and y are bad vertices.

Assume that $\mu(e, c) = b_1 - 1$. Then $[x] \cap [c]$ contains $b_1 - 2$ vertices in $[e]$ and $b_1 - 1$ bad vertices in $[z]$. Since $[c] \cap [z]$ contains the good vertex w, we have $\mu(z, c) = b_1, |\Delta| = b_1$. If u in $[z]$ is a bad vertex other than y, then $[y] \cap [u]$ contains the vertices c and z, $b_1 - 2$ vertices in $[c] \cap [z]$, and all $b_1 - 2$ bad vertices in $[e]$. Thus, $\mu(e, c) = \mu(z, c) = b_1$. This implies that $|\Delta| = b_1 - 1$, $[x] \cap [c]$ contains $b_1 - 1$ vertices in $[e]$ and $b_1 - 2$ bad vertices in $[z]$. If u is a good vertex adjacent to w, then $[u] \cap [w]$ contains z, along with $b_1 - 1$ vertices in Δ, and $b_1 - 2$ vertices in $[c] \cap [z]$. A contradiction. Therefore, a and w are the only good vertices, and $\lambda(d,c) \leq b_1 - 1$ for $d \in \Delta$ other than c. A contradiction with the fact that $b_1 \geq 3$. Lemma 3.4 is proved.

Lemma 3.5. *For each shortest 3-path eacw between vertices e and w, the graph* $[w] \cap \Gamma_3(a)$ *is a subgraph of* $\Gamma_2(e)$.

PROOF. Assume that $[w]$ contains a vertex $z \in \Gamma_3(e) \cap \Gamma_3(a)$. Put $\Delta = c^\perp - ([e] \cup [z])$. First suppose that $\mu(e,c) = b_1$. If $\mu(c,z) = b_1$ then $\Delta = [a] \cap [w]$ for all vertices $a \in [c] \cap [e]$ and $w \in [z] \cap [c]$. Hence, $c^\perp = y^\perp$ for each vertex $y \in \Delta$. A contradiction with $|\Delta| = b_1 - 1$.

Thus, $\mu(c,z) = b_1 - 1$ and $|\Delta| = b_1$. As above, $\Delta \subset [w]$ for each vertex w in $[c] \cap [z]$, and $\Delta \cup ([c] \cap [z])$ is a clique on $2b_1 - 1$ vertices. A contradiction with $\lambda(a,c) = \lambda$.

Therefore, $\mu(e,c) = \mu(c,z) = b_1 - 1$, $|\Delta| = b_1 + 1$, and each vertex in $[e] \cap [c]$ fails to be adjacent to a unique vertex in Δ.

In this case some vertex $x \in \Delta$ other than c is adjacent to all vertices of $[e] \cap [c]$. Due to Lemma 2.4, the graph $[x]$ contains all vertices of Δ, and each of these vertices is adjacent to a vertex in $e] \cap [c]$. If Δ is not a subgraph of $x^\perp$, then $\Delta - x^\perp$ consists of the single vertex y which is adjacent to no vertices of $[e] \cap [c]$. In this case the graph $c^\perp \cap y^\perp$ contains at least $b_1 - 1$ vertices in $[z]$ and all b_1 vertices in $\Delta - \{x\}$. This implies that x is adjacent to no vertices in $[z] \cap [c]$, and $\Delta - \{x, y\} = K(c)$. A contradiction.

Thus $\Delta \subset x^\perp$, and $\lambda(x,c) \geq 2(b_1 - 1)$. Lemma 3.5 is proved.

Lemma 3.6. *If* $x \in \Gamma$ *then* $|\Gamma_2(x)| \leq k + 3$. *If* $|\Gamma_2(x)| = k + 3$ *then* $\mu(x,y) = b_1$ *for a unique vertex* $y \in \Gamma_2(x)$, *and* $\mu(x,z) = b_1 - 1$ *for all other vertices of* $\Gamma_2(x)$.

PROOF. The number of edges between $[x]$ and $\Gamma_2(x)$ is $k(k - \lambda - 1) = (3b_1 - 2)b_1$. Therefore, the number of vertices of $\Gamma_2(x)$ is at most

$$\frac{(3b_1 - 2)b_1}{(b_1 - 1)} = 3b_1 + 1 + \frac{1}{(b_1 - 1)}.$$

Since $k = 3b_1 - 2$, the first assertion of Lemma 3.6 follows.

If $|\Gamma_2(x)| = k + 3$, then the number of edges between $[x]$ and $\Gamma_2(x)$ is $3b_1(b_1 - 1) + b_1$, and the second assertion of Lemma 3.6 holds too. Lemma 3.6 is proved.

Lemma 3.6 provides an upper bound for the number of vertices in a graph of diameter 2.

Lemma 3.7. *For all vertices x and y at distance 2, the inequality $b_2(x, y) \leq 1$ holds.*

PROOF. It is sufficient to prove that $|\Delta(a)| = 1$. Let e and u be distinct vertices in $\Delta(a)$. First suppose that $[w] \cap \Gamma_3(a)$ contains a vertex s. By Lemma 3.5, we have $s \in \Gamma_2(e) \cap \Gamma_2(u)$. Note that $[e] \cap [u]$ contains at most one vertex outside $a^{\perp}$. Therefore, $[s]$ contains at least $2(b_1 - 1) - 1$ vertices in $[e] \cup [u]$ which lie outside $w^{\perp}$. Thus, $2b_1 - 3 \leq b_1$. A contradiction with $b_1 \geq 4$.

Hence, $\Gamma_2(a)$ contains at least $2b_1 - 1$ vertices in $[u] \cup [e]$ and $2b_1$ or $2b_1 - 1$ vertices in $w^{\perp}$. Therefore, $4b_1 - 2 \leq 3b_1 + 1$. A contradiction.

Lemma 3.8. *$\mu(e, c) = \mu(a, w) = b_1 - 1$ for each shortest 3-path $eacw$ between vertices e and w.*

PROOF. Assume that $\mu(e, c) = b_1$. Let f be a unique vertex in $[a]$ lying outside $e^{\perp} \cup c^{\perp}$. By Lemma 3.3, the graph $[f] \cap [w]$ contains a vertex $x \in [a]$, and by Lemma 3.1 applied to the path $eaxw$, taking $\mu(f, c) \geq b_1$ into account, we deduce that $\Gamma_3(f)$ contains a vertex u in $[e] \cup [w]$.

Prove now that $\mu(e, f) < b_1$ and $\mu(f, c) < b_1$. To this end assume that $\mu(e, f) = b_1$. In this case $[f] \cap [c]$ contains $b_1 - 2$ vertices in $[a] - e^{\perp}$. Therefore, $[f]$ contains a unique vertex $z \in [e] \cap [c]$ other than a. Hence, $[a] \cap [z]$ contains e, c, f, and $b_1 - 2$ vertices in each of $[f] \cap [c]$ and $[e] \cap [c]$. A contradiction. The case $\mu(c, f) = b_1$ is considered similarly.

It is now clear that a vertex u in $\Gamma_3(f)$ lies in $[w]$. Since by Lemma 3.8, we have $a \in \Gamma_2(u)$, it follows that either c or e belongs to $[u]$. Since $d(e, w) = 3$, we have $c \in [w]$ and $\mu(c, f) = b_1$. A contradiction.

Let us complete the proof of the Corollary. As follows from Lemmas 3.8 and 3.1, the graph $[a] \cap [c]$ contains a unique vertex f outside $[e] \cup [w]$. If $[f]$ contains no vertex $y \in [a] \cap [w]$, then $\mu(e, y) = b_1$. Then $[f]$ contains $[e] \cap [c]$ and $[a] \cap [w]$; in particular, $[a] \cap [c]$ is contained in $f^{\perp}$. If $x \in [a] \cap [w]$, then $[e] \cap [x] = [e] \cap [c]$, for otherwise $[a] \cap [e] \subset f^{\perp}$, and $\lambda(a, f) > \lambda$. Hence, $x^{\perp} \cap c^{\perp}$ contains $f^{\perp} \cap c^{\perp}$ and the vertex w. A contradiction.

Thus, to complete the proof of the Corollary, it remains to find the number of vertices in the graphs under consideration having the diameter 2. If $k \geq 3b_1$, then by Lemma 1.4.2 in [1], the number v of vertices is at most $2k - 2$. If $k = 3b_1 - 2$ then Lemma 3.6 implies $v \leq 2k + 4$. Let $k = 3b_1 - 1$. If $b_1 = 1$ or $b_1 = 2$, then Γ is either a polygon or the icosahedron. Assume that $b_1 \geq 3$. Then for each vertex $a \in \Gamma$, the number of edges between $[a]$ and $\Gamma_2(a)$ is $kb_1 = (3b_1 - 1)b_1$. Thus the number of vertices of $\Gamma_2(a)$ is at most

$$\frac{(3b_1-1)b_1}{(b_1-1)} = 3b_1 + 2 + \frac{2}{(b_1-1)}.$$

If $|\Gamma_2(a)| > k+3$, then $b_1 = 3$ and $\mu(a,b) = 2$ for all vertices a and b such that $d(a,b) = 2$. Therefore, Γ is a strongly regular graph with parameters (21, 8, 4, 2). However, no graph with such parameters exists since the value of $(\lambda-\mu)^2+(k-\mu)$ fails to be a perfect square. The proof of the Corollary is complete.

Acknowledgments. The author thanks A. D. Korshunov for his suggestions that improved exposition.

References

1. A. E. Brouwer, A. M. Cohen, and A. Neumaier (1989) *Distance-Regular Graphs*, Springer-Verlag, Berlin etc.

Institute of Mathematics and Mechanics
Kovalevskaya ul., 16
Ekaterinburgh, 620219
RUSSIA

TRANSLATED BY O. V. BORODIN

ON CLOSED CLASSES IN k-VALUED LOGIC WHICH CONTAIN A HOMOGENEOUS SWITCHING FUNCTION[†)]

S. S. Marchenkov

It is proved that for every k, $k \geq 2$, in k-valued logic there exist only finitely many closed classes containing a homogeneous switching function $s(x, y, z)$ that equals z at $x = y$, equals y at $x = z$, and equals x in other cases. Every such class consists of all functions preserving a finite set of relations of three explicitly indicated types.

Denote by P_k the set of all functions on $E_k = \{0, 1, \ldots, k-1\}$ taking values in E_k.

Homogeneous functions, i.e. functions self-dual with respect to all permutations on E_k, play an appreciable role in universal algebra and the theory of functions of k-valued logic [1]. A distinctive particularity of these functions consists in the fact that they can be defined, using only variables and the equality relation between variables.

Homogeneous functions and closed classes of homogeneous functions were investigated intensively and at present all closed classes of homogeneous functions in P_k, are described for every k, $k \geq 2$. For $k = 2$ the notions of homogeneous function and self-dual function coincide.

All 6 closed classes of self-dual functions for $k = 2$ were described, for example, in [2]. For $k = 3$ there are 8 closed classes of homogeneous functions; for $k = 4$ there are 14 such classes and for every $k \geq 5$ there are $4k - 3$ closed classes (see [3–6]). Each closed class of homogeneous functions has a finite basis with respect to superposition.

The information on closed classes shows that the presence of nontrivial (non-selector) function in a closed class is a condition that in many cases permits us to establish the existence of a finite basis in this class.

In particular, it was shown in [7] that every closed class in P_k ($k \geq 3$) con-

†) This research was supported by the Russian Foundation for Basic Research (Grant 93–01–16006).

A. D. Korshunov (ed.), Operations Research and Discrete Analysis, 183–195.

taining at least one of the following homogeneous functions

$$d(x,y,z) = \begin{cases} x & \text{for} \quad x = y, \\ z & \text{for} \quad x \neq y; \end{cases}$$

$$l_3(x,y,z) = \begin{cases} x & \text{if} \quad x,\ y\text{, and } z \quad \text{are pairwise distinct,} \\ z & \text{otherwise,} \end{cases}$$

and every closed class in P_3 containing the homogeneous function $r_3(x,y) = 2x+2y$ have finite bases. However, the proofs, proposed in [7] and using results in [8, 9], do not allow us to extract the bases explicitly and to obtain an effective description for all classes under consideration.

In this paper we propose a trick that permits us to obtain an effective description for all closed classes containing prescribed functions (not necessarily homogeneous). This trick is based on Galois theory for Post algebras [10]. We apply the trick to studying closed classes among which there is a multiform generalization of the self-dual linear function $x + y + z$, i.e., the homogeneous function

$$s(x,y,z) = \begin{cases} z & \text{if} \quad x = y, \\ y & \text{if} \quad x = z, \\ x & \text{in the other cases.} \end{cases}$$

For $k = 2$, the function $s(x,y,z)$ coincides obviously with the function $x + y + z$ which forms the basis of the class L_4 of all linear α-functions [2]. For $k \geq 3$, the function s is a basis for the closed class of homogeneous functions which is denoted by $S_k L_4^*$ in [5, 6].

Observe that for any k, $k \geq 3$, the class $S_k L_4^*$ contains the homogeneous function $l_3(x,y,z)$. Hence, in accordance with the results of [7], every closed class containing the function s has finite basis. However [7] does not imply that the number of such closed classes is finite. Below we show that each closed class in P_k, where $k \geq 2$, containing the function s is the class of all functions preserving a suitable finite set of relations on E_k. From this it follows in particular that for every fixed k the number of the indicated classes is finite.

We give necessary definitions. The reader is assumed familiar with the notions of superposition, closure under superposition and a closed class [11]. If $Q \subseteq P_k$, then by $[Q]$ we denote the closure of a set Q with respect to superposition. Unlike [11] we consider only closed classes containing the selector functions $e_n^i(x_1,\ldots,x_i,\ldots,x_n) = x_i$, where $1 \leq i \leq n$, $n = 1,2,\ldots$. By $\mathfrak{A}_k$ we denote the lattice of all closed classes in P_k containing the selector functions. The least element of $\mathfrak{A}_k$ is the closed class comprising only selector functions; the greatest element coincides with P_k.

Along with the functions in P_k we consider relations on E_k. The set of all relations on E_k is denoted by Π_k. If $\rho(x_1, \dots, x_m) \in \Pi_k$ then the relation ρ is identified with the set of all tuples in E_k^m on which ρ is true. We use the respective terminology: the full relation (which is identically true), the empty relation (which is false identically), a relation ρ is an extension ($\sigma \subseteq \rho$) or a restriction ($\rho \subseteq \sigma$) of a relation σ.

Define some operations on the set Π_k (see also [10]). The $(m+n)$-ary relation

$$\rho(x_1, \dots, x_m) \& \sigma(x_{m+1}, \dots, x_{m+n})$$

is called the *conjunction* of the relations $\rho(x_1, \dots, x_m)$ and $\sigma(x_1, \dots, x_n)$. The $(m-1)$-ary relation

$$(\exists x_i)\rho(x_1, \dots, x_i, \dots, x_m),$$

where the domain of the quantifier $\exists x_i$ is the set E_k, is called the *projection* of the relation $\rho(x_1, \dots, x_m)$ to the variable $\rho(x_1, \dots, x_m)$. It is assume that the reader is familiar with the operations of permutation and identification of variables.

The relation $x_i = x_j$ is the *elementary diagonal*; a relation that can be constructed from some elementary diagonals by conjunctions and identifications of variables is called *diagonal*. It is also convenient to regard the empty relation as diagonal.

Let $R \subseteq \Pi_k$. The *closure* of R (denoted by $[R]$) is the least set of relations in Π_k that contains all diagonals and all relations in R and is closed under the operations of conjunction, projection, permutation and identification of variables. By $\mathcal{Z}_k$ we denote the lattice of all closed sets of relations on E_k. The *least element* in $\mathcal{Z}_k$ is the set of all diagonals; the *greatest element* is Π_k.

Let

$$f(x_1, \dots, x_n) \in P_k, \quad \rho(x_1, \dots, x_m) \in \Pi_k.$$

We say that the function f *preserves the relation* ρ if for any n tuples $(a_{11}, \dots, a_{m1}), \dots, (a_{1n}, \dots, a_{mn})$ satisfying the relation ρ the tuple $(f(a_{11}, \dots, a_{1n}), \dots, f(a_{m1}, \dots, a_{mn}))$ also satisfies the relation ρ. If $Q \subseteq P_k$ and $R \subseteq \Pi_k$, we say that Q preserves R if every function in Q preserves every relation in R. The set of all functions in P_k preserving R is denoted by $\operatorname{Pol} R$, and the set of all relations in Π_k preserved by functions in Q, is denoted by $\operatorname{Inv} Q$. If Q is the set of all selector functions in P_k and R is the set of all diagonals in Π_k, then it is easy to see that $\operatorname{Pol} R = \Pi_k$, $\operatorname{Inv} P_k = R$, $\operatorname{Pol} \Pi_k = Q$, and $\operatorname{Inv} Q = \Pi_k$.

It is known [10] that the mappings Pol and Inv define an anti-isomorphism of the lattices $\mathfrak{A}_k$ and $\mathcal{Z}_k$. More precisely, if $Q \subseteq P_k$ and $R \subset \Pi_k$, then $\operatorname{Pol} R$ is a closed class in $\mathfrak{A}_k$, $\operatorname{Inv} Q$ is a closed set in $\mathcal{Z}_k$, $[Q] = \operatorname{Pol}\operatorname{Inv} Q$, and $[R] = \operatorname{Inv}\operatorname{Pol} R$. Thus, the closed classes of functions in P_k can be defined through the corresponding closed sets of relations on E_k.

We apply these results to describing all closed classes in P_k that contain the function s. With this aim, we first define the set of relations $\operatorname{Inv}\{s\}$.

Let C_k denote the collection of the following relations on E_k: all relations of the form $x \in E$, where E is a proper (nonempty and nonfull) subset of E_k; all relations of the form

$$\pi(x_1) = x_2, \tag{1}$$

where π is a permutation on E_k; the relation

$$\lambda_k(x_1, x_2, x_3) \equiv (x_1 \in E_2)\&(x_2 \in E_2)\&(x_3 \in E_2)\&(x_1 + x_2 + x_3 = 0),$$

where addition is taken modulo 2.

Since s is a structural homogeneous function [1], it preserves (by definition) all relations of the form $x \in E$ and (1). In addition, it is easy to see that the function s preserves the relations λ_k. Hence, $s \in \operatorname{Pol} C_k$. However, for $k = 2$ the function s forms a basis for the class L_4, and for $k \geq 3$ it forms a basis for the class $S_k L_4^*$. Therefore, $L_4 \subseteq \operatorname{Pol} C_2$ and $S_k L_4^* \subseteq \operatorname{Pol} C_k$ if $k \geq 3$.

Show that $L_4 = \operatorname{Pol} C_2$. Assume that $L_4 \neq \operatorname{Pol} C_2$. Since C_2 contains all relations (1), the closed class $\operatorname{Pol} C_2$ must comprise only self-dual functions. It is known [2] that the class L_4 is directly contained in a sole closed class of self-dual functions, namely in the class L_5. Therefore, if $L_4 \neq \operatorname{Pol} C_2$, then $L_5 \subseteq \operatorname{Pol} C_2$. However, the class L_5 contains the function $x + 1$ which does not preserve the relation $x = 0$ in C_2. Hence, $L_4 = \operatorname{Pol} C_2$.

The same argument applies to proving the equality $S_k L_4^* = \operatorname{Pol} C_k$ for $k \geq 3$. In [4–6] it was established that the class $S_3 L_4^*$ is contained directly in one closed class of homogeneous functions S_3^*, and for $k \geq 4$ the class $S_k L_4^*$ is contained only in the closed classes S_k^* and $S_k L_4$. Further, for every $k \geq 3$ the class S_k^* contains the homogeneous function

$$p(x, y, z) = \begin{cases} z & \text{if } x = y, \\ x & \text{if } x \neq y, \end{cases}$$

and for $k \geq 4$ the class $S_k L_4$ contains the homogeneous function

$$r_k(x_1, \ldots, x_{k-1}) = \begin{cases} x_k & \text{if } \{x_1, \ldots, x_{k-1}, x_k\} = E_k, \\ x_1 & \text{otherwise.} \end{cases}$$

It is easy to see that the function p does not preserve the relation λ_k and the function r_k does not preserve the relation $x \in E_{k-1}$. Thus, $S_k L_4^* = \operatorname{Pol} C_k$.

By the results of [10] we have $\operatorname{Inv}\{s\} = \operatorname{Inv}[\{s\}]$ and $\operatorname{Pol} C_k = \operatorname{Pol}[C_k]$. Using the equalities $L_4 = [\{s\}] = \operatorname{Pol} C_2$ for $k = 2$, and $S_k L_4^* = [\{s\}] = \operatorname{Pol} C_k$ for $k \geq 3$, we obtain $\operatorname{Inv}\{s\} = [C_k]$.

Describe the relations that lie in $[C_k]$. Show that the set $[C_k]$ contains all relations of the form

$$(x_1 \in F_1)\&\ldots\&(x_n \in F_n)\&(\pi_1(x_1) + \cdots + \pi_n(x_n) = 0), \tag{2}$$

where $n \geq 2$, $|F_1| = \cdots = |F_n| = 2$, $\pi_1, \ldots, \pi_n$ are permutations on E_k such that $\pi_1(F_1) = \cdots = \pi_n(F_n) = E_2$ with addition taken modulo 2 and $|F_i|$ denoting the size of F_i.

Observe that if $[C_k]$ contains the relation

$$\tau_m(x_1, \ldots, x_m) \equiv (x_1 \in E_2)\&\ldots\&(x_m \in E_2)\&(x_1 + \cdots + x_m = 0),$$

where $m \geq 3$, then the relation $\tau_{m+1}(x_1, \ldots, x_{m+1})$ too belongs to $[C_k]$ since

$$\tau_{m+1}(x_1, \ldots, x_{m+1}) \equiv (\exists y)(\tau_m(x_1, \ldots, x_{m-1}, y)\&\lambda_k(x_m, x_{m+1}, y)).$$

In addition, we have

$$\tau_2(x_1, x_2) \equiv (\exists y)((y = 0)\&\lambda_k(x_1, x_2, y)).$$

Now, we obtain (2) from the relation τ_n and the relations $\pi_i(x_1) = x_2$ by the formula

$$(\exists y_1)\ldots(\exists y_n)((\pi_1(x_1) = y_1)\&\ldots\&(\pi_n(x_n) = y_n)\&\tau_n(y_1, \ldots, y_n)).$$

Lemma 1. *Each relation in $[C_k]$ is representable as a conjunction (not necessarily with disjoint sets of variables) of one-place relations, two-place relations of the form* (1) *and the relations of the form* (2).

PROOF. We apply induction on the structure of relations in $[C_k]$. Note that the two-place diagonals take the form (1). Further, to identify the variable x_j with the variable x_i in the relation $\rho(x_1, \ldots, x_n)$ it suffices to consider the relation

$$(\exists x_j)(\rho(x_1, \ldots, x_i, \ldots, x_j, \ldots, x_n)\&(x_i = x_j)).$$

To complete the proof, it remains to establish that the relation

$$(\exists x_i)\rho(x_1, \ldots, x_i, \ldots, x_n), \tag{3}$$

can be transformed as mentioned in the lemma with ρ representable as a conjunction K of one-place relations and (1) and (2). Without loss of generality, we assume that $i = n$.

Let

$$(x_n \in G_1), \dots, (x_n \in G_m)$$

be all one-place relations in K containing the variable x_n. It is evident that

$$(x_n \in G_1)\& \dots \&(x_n \in G_m) \equiv (x_n \in G),$$

where $G = G_1 \cap \cdots \cap G_m$. If x_n occurs only in one-place relations of the conjunction K, then

$$\rho(x_1, \dots, x_n) \equiv (x_n \in G)\&\sigma(x_1, \dots, x_{n-1}),$$

where the relation σ is performed by conjunction of all factors in K not containing the variable x_n. Therefore, in this case the relation (3) coincides with the relation

$$(\exists x_n)(x_n \in G)\&\sigma(x_1, \dots, x_{n-1}).$$

Since the first factor of this relation is either the full relation for $G \neq \varnothing$, or the empty relation for $G = \varnothing$, the relation (3) coincides with either the relation $\sigma(x_1, \dots, x_{n-1})$, or the empty relation, respectively. The empty relation is representable for example as $(x \in \{0\})\&(x \in \{1\})$.

Assume that the variable x_n in K appears in at least one non-one-place relation. Consider first the case in which this relation is of the form (1). Since

$$(\pi(x_1) = x_2) \equiv (x_1 = \pi^{-1}(x_2)),$$

where the π^{-1} on E_k is the inverse permutation of π, we can suppose that the variable x_n enters the relation $\pi(x_j) = x_n$. Therefore, the relation ρ is representable as

$$\rho(x_1, \dots, x_n) \equiv (\pi(x_j) = x_n)\&\sigma(x_1, \dots, x_n),$$

where the relation $\sigma(x_1, \dots, x_n)$ is represented as a conjunction of factors in K different from $\pi(x_j) = x_n$. According to logical rules, we have

$$(\exists x_n)((\pi(x_j) = x_n)\&\sigma(x_1, \dots, x_n)) \equiv \sigma(x_1, \dots, x_{n-1}, \pi(x_j)).$$

It is easy to see that the replacement of x_n by $\pi(x_j)$ in one-place relations and relations of the form (1) produces relations of the same types. Consider the replacement of x_n by $\pi(x_j)$ in (2):

$$(x_1 \in F_1)\& \dots \&(x_{n-1} \in F_{n-1})\&(\pi(x_j) \in F_n)$$
$$\&(\pi_1(x_1) + \cdots + \pi_{n-1}(x_{n-1}) + \pi_n(\pi(x_j)) = 0). \qquad (4)$$

Obviously, there are at most two elements satisfying the relation

$$(x_j \in F_j)\&(\pi(x_j) \in F_n). \tag{5}$$

If the relation (5) is empty, then (4) is also empty. If exactly one element a satisfies (5), then the relation (4) is equivalent to the relation

$$\begin{gathered}(x_1 \in F_1)\&\ldots\&(x_{j-1} \in F_{j-1})\&(x_{j+1} \in F_{j+1}))\\ \&\ldots\&(x_{n-1} \in F_{n-1})\&(\pi_1'(x_1)+\pi_2(x_2)+\cdots+\pi_{j-1}(x_{j-1})+\pi_{j+1}(x_{j+1})\\ +\cdots+\pi_{n-1}(x_{n-1})=0)\&(x_j=a),\end{gathered}$$

where π_1' is defined as follows:

$$\pi_1'(x_1)=\begin{cases}\pi_1(x_1)+\pi_j(a)+\pi_n(\pi(a)) & \text{if} \quad x_1 \in F_1,\\ \pi_1(x_1) & \text{otherwise.}\end{cases}$$

If two elements satisfy the relation (5), then the containments $x_j \in F_j$ and $\pi(x_j) \in F_n$ are equivalent and the functions $\pi_j(x_j)$ and $\pi_n(\pi(x_j))$ on the set F_j either agree, or can be transformed to one another by adding $1 \mod 2$. Therefore, in this case (4) is equivalent to the relation

$$\begin{gathered}(x_1 \in F_1)\&\ldots\&(x_{n-1} \in F_{n-1})\&(\pi_1'(x_1)+\pi_2(x_2)\\ +\cdots+\pi_{j-1}(x_{j-1})+\pi_{j+1}(x_{j+1})+\cdots+\pi_{n-1}(x_{n-1})=0),\end{gathered}$$

where

$$\pi_1'(x_1)=\begin{cases}\pi_1(x_1)+1 & \text{if } x_1 \in F_1 \text{ and the functions } \pi_j(x_j),\ \pi_n(\pi(x_j))\\ & \text{assume distinct values on } F_j,\\ \pi_1(x_1) & \text{in the other cases.}\end{cases}$$

Suppose that the conjunction K contain no relations of the form (1) with the variable x_n. Since in the sequel only the relations in K that contain the variable x_n are essential, we assume for notational simplicity that

$$\rho(x_1,\ldots,x_n) \equiv (x_n \in G)\&\Big(\mathop{\&}_{1\le b\le q} \sigma_b(x_1^b,\ldots,x_{p_b}^b,x_n)\Big),$$

where all relations σ_b are relations of the form (2) and the variables x_t^b belong to the set $\{x_1,\ldots,x_{n-1}\}$. Let in the representations (2) of the relations $\sigma_1,\ldots,\sigma_q$ the permutations $\pi_1,\ldots,\pi_q$ be connected with x_n. If the set

$$G \cap \pi_1^{-1}(E_2) \cap \cdots \cap \pi_q^{-1}(E_2) \tag{6}$$

is empty then the relation (3) is also empty. If (6) consists of a sole element a, then (3) is equivalent to the relation

$$\underset{1\le b\le q}{\&} \sigma_b\big(x_1^b,\dots,x_{p_b}^b,a\big),$$

where for $p_b \ge 2$ we transform the relation $\sigma_b\big(x_1^b,\dots,x_{p_b}^b,a\big)$ to the form (2).

Suppose that (6) consists of two elements (this set can obviously contain at most two elements). Then in the representation of the relation ρ the conjunctive factor $(x_n \in G)$ can be omitted and elimination of the quantifier $\exists x_n$ in (3) can be made in the same manner as in the corresponding case P_2:

$$(\exists x_n)\Big(\underset{1\le b\le q}{\&}\big(x_1^b+\dots+x_{p_b}^b+x_n=0\big)\Big)$$
$$\equiv \underset{1\le b<c\le q}{\&}\big(x_1^b+\dots+x_{p_b}^b+x_1^c+\dots+x_{p_c}^c=0\big).$$

Lemma 1 is proved.

Lemma 2. *Let a relation $\rho(x_1,\dots,x_n)$ in the set $[C_k]$ be presentable in the form of the conjunction of one-place relations and relations of the form (1) and (2). Then ρ is representable as conjunction of one-place relations, relations of the form (2) and relations of type*

$$(y_1 \in E)\&(\pi(y_1)=y_2), \tag{7}$$

where $E \subseteq E_k$ and π is a permutation on E_k with relations of all types entering the conjunction belonging to the set $[\{\rho\}]$.

PROOF. Let a conjunction K of one-place relations and relations of the form (1) and (2) present the relation ρ. Let $\sigma(x_{i_1},\dots,x_{i_m})$ be a relation entering the conjunction K. Denote by $\tau(x_{i_1},\dots,x_{i_m})$ the projection of the relation $\rho(x_1,\dots,x_n)$ over all the variables different from $x_{i_1},\dots,x_{i_m}$. It is obvious that $\tau \in [\{\rho\}]$. By the definition of projection it follows that the relation τ regarded as a relation depending on all the variables $x_1,\dots,x_n$ is an extension of the relation ρ. On the other hand, since the relation $\sigma(x_{i_1},\dots,x_{i_m})$ is a conjunctive factor in the representation K of the relation ρ, the relation τ can be only a restriction of the relation σ. Hence, the conjunction, consisting of all the relations τ corresponding to the relations σ in the conjunction K, realizes the relation ρ.

If the relation σ has the form (1), then the relation τ, as a restriction of σ, takes the form (7). Let the relation σ have form (2). If τ coincides with σ, then τ has form (2). If τ differs from σ according to the inclusions $\tau \in [\{\rho\}]$, $\rho \in [C_k]$ and Lemma 1, the relation τ is representable as conjunction of one-place relations and relations of the form (1) and (2). Since $\tau \in [C_k]$, we use induction on the number of variables in the relation τ.

It is clear that in the case $m = 2$ the relation σ is representable as (7) and, therefore, the relation τ is also of the form (7). Now, the only obstacle to performing induction is the case $m = n$. Since we suppose that τ differs from σ, the conjunction K must contain at least one relation $\sigma'(x_{j_1}, \ldots, x_{j_q})$. To simplify presentation, we suppose that $j_1 = 1, \ldots, j_q = q$.

Let $q = 1$, $\sigma'(x_1) \equiv (x_1 \in F)$, and let F_1 be the set corresponding to the variable x_1 in the representation (2) of the relation σ. If $F_1 \subseteq F$, then the relation σ' in the conjunction K can be omitted. In case $F_1 \cap F = \varnothing$, ρ is empty and consequently τ is also the empty relation. Assume that $F_1 \cap F = \{a\}$. Then as well as in Lemma 1, the relation $\sigma(x_1, \ldots, x_n)$ is equivalent to conjunction of the one-place relation $x_1 = a$ and the $(n-1)$-ary relation

$$(x_2 \in F_2)\&\ldots\&(x_n \in F_n)\&(\pi_2'(x_2) + \pi_3(x_3) + \cdots + \pi_n(x_n) = 0),$$

where

$$\pi_2'(x_2) = \begin{cases} \pi_2(x_2) + 1 & \text{if } \pi_1(a) = 1 \text{ and } x_2 \in F_2, \\ \pi_2(x_2) & \text{in other cases.} \end{cases}$$

Thus, the number of variables in σ decreases by one.

Let $q \geq 2$. If $q = 2$ and the relation σ' has form (1), i.e.,

$$\sigma'(x_1, x_2) \equiv (\pi(x_1) = x_2),$$

then the conjunction of the relations σ and σ' is equivalent to the conjunction of the relation σ' and the relation

$$(x_1 \in F_1)\&(\pi(x_1) \in F_2)\&(x_3 \in F_3)\&\ldots\&(x_n \in F_n)$$
$$\&(\pi_1(x_1) + \pi_2(\pi(x_1)) + \pi_3(x_3) + \cdots + \pi_n(x_n) = 0),$$

that is obtained from the relation σ on replacing x_2 with $\pi(x_1)$. As well as in Lemma 1, the latter relation gives either the empty relation or the conjunction of a one-place relation of the variable x_1 and an $(n-2)$-ary relation of the form (2) of the variables $x_3, \ldots, x_n$.

Suppose that the relation σ' is of form (2), i.e.,

$$\sigma'(x_1, \ldots, x_q) \equiv (x_1 \in G_1)\&\ldots\&(x_q \in G_q)\&(\varphi_1(x_1) + \cdots + \varphi_q(x_q) = 0),$$

where $\varphi_1, \ldots, \varphi_q$ are permutations on E_k. Taking into account the arguments for the case $q = 1$, we assume that

$$G_1 = F_1, \ldots, G_q = F_q.$$

By the properties of linear Boolean functions for $x_1 \in F_1, \ldots, x_q \in F_q$ we have

$$\varphi_1(x_1) + \cdots + \varphi_q(x_q) = \pi_1(x_1) + \cdots + \pi_q(x_q) + b,$$

where $b \in \{0,1\}$. Therefore, for $q = n$ the conjunction of the relations σ and σ' either is equivalent to the relation σ (for $b = 0$) or is the empty relation (for $b = 1$). If $q < n$, then the conjunction of the relations σ and σ' is equivalent to the conjunction of the relation σ' and the relation

$$(x_{q+1} \in F_{q+1}) \& \ldots \& (x_n \in F_n) \& (b + \pi_{q+1}(x_{q+1}) + \cdots + \pi_n(x_n) = 0).$$

The latter relation in the case $b = 1$ is reduced to the form (2) as above. Lemma 2 is proved.

Lemma 3. *Let $\rho(x_1, \ldots, x_n)$ be a relation of (2), $m = \max(\lfloor k/2 \rfloor, 4)$, and let R be a set of all at most m-ary relations in $[\{\rho\}]$. Then $[\{\rho\}] = [R]$.*

PROOF. The inclusion $[R] \subseteq [\{\rho\}]$ is evident. Therefore, we prove the inclusion $[\{\rho\}] \subseteq [R]$. Since the relation $[\{\rho\}] \subseteq [R]$ is valid for $n \leq m$, we assume that $n > m$.

Assume that in the representation (2) of the relation ρ among the sets F_1, $\ldots$, F_n at least two are equal. Let, for instance, $F_1 = F_2$. Put $\sigma_1(x_3, \ldots, x_n) \equiv (\exists x_2)(\rho(x_2, x_2, x_3, \ldots, x_n))$. By the definition of ρ we have

$$\sigma_1(x_3, \ldots, x_n) \equiv (x_3 \in F_3) \\ \& \ldots \& (x_n \in F_n) \& (a + \pi_3(x_3) + \cdots + \pi_n(x_n) = 0), \quad (8)$$

where $a = 0$ if the permutations π_1 and π_2 coincide on F_1, and $a = 1$ otherwise. Let

$$\sigma_2(x_1, x_2, x_3, y_3) \equiv (\exists x_4) \ldots (\exists x_n)(\rho(x_1, \ldots, x_n) \& \rho(x_1, x_1, y_3, x_4, \ldots, x_n)).$$

Then

$$\sigma_2(x_1, x_2, x_3, y_3) \equiv (x_1 \in F_1) \& (x_2 \in F_2) \& (x_3 \in F_3) \\ \& (y_3 \in F_3) \& (a + \pi_1(x_1) + \pi_2(x_2) + \pi_3(x_3) + \pi_3(y_3) = 0).$$

Hence,

$$(\exists y_3)(\sigma_1(y_3, x_4, \ldots, x_n) \& \sigma_2(x_1, x_2, x_3, y_3)) \equiv \rho(x_1, \ldots, x_n).$$

Thus, $\rho \in [\{\sigma_1, \sigma_2\}]$, where $\{\sigma_1, \sigma_2\} \subseteq [\{\rho\}]$, the relation σ_1 depends on $n - 2$ variables, and the relation σ_2 depends on four variables.

Suppose now that among the sets $F_1, \ldots, F_n$ at least two are distinct and intersecting sets, for example, the sets F_1 and F_2. If $F_1 \cap F_2 = \{b\}$, then the relation σ_1 defined in the beginning of the proof satisfies (8), where $a = 0$ if $\pi_1(b) = \pi_2(b)$, and $a = 1$ otherwise.

Let

$$\sigma_3(x_1, x_2) \equiv (\exists x_3) \ldots (\exists x_n)(\rho(x_1, \ldots, x_n) \& \sigma_1(x_3, \ldots, x_n)).$$

It is easy to see that

$$\sigma_3(x_1, x_2) \equiv (x_1 \in F_1) \& (x_2 \in F_2) \& (a + \pi_1(x_1) + \pi_2(x_2) = 0).$$

The relation σ_3 is also presentable in the form (7), i.e.,

$$\sigma_3(x_1, x_2) \equiv (x_1 \in F_1) \& (\pi(x_1) = x_2),$$

where π is a permutation on E_k and $\pi(F_1) = F_2$. Using this representation of σ_3, the set F_2 in the relation ρ can be replaced by the set F_1. In fact, let

$$\sigma_4(x_1, \ldots, x_n) \equiv (\exists y_2)(\sigma_3(x_2, y_2) \& \rho(x_1, y_2, x_3, \ldots, x_n)).$$

Then σ_4 is representable as

$$\sigma_4(x_1, \ldots, x_n) \equiv (x_1 \in F_1) \& (x_2 \in F_1) \& (x_3 \in F_3)$$
$$\& \ldots \& (x_n \in F_n) \& (\pi_1(x_1) + \pi_2(\pi(x_2)) + \pi_3(x_3) + \cdots + \pi_n(x_n) = 0).$$

Moreover,

$$(\exists y_2)(\sigma_3(y_2, x_2) \& \sigma_4(x_1, y_2, x_3, \ldots, x_n)) \equiv \rho(x_1, \ldots, x_n).$$

Hence, $[\{\rho\}] \subseteq [\{\sigma_3, \sigma_4\}]$. But in the representation (2) of the relation σ_4 the sets F_1 and F_2 coincide. Therefore, we can use the argument of the beginning of the proof.

To complete the proof of Lemma 3 we note that the set E_k can have at most $\lfloor k/2 \rfloor$ mutually disjoint two-element subsets.

Theorem. *For every k, $k \geq 2$, P_k has finitely many closed classes containing the homogeneous function s. Every such class is representable as* Pol R, *where R is a finite set of relations on E_k containing one-place relations, two-place relations of the form* (7), *and m-ary relations of the form* (2), *where $m \leq \max(\lfloor k/2 \rfloor, 4)$.*

PROOF. As was established prior to Lemma 1, $[\{s\}] = L_4 = \mathrm{Pol}\, C_2 = \mathrm{Pol}[C_2]$ in Boolean case, and $[\{s\}] = S_k L_4^* = \mathrm{Pol}\, C_k = \mathrm{Pol}[C_k]$ for $k \geq 3$. According to [10] every closed class in P_k containing the function s (i.e., containing the whole class L_4 or the class $S_k L_4^*$ for $k \geq 3$) is representable as $\mathrm{Pol}\, R$, where R is a closed set of relations in $[C_k]$. By Lemmas 1 and 2, each relation ρ in R is representable as conjunction of one-place relations and relations of the form (2) and (7), where all such relations are contained in the set $[\{\rho\}]$. If we denote this finite set of relations by R_1, then the inclusion $R_1 \subseteq [\{\rho\}]$ implies $\mathrm{Pol}\, \rho \subseteq \mathrm{Pol}\, R_1$. At the same time, the conjunction of the relations in R_1 defines the relation ρ. Therefore, $\rho \in [R_1]$ and hence $\mathrm{Pol}\, R_1 \subseteq \mathrm{Pol}\, \rho$. Thus, $\mathrm{Pol}\, \rho = \mathrm{Pol}\, R_1$.

Using Lemma 3, we choose in $[R_1]$ a finite subset R_2 of relations whose arity is not greater than m ($m = \max(\lfloor k/2 \rfloor, 4)$) so that equality $\mathrm{Pol}\, \rho = \mathrm{Pol}\, R_2$ holds. Obviously, to obtain the class $\mathrm{Pol}\, R$ instead of the closed set of relations R it is sufficient to take the finite set of relations $\bigcup R_2$, where the union is taken over all the finite sets R_2 constructed in the above-indicated manner for the relations ρ in R. This implies finiteness of the number of closed classes containing the function s. The theorem is proved.

From this theorem it follows that every closed class containing the function s is an intersection of finitely many closed classes each of which is defined by a unique relation of the form $x \in E$, (7) or (2), where in the latter relation $n \leq \max(\lfloor k/2 \rfloor, 4)$, and the sets $F_1, \ldots, F_n$ are mutually disjoint. The relations $x \in E$ determine precomplete classes in P_k. Relations of the form (7) determine extensions of classes of self-dual functions, and relations of the form (2) determine extensions of classes of linear Boolean functions, where linearity reveals itself on the set $F_1 \times \cdots \times F_n$.

References

1. E. Marczewski (1964) Homogeneous operations and homogeneous algebras, *Fund. Math.* **56**, No. 2, 81–103.
2. S. V. Yablonskiĭ, G. P. Gavrilov, and V. B. Kudryavtsev (1966) *Functions of the Algebra of Logic and Post Classes* (in Russian), Nauka, Moscow.
3. S. S. Marchenkov (1979) On closed classes of self-dual functions of many-valued logic (in Russian), in: *Problemy Kibernetiki.* Vol. 36, Nauka, Moscow, pp. 5–22.
4. B. Csakany and T. Gavalcova (1980) Finite homogeneous algebras. I, *Acta Sci. Math.* **42**, No. 1–2, 57–65.
5. S. S. Marchenkov (1981) On homogeneous algebras (in Russian), *Dokl. Akad. Nauk SSSR* **256**, No. 4, 787–790.
6. S. S. Marchenkov (1982) Homogeneous algebras (in Russian), in: *Problemy Kibernetiki.* Vol. 39, Nauka, Moscow, pp. 85-106.

7. S. S. Marchenkov (1984) Closed classes in P_k containing homogeneous functions (in Russian), Akad. Nauk SSSR, Inst. Prikl. Mat., Preprint, No. 35.

8. K. A. Baker and A. F. Pixley (1975) Polynomial interpolation and the Chinese remainder theorem for algebraic systems, *Math. Z.* **143**, No. 3, 165–174.

9. V. A. Taĭmanov (1983) Cartesian powers of P_2 (in Russian), *Dokl. Akad. Nauk SSSR* **270**, No.6, 1327–1330.

10. V. G. Bodnarchuk, L. A. Kaluzhnin, V. N. Kotov, and B. A. Romov (1969) Galois theory for Post algebras. I and II (in Russian), *Kibernetika* (Kiev), I: No. 3, 1–10; II: No. 5, 1–9.

11. S. V. Yablonskiĭ (1974) Introduction to the theory of functions of k-valued logic, in: *Mathematics and Mathematical Questions of Cybernetics* (in Russian), Nauka, Moscow, pp. 9–66.

Keldysh Institute
for Applied Mathematics
Miusskaya Pl., 4
Moscow 125047
RUSSIA

TRANSLATED BY A. V. RYAZANOV

STRONGLY ESSENTIAL VARIABLES OF THE FUNCTIONS OF k-VALUED LOGIC

I. Mirchev

We establish some conditions that are sufficient for existence of strongly essential and c-strongly essential variables of the functions of k-valued logic.

Introduction

Several researchers studied problems about strongly essential variables of discrete functions. O. B. Lupanov [1] has shown that an arbitrary Boolean function depending essentially on at least two variables has at least one strongly essential variable. This fact was also proved by N. A. Solov'ëv [2]. A similar result for arbitrary functions was established by A. Salomaa [3]. Ju. Ya. Breitbart [4] proved that an arbitrary Boolean function depending essentially of at least two variables has at least two strongly essential variables. Related problems for the functions of k-valued logic were investigated by K. Chimev [5] (see also [6]). In this paper we propose some conditions that are sufficient for existence of strongly essential and c-strongly essential variables of the functions of k-valued logic.

§ 1. Basic Notions and Definitions

DEFINITION 1 [5]. A function of k-valued logic $f(x_1, ..., x_n)$ *depends essentially on* $x_i, 1 \leq i \leq n$, if, for some constants $c_1, \dots, c_{i-1}, c_{i+1}, \dots, c_n$, the function $f(x_1 = c_1, \dots, x_{i-1} = c_{i-1}, x_i, x_{i+1} = c_{i+1}, \dots, x_n = c_n)$ has at least two distinct values.

An inessential variable is called a *fictitious* variable.

$Ess(f)$ denotes the set of all essential variables of a function f, and $F(n)$ stands for the set of all functions depending essentially on n variables.

DEFINITION 2 [7]. Let $N \subseteq Ess(f)$. The set $M = \{x_1, \dots, x_m\} \subseteq Ess(f)$ is called the *dominant of the set* N (for f) if there exist constants $c_1, \dots, c_m$ such that

$$N \cap Ess(f(x_1 = c_1, \dots, x_m = c_m)) = \varnothing,$$

and M is the least set satisfying this property (i.e., there is no other $M_1 \subset M$ with the same property).

The notation $M \xrightarrow{d} N$ means that N is dominated by M.

A. D. Korshunov (ed.), Operations Research and Discrete Analysis, 197–203.

DEFINITION 3 [8]. Let $Q \subseteq Ess(f)$. The set $P = \{x_1, \ldots, x_p\} \subseteq Ess(f)$ is called the *strong dominant of* Q if there exists a collection $\widetilde{P}^* = (c_1^*, \ldots, c_p^*)$ such that P is the dominant of Q with respect to $\widetilde{P}^*$ and

$$\bigcup_{i=1}^{p} Zx_i = Q, \quad \text{where} \quad Zx_i = Q \setminus Ess(f(x_i = c_i^*)).$$

The notation $P \xrightarrow{\text{sd}} Q$ means that Q is strongly dominated by P.

The set Zx_i is called the (*active*) *region of* x_i in Q.

Consider two examples illustrating this definition.

EXAMPLE 1. Let $f = x_1x_5 + x_2x_5 + x_3x_5 + x_3x_5\bar{x}_6 + x_4x_6 \pmod 2$ be a function on $A = \{0,1\}$, $P = \{x_5, x_6\}$ and $Q = \{x_1, x_2, x_3, x_4\}$. It is clear that $P \xrightarrow{\text{sd}} Q$ with respect to $\widetilde{P}^* = (0,0)$, where $Zx_5 = \{x_1, x_2, x_3\}$ and $Zx_6 = \{x_4\}$.

EXAMPLE 2. Let $f = x_1x_3 + x_2x_3 \pmod 2$ be a function on $A = \{0,1\}$, $P = \{x_1, x_2\}$ and $Q = \{x_3\}$. It is obvious that $P \xrightarrow{\text{d}} Q$ for f, but P is not the strong dominant of Q (for f).

Some results about strongly dominating sets of variables of discrete functions can be found in [8].

A function f and all the functions which can be obtained from f by substitution of some variables with constants are called *subfunctions of* f.

DEFINITION 4 [5]. Let $f \in F(n)$, $n \geq 1$, and $\varnothing \neq M \subseteq Ess(f)$. A variable $x \in M$ is called *strongly essential for* f *with respect to* M if for some constant c the inclusion holds:

$$M \setminus \{x\} \subseteq Ess(f(x = c)). \tag{1}$$

If x is a strongly essential variable of f with respect to the whole set $Ess(f)$, then x is called a *strongly essential variable of* f.

By $Ess^*(f)$ we denote the set of all strongly essential variables of the function f.

If (1) holds for an arbitrary constant c, then the variable x is called a *c-strongly essential with respect to* M.

If x is a c-strongly essential variable of f with respect to $Ess(f)$, then x is called a *c-strongly essential variable of* f.

§ 2. Main Results

We will assume that P and Q are nonempty sets of essential variables of a k-valued function $f \in F(n)$ such that $P \cap Q = \varnothing$ and $P \cup Q = Ess(f)$.

The following two propositions are given without proof.

Proposition 1. *Every essential variable of an arbitrary subfunction f_1 of a function f is an essential variable of f.*

Proposition 2. *If x_i is a fictitious variable for f, then, for an arbitrary constant c_i,*

$$f(x_1,\dots,x_{i-1},x_i,x_{i+1},\dots,x_n) = f(x_1,\dots,x_{i-1},c_i,x_{i+1},\dots,x_n).$$

Theorem 1 [8]. *If $P \xrightarrow{\text{sd}} Q$ for f, then, for every variable $x_i \in P$, the following relation holds: $Zx_i \not\subset M$, where*

$$M = \bigcup_{\substack{x_j \in P \\ x_j \neq x_i}} Zx_j.$$

Corollary 1. *If $P \xrightarrow{\text{sd}} Q$ for f, then $Zx_i \not\subset Zx_j$, where $x_i \in P$, $x_j \in P$, and $x_i \neq x_j$.*

Corollary 2. *If $P \xrightarrow{\text{sd}} Q$ for f, then $Zx_i \neq \varnothing$ for every $x_i \in P$.*

Corollary 3. *If $P \xrightarrow{\text{sd}} Q$ for f, then $|P| \leq |Q|$.*

Theorem 2. *If $P \xrightarrow{\text{sd}} Q$ for f with respect to $\widetilde{P}^*$ and $|P| = |Q|$, then*
(1) all variables in Q are strongly essential of f;
(2) for any $P_1 \subset P$, $P_1 \neq P$, in Q there exist at least $k = |P| - |P_1|$ variables c-strongly essential for f with respect to $P_1 \cup Q$.

PROOF. 1. By Theorem 1 and its corollaries, we conclude that each variable in Q is an active region for the only variable of the set P. Without loss of generality, we suppose $P = \{x_1,\dots,x_p\}$, $Q = \{x_{p+1},\dots,x_{2p}\}$, and $Zx_i = \{x_{p+i}\}$, $i = 1,\dots,p$. Prove that an arbitrary variable in Q, for example, x_{2p}, is strongly essential for f.

Since the variable x_p is essential for f, Definition 1 and Proposition 1 imply that there exists a constant α_{2p} such that the function $f(x_{2p} = \alpha_{2p})$ depends essentially on x_p, i.e.,

$$x_p \in Ess(f(x_{2p} = \alpha_{2p})). \tag{2}$$

We provide that after replacing the variable x_{2p} by any constant c_{2p} the resulting subfunction $f_1 = f(x_{2p} = c_{2p})$ depends essentially on all the variables in the set $(Q \setminus \{x_{2p}\}) \cup (P \setminus \{x_p\})$.

Assume that at least one variable in $Q \setminus \{x_{2p}\}$ is fictitious for $f_1 = f(x_{2p} = c_{2p})$. Without loss of generality, we can suppose that $x_{p+1} \notin Ess(f_1)$. Then, for any subfunction of the function $f_1 = f(x_{2p} = c_{2p})$, the variable x_{p+1} is also fictitious. Therefore, for the subfunction $f_2 = f\left(x_p = c_p^*,\, x_{2p} = c_{2p}\right)$, where c_p^* is a constant in the collection $\widetilde{P}^* = \left(c_1^*, \ldots, c_p^*\right)$, the variable x_{p+1} is also fictitious. By Proposition 2, we obtain

$$f\left(x_p = c_p^*\right) = f\left(x_p = c_p^*,\, x_{2p} = c_{2p}\right).$$

Hence, x_{p+1} is fictitious variable of $f\left(x_p = c_p^*\right)$, i.e., $x_{p+1} \in Zx_p$. This contradicts the relation $Zx_p = \{x_{2p}\}$. Therefore, for an arbitrary variable x_j in the set $Q \setminus \{x_{2p}\}$, we have

$$x_j \in Ess(f_1). \tag{3}$$

Suppose now that at least one variable in the set $P \setminus \{x_p\}$ is a fictitious variable of the function $f_1 = f(x_{2p} = c_{2p})$. Assume it x_1. Then, by Proposition 2, for any constant β_1,

$$f_1 = f(x_{2p} = c_{2p}) = f(x_{2p} = c_{2p},\, x_1 = \beta_1).$$

Hence, in the function $f_1 = f(x_{2p} = c_{2p}) = f(x_{2p} = c_{2p},\, x_1 = c_1^*)$, where c_1^* is a constant in the collection $\widetilde{P}^* = \left(c_1^*, \ldots, c_p^*\right)$, the variable x_{p+1} is fictitious. This contradicts (3). Therefore, for any variable x_k in $P \setminus \{x_p\}$, we obtain

$$x_k \in Ess(f_1). \tag{4}$$

It follows from (2), (3), and (4) that all variables of the function $f(x_{2p} = 2\alpha)$ are essential, i.e., $x_{2p} \in Ess^*(f)$. Since x_{2p} was chosen arbitrarily, all variables in Q are strongly essential for f.

2. If $P_1 \subset P$ and $P_1 \neq P$, then from the proof of Subsection 1 and the definition of c-strongly essential variable we conclude that variables in the set $Q \setminus \left(\bigcup_{x_1 \in P} Zx_i\right)$ are c-strongly essential for f with respect to $P_1 \cup Q$. Since $|P| = |Q|$, we have

$$\left|Q \setminus \left(\bigcup_{x_i \in P} Zx_i\right)\right| = |P| - |P_1| \geq 1.$$

Theorem 2 is proved.

Corollary 4. *If $P \xrightarrow{\text{sd}} Q$ for f and $Q \xrightarrow{\text{sd}} P$ for f, then all variables of the function f are strongly essential.* (Example 3.)

Corollary 5. *If $P \xrightarrow{\text{sd}} Q$ for f, then the number of strongly essential variables of the function f, belonging to Q, is not less than the number of one-element regions of the set Q.*

Corollary 6. *If $P \xrightarrow{\text{sd}} Q$ for f, then each one-element region Zx_i, $x_i \in P$, is c-strongly essential for f with respect to $Ess(f) \setminus \{x_i\}$.*

EXAMPLE 3. Let $f = x_1x_3 + x_2x_4 \pmod 2$. If $P = \{x_1, x_2\}$ and $Q = \{x_3, x_4\}$, then $P \xrightarrow{\text{sd}} P$ for f with respect to the collection $\widetilde{P}^* = (0,0)$, and $P \cup Q = Ess(f) = Ess^*(f)$.

Example 3 shows that $P \xrightarrow{\text{sd}} Q$ and $Q \xrightarrow{\text{sd}} P$ for f, but there is no c-strongly essential variable of f.

The following Example 4 demonstrates that under the conditions of Theorem 2 there can exist a not strongly essential variable in the set P.

EXAMPLE 4. Let $f = x_1x_2x_3 + \bar{x}_1x_4 \pmod 2$. If $P = \{x_1, x_2\}$ and $Q = \{x_3, x_4\}$, then $P \xrightarrow{\text{sd}} Q$ for f with respect to the collection $\widetilde{P}^* = (1,0)$, but $x_1 \notin Ess^*(f)$.

If $P \xrightarrow{\text{d}} P$ for f, then P is a *self-dominating set* of the function f.

Theorem 3. *If Q is a self-dominating set of the function f and $Q \subseteq Ess(f)$, then each variable in Q is c-strongly essential for f with respect to Q.*

PROOF. Let $Q = \{x_1, \ldots, x_q\}$, $q \leq |Ess(f)|$. Without loss of generality, we can assume that there exists at least one variable x_i, $1 \leq i \leq q$, that is not c-strongly essential for f with respect to Q. Then there exists a constant α_i such that $f_1 = f(x_i = \alpha_i)$ has at least one variable x_j in the set $Q \setminus \{x_i\}$ which is fictitious. Hence, for all subfunctions of the function $f_1 = f(x_i = \alpha_i)$ the variable x_j is also fictitious, i.e., in the function g resulting from replacing the variables in $Q \setminus \{x_j\}$ by constants in f_1, the variable x_j is fictitious. Hence, when replacing variables in $Q \setminus \{x_j\}$ by some constants, we obtain a subfunction whose variables pertaining to Q are all fictitious. This contradicts the relation $Q \xrightarrow{\text{d}} Q$ (the minimality condition). Theorem 3 is proved.

The following example demonstrates that there exist sets that are not self-dominating and the converse of Theorem 3 is invalid.

EXAMPLE 5. Let $f = x_1x_2 + x_1x_3 + x_2x_3 \pmod 2$. Every variable in the set $Q = \{x_1, x_2, x_3\}$, $Q = Ess(f)$, is c-strongly essential for f, but the set Q is not self-dominating since the proper subset $Q_1 = \{x_1, x_2\}$ of Q dominates Q.

Theorem 4. *Assume that $P \xrightarrow{\text{sd}} Q$ for f with respect to the collection $\widetilde{P}^*$ and the variable $x_i \in P$ has an active region Zx_i such that $|Zx_i| \geq 2$. Then any c-strongly essential variable x_j for the function f with respect to Zx_i is c-strongly essential for f.*

PROOF. Since x_j is c-strongly essential for the function f with respect to Zx_i, after replacing the variable x_j by any constant c_j, the other variables in the region Zx_i remain essential for $f_1 = f(x_j = c_j)$.

Next, from the proof of Theorem 2 it follows that all variables of the set

$$(P \setminus \{x_i\}) \cup (Q \setminus Zx_i)$$

are essential variables of the function f_1.

We show that $x_i \in Ess(f_1)$. Assume that this is not so, i.e., $x_i \notin Ess(f_1)$. Then

$$f_1 = f(x_j = c_j) = f\big(x_j = c_j, x_i = c_i^*\big), \quad \text{where } c_i^* \in \widetilde{P}^*.$$

Hence, at least one variable in Zx_i must be both essential and fictitious for f_1. We have arrived at contradiction. Theorem 4 is proved.

Corollary 7. *Let $P \xrightarrow{\text{sd}} Q$ for f, and let the active regions $Zx_{i_1}, \ldots, Zx_{i_k}$ be self-dominating sets, where $|Zx_{i_j}| \geq 2$, $j = 1, \ldots, k$. Then the number of c-strongly essential variables of f is not less than $|\bigcup_{j=1}^{k} Zx_{i_j}|$.*

Example 3 shows that the condition $|Zx_{i_j}| \geq 2$ cannot be omitted.

The following example demonstrates that the objects satisfy the conditions of Theorem 4 exist.

EXAMPLE 6. Let $f = x_1x_3 + x_2(x_4 + x_5) \pmod 2$, $P = \{x_1, x_2\}$, $Q = \{x_3, x_4, x_5\}$, and $P \xrightarrow{\text{sd}} Q$ for f with respect to the collection $(0,0)$. Then $Zx_2 = \{x_4, x_5\}$ and the variables x_4, x_5 are c-strongly essential for f with respect to Zx_2 and c-strongly essential for f.

References

1. O. B. Lupanov (1962) On a class of circuits of functional gates (in Russian), in: *Problemy Kibernetiki.* Vol. 7, Gosudarstv. Izdat. Fiz.-Mat. Lit., Moscow, pp. 61–144.
2. N. A. Solov'ëv, (1963) On a essential dependence of functions of the algebra of logic (in Russian), in: *Problemy Kibernetiki.* Vol. 7, Gosudarstv. Izdat. Fiz.-Mat. Lit., Moscow, pp. 333–335.
3. A. Salomaa (1963) On essential variables of functions, especially in the algebra of logic, *Ann. Acad. Sci. Fenn. Ser. AI Math.* No. 339, 1–11.
4. Yu. Ya. Breĭtbart (1967) On essential variables of functions of the algebra of logic (in Russian), *Dokl. Akad. Nauk SSSR* **172**, No. 1, 9–10.
5. K. N. Chimev and I. Mirchev (1993) Separable and Dominating Sets of Variables, S. W. University, Blagoevgrad.

6. A. Ehrenfeucht, J. Kahn, R. Maddux, and J. Mucielski (1982) On the dependence of functions on their variables, *J. Combin. Theory Ser. A* **33**, No. 1, 106–108.
7. S. Shtrakov (1987) Dominating and Annulling Sets of Variables for the Functions, Blagoevgrad.
8. I. Mirchev and S. Shtrakov (1988) Strongly dominating sets of variables, *Közl.-MTA Számitástech. Automat. Kutató Inst.* Budapest, No. 39, 121–129.

Yugo-Zapadnyĭ Universitet
im. N. Rilskogo
ul. Al. Velichkova, 66
Blagoevgrad 2700
BULGARIA

TRANSLATED BY A. V. RYAZANOV

ON COMPARISON BETWEEN THE SIZES OF READ-k-TIMES BRANCHING PROGRAMS†)

E. A. Okol'nishnikova

We compare the complexities of Boolean functions for read-k-times branching programs at various values of k. We show that for each natural number k, $k \geq 2$, there exist a natural number s_k and a sequence of Boolean functions such that the complexity of computation of every function of this sequence by read-k-times branching programs in an exponential number of times exceeds the complexity of computation of the same function by read-s_k-times branching programs (with respect to the number of variables of the Boolean function). We can take k^2 as s_k. A more precise estimate for s_k will be given before Lemma 5.

Introduction

The problem of finding nontrivial lower bounds for the complexity of "explicitly defined" Boolean functions is one of the most important problems in the theory of synthesis and complexity of logical schemes. To obtain such bounds is a complicated problem. At present nontrivial lower bounds for the complexity of Boolean functions are known for a few functions only. Therefore, it is of interest to reveal the factors that somehow influence the complexity of Boolean functions.

In the present paper we study factors that influence the complexity of branching programs computing Boolean functions. Branching programs are logical schemes which well simulate calculations by means of a single processor reading at most one input bit in unit time. (The definition of a branching program will be given in § 2, see also [1].) Branching programs were first studied by V. A. Kuz'min [2]. He obtained an asymptotic formula for Shannon's function for the complexity of branching programs computing Boolean functions. In the last two decades some interesting results on the complexity of such programs were obtained. The detailed survey of results on lower bounds for complexity of branching programs can be found in [3, 4].

A *read-k-times branching program* is a branching program such that each variable occurs at most k times in any computation path (whether or not consistent). These programs are also called *syntactic* read-k-times branching programs [5]. A possible alternative definition of a read-k-times branching program is

†) This research was supported by the Russian Foundation for Basic Research (Grant 93–01–16009).

A. D. Korshunov (ed.), Operations Research and Discrete Analysis, 205–225.

as follows: we assume that there are restrictions on the number of occurrences of each variable only in consistent paths rather than in all computation paths. In [5, 6], exponential lower bounds for the complexity of deterministic syntactic read-k-times branching programs [6] and for nondeterministic syntactic read-k-times branching programs [5] computing "explicitly defined" Boolean functions were obtained. At present nontrivial lower bounds for the complexity of nonsyntactic read-k-times branching programs are unknown for $k > 1$. In the present paper we discuss only syntactic read-k-times branching programs.

Each Boolean function can be computed by a read-k-times branching program for various values of k. Hence, the question arises about comparison between the complexities of computation of the Boolean function by read-k_1-times branching programs and by read-k_2-times branching programs. It is shown in [5, 7–10] that the complexity of computation of some sequences of Boolean functions by read-once branching programs in an exponential number of times exceeds the complexity of computation of the same functions by read-2-times branching programs (with respect to the number of variables of Boolean functions).

In the present paper we show that to each natural number k, $k \geq 2$, there exist a natural number s_k and a sequence of "explicitly defined" Boolean functions such that the complexity of realization of each function of this sequence by read-k-times branching programs in an exponential number of times exceeds the complexity of realization of the same function by read-s_k-times branching programs (with respect to the number of variables of the Boolean function). We can take k^2 as s_k. A more exact bounds for s_k will be given before Lemma 5. In [6] it is shown that the lower bounds for the complexity of read-k-times branching programs can be used to obtain lower bounds for the complexity of branching programs without constraints. Thus, the complexity of read-k-times branching programs is connected with the complexity of usual branching programs.

We suggest some modification of the method applied in [6] to obtaining lower bounds. This method is based on the information-theoretical idea of entropy consideration with respect to the parts of the program and the function. This method we apply to a function satisfying the following property: its subfunctions are in general "badly" representable as

$$\bigvee_i f_i(X)\&g_i(Y),$$

where X and Y are some disjoint subsets of variables with $|X|$ and $|Y|$ not small.

This paper is organized as follows. In § 1 we give some definitions concerning branching programs and some elementary properties of such programs. In § 2 we define a sequence of Boolean functions $F_{n,s}$. Then we prove that there is an exponential increase in the complexity of computation of the Boolean function $F_{n,s}$

when we pass from read-s_k-times branching programs to read-k-times branching programs. In § 3 we define some mapping Ψ that takes each conjunction K, $K \in \mathcal{K}_0$, to a sequence $\Psi(K)$ of nodes of a branching program. In § 4 we estimate the size of a set $\Psi^{-1}(B)$ for a sequence of nodes B of a branching program. We show that $|\Psi^{-1}(B)| \ll |\mathcal{K}_0|$. This fact allows us to prove the main theorem of § 5 in which we prove that to every natural number k, $k \geq 2$, there exist a natural number s_k and a sequence of Boolean functions such that the complexity of computation of each function of this sequence by read-k-times branching programs exceeds an exponential number of times the complexity of computation of the same function by read-s_k-times branching programs (in regard to the number of variables of the Boolean function).

§ 1. Definitions and Preliminaries

Recall that a (deterministic) *branching program* is a directed acyclic graph consisting of one source (no predecessor, there are two arcs leaving this node), *inner nodes* (there is a predecessor and there are two arcs leaving this node), and outputs nodes, i.e. exit nodes (there are no arcs leaving this node). Each inner node (as well as the source) is labelled with a Boolean variable in $X = \{x_1, \ldots, x_n\}$. Each exit is labelled with 0 or 1. One arc leaving the source or an inner node is labelled with 0, the other is labelled with 1.

Let $\tilde{a} = (a_1, \ldots, a_n)$, $a_i \in \{0,1\}$ and $i = 1, 2, \ldots, n$, be an input tuple of P. A branching program $\mathcal{P}$ computes the value $P(\tilde{a})$ as follows. The computation starts at the source. If we reach a node labelled with x_i, then we move to the successor along the arc labelled with 1 if $a_i = 1$, and by the arc labelled with 0 if $a_i = 0$. Finally, we reach the exit labelled with 0 or 1. The resulting value is denoted by $P(\tilde{a})$.

A branching program $\mathcal{P}$ is said to *compute a Boolean function* $f(x_1, \ldots, x_n)$ if for every input tuple $\tilde{a} = (a_1, \ldots, a_n)$ the equality $P(\tilde{a}) = f(\tilde{a})$ holds.

A (*syntactic*) *read-k-times branching program* (or a *k-program*, for short) is a branching program such that each variable occurs at most k times in any computation path (whether or not consistent).

A read-k-times branching program is said to be *uniform* if, for every node a of this program and for every variable x, the number of occurrences of x in each path from the source to a node a depends only on a and x. Moreover, every variable occurs in every path from the source to an exit exactly k times.

The number of nodes in $\mathcal{P}$ is said to be the *size* $B(\mathcal{P})$ of a branching program $\mathcal{P}$. Denote by $Bk(f)$ the size of a minimal k-program, and by $UBk(f)$, the size of a minimal uniform k-program computing a Boolean function f.

Let a branching program $\mathcal{P}_f$ compute a Boolean function f. For any path α from the source to the exit labelled with 1 there is a conjunction $K(\alpha)$ that corresponds to this path. This correspondence is defined as follows. The conjunction K contains x_i if and only if we have chosen an arc labelled with 1 after traversing a node labelled with x_i in this path. The conjunction K contains $\bar{x}_i$ if and only if we have chosen an arc labelled with 0 after traversing a node labelled with x_i in this path. In doing so some resulting conjunctions contain both a variable and its negation, i.e., these conjunctions are equal to zero.

A node a_i is said to *precede* a node a_j if and only if there is a path in $\mathcal{P}$ from a_i to a_j.

Let us consider a subset of nodes $a_1, \ldots, a_t$ in a branching program $\mathcal{P}$ (these nodes need not differ). This subset is called a *sequence of nodes* if there is a path in $\mathcal{P}$ from the source to the exit labelled with 1 such that this path includes nodes $a_1, \ldots, a_t$; a_i precedes a_j or a_i coincides with a_j for $i < j$.

It is clear that a conjunction K corresponds to some path from the source to the exit labelled with 1 in a branching program $\mathcal{P}_f$ if and only if K is *admissible* to the Boolean function f, i.e., $N_K \subseteq N_f$. (As usual, N_f is the set of all "ones" of a Boolean function f.)

§ 2. Definition and Properties of the Boolean Function $F_{n,s}$

In this section we define a sequence of Boolean functions $F_{n,s}$. In § 5 we prove that there is an exponential increase in the complexity of the Boolean function $F_{n,s}$ when we pass from read-s_k-times branching programs to read-k-times branching programs (with respect to the number of variables of the Boolean function).

Let n and s be integers, n divisible by s. Let $X_{n,s} = \{x_{i_1,i_2,\ldots,i_s} \mid 1 \le i_1 < i_2 < \cdots < i_s \le n\}$ be a set of variables. It is clear that

$$|X_{n,s}| = \binom{n}{s} \overset{\text{def}}{\equiv} N. \tag{1}$$

We define a subset W_i of the set $X_{n,s}$. A variable $x_{i_1,i_2,\ldots,i_s}$ belongs to W_i if and only if this variable has i among its indices. A Boolean function $F_{n,s}(X_{n,s})$ is defined as follows:

$$F_{n,s} = \bigwedge_{i=1}^{n} \Big(\bigvee_{x_{i_1,\ldots,i_s} \in W_i} x_{i_1,\ldots,i_s} \Big). \tag{2}$$

EXAMPLE 1. For $n = 4$ and $s = 2$ we have $X_{4,2} = \{x_{1,2}, x_{1,3}, x_{1,4}, x_{2,3}, x_{2,4}, x_{3,4}\}$; $W_1 = \{x_{1,2}, x_{1,3}, x_{1,4}\}$, $W_2 = \{x_{1,2}, x_{2,3}, x_{2,4}\}$, $W_3 = \{x_{1,3}, x_{2,3}, x_{3,4}\}$, and $W_4 = \{x_{1,4}, x_{2,4}, x_{3,4}\}$. Then

$$F_{4,2} = (x_{1,2} \vee x_{1,3} \vee x_{1,4}) \& (x_{1,2} \vee x_{2,3} \vee x_{2,4}) \& (x_{1,3} \vee x_{2,3} \vee x_{3,4}) \& (x_{1,4} \vee x_{2,4} \vee x_{3,4}).$$

It is easy to see that the Boolean function $F_{n,s}$ can be computed by a s-program of the size

$$n|W_i| = n\binom{n-1}{s-1} = s\binom{n}{s} = s|X_{n.s}|. \tag{3}$$

The family of the functions $F_{n,2}$ was used in [10] for comparing complexities of Boolean functions for read-1-time branching programs and for read-2-times branching programs and in [11, 12] for comparing complexities of Boolean functions for circuits of functional elements with or without constraints.

Let us give some definitions.

Let $D \subset X_{n,s}$. Denote by $V(D)$ the set of all indices (without regard for multiplicity) of all variables in D. In particular, $V(X_{n,s}) = \{1, 2, \ldots, n\}$.

Let a conjunction K depend on some variables of the set $X_{n,s}$. Denote by $R(K)$ the set of variables that are contained without negation in K, and by $V(K)$, the set of all indices (without regard to multiplicity) of all variables in $R(K)$. It is clear that

$$V(K) = V(R(K)).$$

EXAMPLE 2. For the conjunction $K_1 = x_{1,2} \,\&\, \bar{x}_{1,3} \,\&\, x_{2,4} \,\&\, \bar{x}_{3,5}$ we have $V(K_1) = V(x_{1,2},\ x_{2,4}) = \{1, 2, 4\}$.

A conjunction K is said to be *nonintersecting* if k depends on all variables in $X_{n,s}$ and if for every integer i, $1 \leq i \leq n$, there is a unique variable in $R(K)$ that has this integer among its indices. It is clear that $V(K) = \{1, 2, \ldots, n\}$ for every nonintersecting conjunction K.

The number of nonintersecting conjunctions depending on the variables of $X_{n,s}$ is denoted by $\Phi(n, s)$. It can easily be checked that

$$\Phi(n, s) = \frac{n!}{(s!)^{n/s}(n/s)!}. \tag{4}$$

Consider some properties of the Boolean function $F_{n,s}$ and of conjunctions realized by a branching program computing the Boolean function $F_{n,s}$.

From (2) we have

Lemma 1. *If a conjunction K depends on some variables in $X_{n,s}$, then K is admissible to the Boolean function $F_{n,s}$ if and only if $V(K) = \{1, 2, \ldots, n\}$.*

EXAMPLE 3. Consider the Boolean function $F_{4,2}$. The conjunctions $x_{1,2} \,\&\, x_{1,3} \,\&\, \bar{x}_{1,4} \,\&\, x_{2,3} \,\&\, \bar{x}_{2,4} \,\&\, \bar{x}_{3,4}$ and $x_{1,2} \,\&\, x_{1,3} \,\&\, \bar{x}_{2,4}$ are inadmissible for $F_{4,2}$, the conjunction $x_{1,2} \,\&\, x_{1,3} \,\&\, \bar{x}_{1,4} \& \, x_{3,4}$ is admissible for the Boolean function $F_{4,2}$, the conjunction $x_{1,2} \,\&\, \bar{x}_{1,3} \,\&\, \bar{x}_{1,4} \,\&\, \bar{x}_{2,3} \,\&\, \bar{x}_{2,4} \,\&\, x_{3,4}$ is admissible and nonintersecting for $F_{4,2}$.

From Lemma 1 we obtain

Corollary 1. *Let $\mathcal{P}_0$ be a uniform k-program computing the Boolean function $F_{n,s}$. Then, to every nonintersecting conjunction K, there exists a path (from the source to the exit labelled with* 1*) realizing the conjunction K in $\mathcal{P}_0$.*

PROOF. By the definition of nonintersecting conjunction and Lemma 1 we obtain that the elementary conjunction K is admissible for the Boolean function $F_{n,s}$. Therefore, there exists a path (from the source to the exit labelled with 1) realizing K in $\mathcal{P}_0$. This proves Corollary 1.

REMARK 1. For describing the Boolean function $F_{n,s}$, we may use hypergraphs on n vertices. (The definition of hypergraph can be found, for example, in [13, Chapter XI].) To each conjunction $K(X_{n,s})$ we assign a hypergraph $G(K)$ on n vertices $1, 2, \dots, n$ as follows: the hypergraph $G(K)$ contains an s-arc $(i_1, \dots, i_s)$ if and only if $x_{i_1,\dots,i_s} \in R(K)$. Then the conjunction $K(X_{n,s})$ is admissible to the Boolean function $F_{n,s}$ if and only if there exists a hypergraph without isolated vertices corresponding to this conjunction. Every hypergraph corresponding to a nonintersecting conjunction has no isolated vertices, every vertex is incident to one and only one s-arc. These hypergraphs are analogous to the perfect matchings in the graph theory.

In these terms the set $V(K)$ can be defined as the set of vertices of the hypergraph $G(K)$ which are incident to some s-arc of $G(K)$.

§ 3. Construction of the Sequence $\Psi(K)$

By Lemma 1 in [9], each k-program $\mathcal{P}$ computing a Boolean function f of N variables can be transformed into a uniform k-program $\mathcal{P}_0$ that computes f and

$$Bk(\mathcal{P}_0) \le (2kN)Bk(\mathcal{P}).$$

From this it follows that, for every Boolean function f,

$$UBk(f) \le (2kN)Bk(f). \tag{5}$$

Let $\mathcal{P}_0$ be a uniform k-program computing a Boolean function f. It is clear that each path in $\mathcal{P}_0$ realizes an elementary conjunction or a conjunction that is equal to zero. Let α be a consistent path in $\mathcal{P}_0$ realizing a conjunction K, where K is not equal to zero. We say that a variable x *occurs in an interval* $[b, c)$ of a path α if there is an arc leaving a node labelled with x in the interval $[b, c)$ of the path α. Similarly, we say that a variable y *occurs without negation in an interval* $[b, c)$ of a path α if there is an arc labelled with 1 and issuing from a node labelled with y in the interval $[b, c)$ of the path α. It is clear that $y \in R(K)$.

Let $a_1, a_2, \dots, a_{2m}$ be a sequence of nodes of a uniform k-program $\mathcal{P}_0$, and let all nodes of this sequence belong to the path α.

We introduce the following notation:

- $R^1_\alpha(a_1, a_2, \ldots, a_{2m})$ ($Q^1_\alpha(a_1, a_2, \ldots, a_{2m})$) is the set of variables without negation (the set of variables with or without negation respectively) that occur only in the intervals $[a_1, a_2), \ldots, [a_{2m-1}, a_{2m})$ of the path α;
- $R^2_\alpha(a_1, a_2, \ldots, a_{2m})$ ($Q^2_\alpha(a_1, a_2, \ldots, a_{2m})$) is the set of variables without negation (a set of variables with or without negation respectively) that occur only outside the intervals above;
- $R^0_\alpha(a_1, a_2, \ldots, a_{2m})$ ($Q^0_\alpha(a_1, a_2, \ldots, a_{2m})$) is the set of variables without negation (a set of variables with or without negation respectively) that occur in both intervals above and outside them.

It is clear that sets $R^j_\alpha(a_1, a_2, \ldots, a_{2m})$, $j = 0, 1, 2$, are pairwise disjoint and the union of these sets is the set of all variables without negation that occur in the path α. Each set R^j_α depends both on the sequence $a_1, a_2, \ldots, a_{2m}$ and on the path α to which this sequence belongs.

A set of variables that occur in an interval $[b, c)$ in a uniform k-program $\mathcal{P}_0$ does not depend on a path to which nodes b and c belong. Therefore, the sets $Q^j_\alpha(a_1, a_2, \ldots, a_{2m})$, $j = 0, 1, 2$, depend only on the sequence of nodes $a_1, a_2, \ldots, a_{2m}$ and do not depend on the path to which these nodes belong. Hence, we may omit the index α in Q^j_α. It is clear that the sets $Q^j(a_1, a_2, \ldots, a_{2m})$, $j = 0, 1, 2$, are pairwise disjoint and the union of these sets is the set of all variables of the Boolean function f. It is obvious that $R^j_\alpha(a_1, a_2, \ldots, a_{2m}) \subseteq Q^j(a_1, a_2, \ldots, a_{2m})$, $j = 0, 1, 2$, for every path α containing the nodes $a_1, a_2, \ldots, a_{2m}$.

Lemma 2. *Let $\mathcal{P}_0$ be a uniform k-program computing the Boolean function $F_{n,s}$; let K be an nonintersecting conjunction; let m and t be natural numbers, $k \le m \le t$. Then there exists a sequence $\Psi(K)$ of $2m$ nodes such that*

(a) *all nodes of the sequence $\Psi(K)$ belong to a path $\alpha(K)$ realizing the conjunction K;*

(b) $|R^1_\alpha(\Psi(K))| = \left\lceil (n/s)\binom{t-k}{m-k} \Big/ \binom{t}{m} \right\rceil$;

$$|R^2_\alpha(\Psi(K))| \ge n/s - m\lceil kn/(ts)\rceil + (k-1)\left\lceil (n/s)\binom{t-k}{m-k} \Big/ \binom{t}{m} \right\rceil;$$

$$|R^0_\alpha(\Psi(K))| \le m\lceil kn/(ts)\rceil - k\left\lceil (n/s)\binom{t-k}{m-k} \Big/ \binom{t}{m} \right\rceil.$$

PROOF. By Corollary 1 of Lemma 1, the uniform k-program $\mathcal{P}_0$ contains a path $\alpha(K)$ (from the source to the exit labelled with 1) realizing the conjunction K. Since $\mathcal{P}_0$ is a uniform k-program, each variable occurs exactly k times in each computation path. Therefore, the path $\alpha(K)$ contains exactly kn/s arcs labelled with 1. We can take $t+1$ nodes $a_0, a_1, \ldots, a_t$ of the path $\alpha(K)$ such that a_0 is the source, a_i precedes a_j, $0 \le i < j \le t$, and the number of arcs labelled with 1 in the path $\alpha(K)$ between a_i and a_{i+1} ($i = 0, 1, \ldots, t-1$) does not exceed $\lceil kn/(ts)\rceil$.

Any m nodes $a_{i_1}, \ldots, a_{i_m}$, $0 \leq i_1 < i_2 < \cdots < i_m \leq t-1$, in the set $\{a_0, a_1, \ldots, a_{t-1}\}$ define m intervals in the path $\alpha(K)$: $[a_{i_1}, a_{i_1+1})$, $[a_{i_2}, a_{i_2+1})$, $\ldots, [a_{i_m}, a_{i_m+1})$, where a_0 is the source and a_{i_m+1} is the exit labelled with 1. The set of the ends of these intervals we denote $A_{i_1,i_2,\ldots,i_m}$, i.e., $A_{i_1,i_2,\ldots,i_m} = \{a_{i_1}, a_{i_1+1}, a_{i_2}, a_{i_2+1}, \ldots, a_{i_m}, a_{i_m+1}\}$. The set $A_{i_1,i_2,\ldots,i_m}$ is a sequence of nodes in the branching program $\mathcal{P}_0$.

In the end of the proof of Lemma 2 we show that there exists an insignificant modification of one of the sequences $A_{i_1,i_2,\ldots,i_m}$ (a sequence B) that satisfies the conditions of Lemma 2.

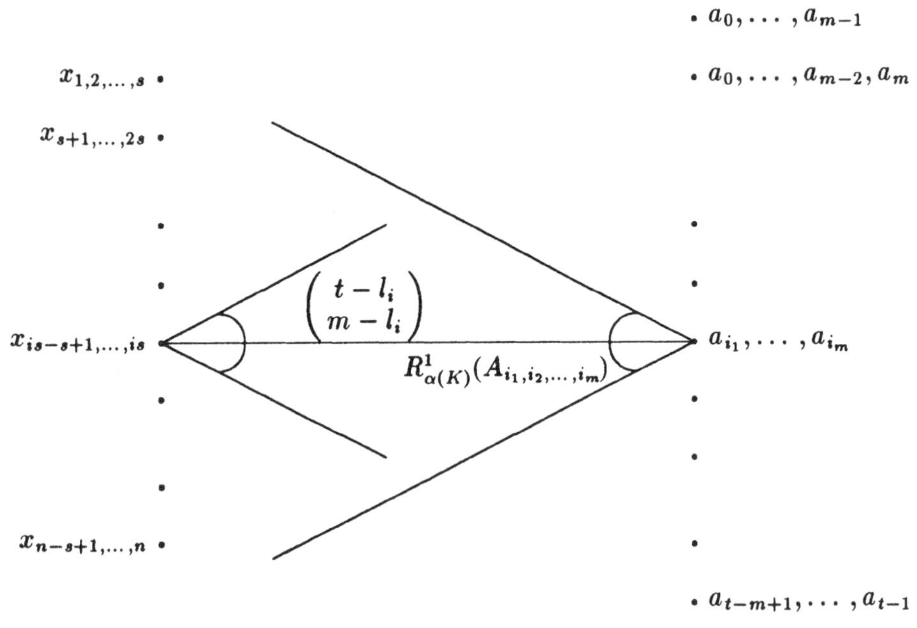

Fig. 1

Without loss of generality, we may assume that

$$K = x_{1,\ldots,s} \,\&\, x_{s+1,\ldots,2s} \,\&\, \ldots \,\&\, x_{n-s+1,\ldots,n}.$$

Consider the graph G (see Fig. 1) containing two sets of vertices: the n/s vertices corresponding to the variables $x_{1,\ldots,s}$, $x_{s+1,\ldots,2s}, \ldots,$ $x_{n-s+1,\ldots,n}$ and the $\binom{t}{m}$ vertices corresponding to all m-subsets $\{a_{i_1}, \ldots, a_{i_m}\}$ of the set $\{a_0, a_1, \ldots,$

$a_{t-1}\}$, and hence to all choices of the sequences $A_{i_1,i_2,\ldots,i_m}$. The vertex corresponding to the variable $x_{(i-1)s+1,\ldots,is}$ is adjacent to the vertex corresponding to the subset $\{a_{i_1}, a_{i_2}, \ldots a_{i_m}\}$, $0 \le i_1 < i_2 < \cdots < i_m \le t-1$, if and only if the variable $x_{(i-1)s+1,\ldots,is}$ belongs to $R^1_{\alpha(K)}(A_{i_1,i_2,\ldots,i_m})$, i.e., if and only if this variable occurs without negation in some intervals $[a_{i_1}, a_{i_1+1})$, $[a_{i_2}, a_{i_2+1})$, $\ldots$, $[a_{i_m}, a_{i_m+1})$ and does not occur outside them. Let l_i denote the number of the intervals $[a_{i_j}, a_{i_j+1})$, $0 \le j \le t-1$, to which the variable $x_{(i-1)s+1,\ldots,is}$ belongs. Then in the graph G the vertex corresponding to the variable $x_{(i-1)s+1,\ldots,is}$ is connected with all vertices corresponding to the subsets containing the intervals above. The number of such intervals is $\binom{t-l_i}{m-l_i}$. It is clear that the vertex corresponding to the subset $\{a_{i_1}, \ldots, a_{i_m}\}$ is connected in G with $|R^1_{\alpha(K)}(A_{i_1,i_2,\ldots,i_m})|$ vertices corresponding to variables, and the number of arcs leaving the left vertices of the bichromatic graph G (see Fig. 1) is equal to the number of arcs leaving the right vertices of the graph G. Therefore,

$$\sum_{i_1,\ldots,i_m} |R^1_{\alpha(K)}(A_{i_1,i_2,\ldots,i_m})| = \sum_{i=1}^{n/s} \binom{t-l_i}{m-l_i}. \tag{6}$$

Since $\mathcal{P}_0$ is a uniform k-program $\mathcal{P}_0$, we have $l_i \le k$. Hence, $\binom{t-l_i}{m-l_i} \ge \binom{t-k}{m-k}$. It follows from this and (6) that

$$\sum_{i_1,\ldots,i_m} |R^1_{\alpha(K)}(A_{i_1,i_2,\ldots,i_m})| \ge (n/s)\binom{t-k}{m-k}.$$

Hence, there exist nodes $a_{i_1}, \ldots, a_{i_m}$ such that the inequality

$$|R^1_{\alpha(K)}(A_{i_1,i_2,\ldots,i_m})| \ge (n/s)\binom{t-k}{m-k} \Big/ \binom{t}{m}$$

holds. Since $|R^1_{\alpha(K)}(A_{i_1,i_2,\ldots,i_m})|$ is a natural number, we obtain

$$|R^1_{\alpha(K)}(A_{i_1,i_2,\ldots,i_m})| \ge \left\lceil (n/s)\binom{t-k}{m-k} \Big/ \binom{t}{m} \right\rceil. \tag{7}$$

If (7) is a strict inequality, then we may gradually diminish intervals (bringing closer to each other the ends of the intervals $[a_{i_j}, a_{i_j+1})$ in the path $\alpha(K)$). Finally, we move from the sequence $A_{i_1,i_2,\ldots,i_m}$ to a sequence $B = (b_1, \ldots, b_{2m})$ such that

$$|R^1_{\alpha(K)}(B)| = \left\lceil (n/s)\binom{t-k}{m-k} \Big/ \binom{t}{m} \right\rceil. \tag{8}$$

The number of variables without negation lying in the intervals $[b_1, b_2)$, $[b_3, b_4)$, $\ldots$, $[b_{2m-1}, b_{2m})$, i.e. the value $|R^1_{\alpha(K)}(B) \cup R^0_{\alpha(K)}(B)|$, satisfies the inequality

$$|R^1_{\alpha(K)}(B) \cup R^0_{\alpha(K)}(B)| \le \lceil kn/(ts) \rceil m - (k-1)|R^1_{\alpha(K)}(B)|. \tag{9}$$

Since

$$|R^2_{\alpha(K)}(B)| = n/s - |R^1_{\alpha(K)}(B) \cup R^0_{\alpha(K)}(B)|$$

it follows from (8) and (9) that

$$\begin{aligned}|R^2_{\alpha(K)}(B)| &\ge n/s - \lceil kn/(ts) \rceil m + (k-1)|R^1_{\alpha(K)}(B)| \\ &\ge n/s - \lceil kn/(ts) \rceil m + (k-1)\left\lceil (n/s)\binom{t-k}{m-k} \Big/ \binom{t}{m} \right\rceil.\end{aligned}$$

So $|R^0_{\alpha(K)}(B)| + |R^1_{\alpha(K)}(B)| + |R^2_{\alpha(K)}(B)| = n/s$, which implies the needed bound for $|R^0_{\alpha(K)}(B)|$.

Thus, the sequence of nodes $\Psi(K) = B$ satisfies (b) of Lemma 2. The sequence B satisfies (a) of Lemma 2 by construction. Lemma 2 is proved.

§ 4. Estimation of the Size of the Set $\Psi^{-1}(B)$

Let ψ be a sequence of nodes of a branching program realizing the Boolean function $F_{n,s}$. From the definition of $Q^j(\psi)$, $j = 0, 1, 2$, it follows that $X_{n,s} = Q^0(\psi) \cup Q^1(\psi) \cup Q^2(\psi)$ and $Q^i(\psi) \cap Q^j(\psi) = \varnothing$ for $i \ne j$. Let $\mathcal{K}$ be a conjunction depending essentially on all variables of $X_{n,s}$. Then $\mathcal{K}$ can be uniquely represented as

$$\mathcal{K} = \mathcal{K}^0(Q^0(\psi)) \& \mathcal{K}^1(Q^1(\psi)) \& \mathcal{K}^2(Q^2(\psi)),$$

where the conjunction $\mathcal{K}^i(Q^i(\psi))$ depends essentially on all variables in $Q^i(\psi)$. Therefore, the conjunctions $\mathcal{K}^i(Q^i(\psi))$ are uniquely determined from the conjunction $\mathcal{K}$ and the sequence ψ.

Let ψ be a sequence of nodes of a branching program. Denote by $T(\psi)$ the set of distinct conjunctions depending essentially on all variables in $X_{n,s}$ such that there exists a path (from the source to the exit labelled with 1) containing the vertices of the sequence ψ.

Lemma 3. *Let $\psi = (a_1, a_2, \ldots, a_{2m})$ be a sequence of nodes of a uniform k-program $\mathcal{P}_0$, and let conjunctions $\mathcal{A}$ and $\mathcal{B}$ belong to the set $T(\psi)$. Suppose that $\mathcal{A}$ and $\mathcal{B}$ are represented as*

$$\mathcal{A} = \mathcal{A}^0(Q^0(\psi)) \ \& \ \mathcal{A}^1(Q^1(\psi)) \ \& \ \mathcal{A}^2(Q^2(\psi))$$

and

$$\mathcal{B} = \mathcal{B}^0(Q^0(\psi)) \ \& \ \mathcal{B}^1(Q^1(\psi)) \ \& \ \mathcal{B}^2(Q^2(\psi)).$$

If the identity $\mathcal{A}^0(Q^0(\psi)) = \mathcal{B}^0(Q^0(\psi))$ *holds, then the identities* $V(\mathcal{A}^1(Q^1(\psi))) = V(\mathcal{B}^1(Q^1(\psi)))$ *and* $V(\mathcal{A}^2(Q^2(\psi))) = V(\mathcal{B}^2(Q^2(\psi)))$ *also hold.*

PROOF. Let α and β be paths in $\mathcal{P}_0$ (from the source to the exit labelled with 1) each of which contains the sequence of nodes in $\psi = (a_1, a_2, \dots, a_{2m})$. Also, let the path α realize the conjunction in $\mathcal{A}$, and let the path β realize the conjunction in $\mathcal{B}$.

Denote by α_1 (β_1) the part of the path α (β) containing the intervals $[a_1, a_2)$, $[a_3, a_4)$, ... , $[a_{2m-1}, a_{2m})$ by α_1 (β_1, respectively). By α_2 (β_2) we denote the remaining part of the path α (β, respectively).

The conjunction $\mathcal{A}$ is a nonintersecting conjunction (by the definition of $T(\psi)$). Hence, the sets $V(\mathcal{A}^0(Q^0(\psi)))$, $V(\mathcal{A}^1(Q^1(\psi)))$, and $V(\mathcal{A}^2(Q^2(\psi)))$ are pairwise disjoint. Therefore, the union of these sets coincides with $\{1, 2, \dots, n\}$. A similar statement for the sequence of nodes $\mathcal{B}$ holds too. If $\mathcal{A}^0(Q^0(\psi)) = \mathcal{B}^0(Q^0(\psi))$, then $V(\mathcal{A}^0(Q^0(\psi))) = V(\mathcal{B}^0(Q^0(\psi)))$. Therefore,

$$V(\mathcal{A}^1(Q^1(\psi))) \cup V(\mathcal{A}^2(Q^2(\psi))) = V(\mathcal{B}^1(Q^1(\psi))) \cup V(\mathcal{B}^2(Q^2(\psi))).$$

Suppose that $V(\mathcal{A}^1(Q^1(\psi))) \neq V(\mathcal{B}^1(Q^1(\psi)))$. Without loss of generality, we may assume that $V(\mathcal{A}^1(Q^1(\psi))) \not\subseteq V(\mathcal{B}^1(Q^1(\psi)))$, i.e., there exists a index i such that $i \in V(\mathcal{A}^1(Q^1(\psi)))$ and $i \notin V(\mathcal{B}^1(Q^1(\psi)))$. Consider a path γ (it is indicated by dashes in Fig. 2)

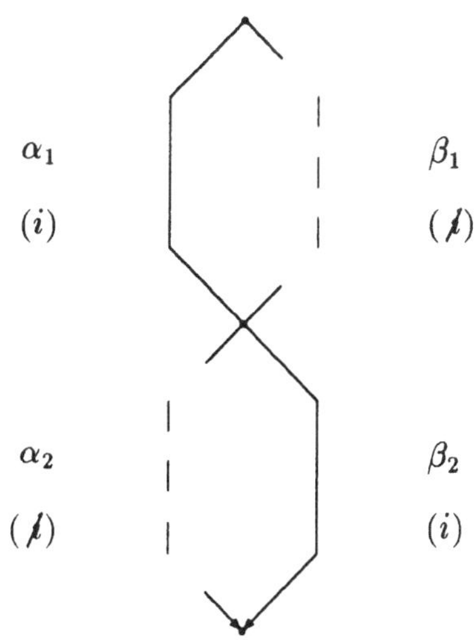

Fig. 2

traversing the arcs of β_1 of the path β and traversing the arcs of α_2 of the path α. We have $\mathcal{A}^0(Q^0(\psi)) = \mathcal{B}^0(Q^0(\psi))$. Therefore, the values of the variables that occur both in the interval β_1 and in the interval α_2 coincide. Hence, the path γ realizes a nonzero conjunction

$$\mathcal{C} = \mathcal{A}^0(Q^0(\psi)) \,\&\, \mathcal{B}^1(Q^1(\psi)) \,\&\, \mathcal{A}^2(Q^2(\psi)).$$

Since the conjunction $\mathcal{A}$ is a nonintersecting conjunction, it follows from $i \in V(\mathcal{A}^1(Q^1(\psi)))$ that $i \notin V(\mathcal{A}^2(Q^2(\psi)))$ and $i \notin V(\mathcal{A}^0(Q^0(\psi)))$. By assumption, $i \notin V(\mathcal{B}^1(Q^1(\psi)))$. Therefore, $i \notin V(\mathcal{C})$. By Lemma 1, the conjunction $\mathcal{C}$ is not admissible for the Boolean function $F_{n,s}$. This contradiction proves Lemma 3.

Below we assume that

$$n_1 = s\left\lceil (n/s)\binom{t-k}{m-k} \Big/ \binom{t}{m} \right\rceil; \tag{10}$$

$$n_0 = sm\lceil kn/(ts)\rceil - sk\left\lceil (n/s)\binom{t-k}{m-k} \Big/ \binom{t}{m} \right\rceil; \tag{11}$$

$$n_2 = n - n_0 - n_1. \tag{12}$$

Lemma 4. *Let $B = (b_1, b_2, \dots, b_{2m})$ be a sequence of nodes of a uniform k-program $\mathcal{P}_0$ computing the Boolean function $F_{n,s}$. Then for all fixed k, m, and t ($k \le m$, $km \le t$), the inequality*

$$\Psi^{-1}(B) \le \binom{n}{n_0}\Phi(n_0, s)\Phi(n_1, s)\Phi(n_2, s)$$

holds (Φ is given by (4)).

PROOF. The sequence of nodes B determines a partition of the set $X_{n,s}$ into three pairwise disjoint subsets $Q^j(B)$, $j = 0, 1, 2$. The mapping Ψ (see Lemma 2) is a mapping of the set of all nonintersecting conjunctions of variables in $X_{n,s}$ into the set of all sequences of nodes in the branching program $\mathcal{P}_0$. Furthermore, if $K \in \Psi^{-1}(B)$, then by the construction of the mapping Ψ, the relations

$$|R^1_{\alpha(K)}(B)| = |R(K) \cap Q^1(B)| = n_1/s,$$

$$|R^0_{\alpha(K)}(B)| = |R(K) \cap Q^0(B)| \le n_0/s,$$

and

$$|R^2_{\alpha(K)}(B)| = |R(K) \cap Q^2(B)| \ge n_2/s$$

hold for the path $\alpha(K)$ containing the sequence of nodes B and realizing the nonintersecting conjunction K.

By the definition of $T(B)$ and Lemma 2 it follows that $\Psi^{-1}(B) \subseteq T(B)$. Further, if the *sets of variables* $(R(K) \cap Q^0(B))$ coincide for some conjunctions K belonging to the set $\Psi^{-1}(B)$, then the *sets of indices* $(V(R(K) \cap Q^1(B)))$ (similarly, $(V(R(K) \cap Q^2(B))))$ also coincide. This, for the conjunctions belonging to $\Psi^{-1}(B)$ the set of indices $\{1, 2, \ldots, n\} \setminus (V(R(K) \cap Q^0(B)))$ is divided into two disjoint subsets $V_1 = V(R(K) \cap Q^1(B))$ and $V_2 = V(R(K) \cap Q^2(B))$. In addition, we have $|V(R(K) \cap Q^0(B))| \leq n_0$, $|V_1| = n_1$, and $|V_2| = n - n_1 - |V(R(K) \cap Q^0(B))|$.

Each conjunction K depending essentially on all variables in $X_{n,s}$ is uniquely determined by the set $R(K)$. Since all conjunctions in $\Psi^{-1}(B)$ are nonintersecting conjunctions, variables in the set $R(K)$ have no common indices for any conjunction $K \in \Psi^{-1}(B)$. We will estimate from above the number of ways to construct the set $R(K)$ for conjunctions in $\Psi^{-1}(B)$. This can be done as follows:

1. We choose a subset U of the set of variables $Q^0(B) \cup Q^2(B)$ such that

(a) the variables in U have no common indices;

(b) $|U| = n_0/s$.

The number of ways to choose a subset of $X_{n,s}$ that contains n_0/s variables without common indices is equal to $\binom{n}{n_0}\Phi(n_0, s)$. Hence, the number of ways to choose the subset U of $Q^0(B) \cup Q^2(B)$ does not exceed $\binom{n}{n_0}\Phi(n_0, s)$.

The number of ways to choose the subset U of $Q^0(B) \cup Q^2(B)$ does not exceed the number of ways to choose a subset of $X_{n,s}$ such that this subset contains n_0/s variables without common indices. Hence, this number does not exceed $\binom{n}{n_0}\Phi(n_0, s)$.

Let $R_0 = U \cap Q^0(B)$. It is clear that $|R_0| \leq n_0/s$. Suppose that K is a conjunction in $\Psi^{-1}(B)$ such that $R(K) \cap Q_0(K) = R_0$. By Lemma 3, the sets $V_1(R_0) = V(Q^1(B) \cap R(K))$ and $V_2(R_0) = V(Q^2(B) \cap R(K))$ are uniquely determined.

2. If $|V_1(R_0)| \neq n_1$, then the process of constructing the set $R(K)$ terminates. If $|V_1(R_0)| = n_1$, then we choose a subset U_1 from $Q^1(B)$ such that

(a) the variables in U_1 have no common indices;

(b) $V(R_0) = V_1(R_0)$.

The number of possibilities of choosing U_1 does not exceed $\Phi(n_1, s)$.

3. If $V(U \cap Q^2(B)) \not\subseteq V_2(R_0)$, then the process of constructing the set $R(K)$ terminates. If $V(U \cap Q^2(B)) \subseteq V_2(R_0)$, then we choose a subset U_2 in $Q^2(B)$ such that

(a) the variables in U_2 have no common indices;

(b) $|U_2| = n_2$;

(c) $V(U_2) = V_2(R_0) \setminus V(U \cap Q^2(B))$.

The number of possibilities of choosing U_2 does not exceed $\Phi(n_2, s)$.

It is easy to check that each conjunction in $\Psi^{-1}(B)$ is taken into account in items 1–3. Hence,

$$\Psi^{-1}(B) \leq \binom{n}{n_0} \Phi(n_0, s)\Phi(n_1, s)\Phi(n_2, s).$$

This proves Lemma 4.

Let k be a natural number, $k \geq 2$. Put

$$s_k^0 = \frac{k^2 \log k - (k^k - k)\log(1 - k^{-k+1}) - (k-1)\log(k-1)}{k \log k - (k-1)\log(k-1)}; \tag{13}$$

$$s_k = \lfloor s_k^0 \rfloor + 1;$$

$$\varepsilon_k = s_k - s_k^0.$$

It is clear that

$$0 < \varepsilon_k \leq 1. \tag{14}$$

Lemma 5. *For any k, $k \geq 2$, the inequality*

$$s_k \leq k^2$$

holds.

PROOF. Since $s_2 = 4$, Lemma 5 is valid for $k = 2$. For any k, $k \geq 3$, we have

$$\begin{aligned} -(k-1)\log(k-1) &= -(k-1)\log k - (k-1)\log(1 - 1/k) \\ &= -(k-1)\log k - (k-1)\sum_{i=1}^{\infty} \frac{1}{ik^i} \\ &= -(k-1)\log k + 1 - \sum_{i=1}^{\infty} \frac{1}{i(i+1)k^i} < -(k-1)\log k + 1 \end{aligned} \tag{15}$$

and

$$-(k-1)\log(k-1) > -(k-1)\log k + 1 - 1/k. \tag{16}$$

Similarly, we can verify that

$$-(k^k - k)\log(1 - 1/k^{k-1}) < k. \tag{17}$$

It follows from (14)–(17) that, for $k \geq 3$,

$$s_k \leq \frac{k^2 \log k + k - (k-1)\log k + 1}{\log k + 1 - 1/k} + 1 \leq \frac{(k^2 - k + 1)\log k + k + 2}{\log k + 1 - 1/k} < k^2. \tag{18}$$

Lemma 5 is proved.

Let $\mathcal{P}_0$ be a uniform k-program realizing the Boolean function $F_{n,s}$. Denote by $\omega_{\mathcal{P}_0}(n, s, k, 2m)$ the number of sequences consisting of $2m$ nodes in $\mathcal{P}_0$.

Put

$$m_k = \lceil k^3/\varepsilon_k \rceil. \tag{19}$$

Lemma 6. *Suppose that $k \geq 2$, $n \geq 3k^{k+5}/\varepsilon_k$, and $\mathcal{P}_0$ is a uniform k-program realizing the Boolean function $F_{n,s}$. Then the inequality*

$$\omega_{\mathcal{P}_0}(n, s_k, k, 2m_k) \geq \frac{\exp(n\varepsilon_k/k^2)}{(2n)^{m_k k^2}}$$

holds.

Proof. Since to each nonintersecting conjunction K we assign a sequence of $2m$ nodes $\Psi(K)$, it follows from Lemma 4 and (4) that the inequality

$$\begin{aligned}\omega_{\mathcal{P}_0}(n, s_k, k, 2m) &\geq \frac{\Phi(n, s)}{\binom{n}{n_0}\Phi(n_0, s)\Phi(n_1, s)\Phi(n_2, s)} \\ &= \frac{(n_0/s)!\,(n_1/s)!\,(n_2/s)!\,(n_1+n_2)!}{(n/s)!\,n_1!\,n_2!},\end{aligned}$$

holds for fixed values of n, k, t, m and for $s = s_k$, where n_0, n_1, and n_2 are defined by (10)–(12).

By using this and Stirling's formula, we obtain

$$\omega_{\mathcal{P}_0}(n, s_k, k, 2m) \geq \sqrt{\frac{2\pi n_0(n_1+n_2)}{ns^2}}\;\frac{n_0^{n_0/s} n_1^{n_1/s} n_2^{n_2/s}(n_1+n_2)^{n_1+n_2}}{n^{n/s} n_1^{n_1} n_2^{n_2} e^{s/(12n)+1/(12n_1)+1/(12n_2)}}. \tag{20}$$

We set

$$m = m_k, \quad t = m_k k, \tag{21}$$

where m_k is taken in (19).

We represent n_1 in the form

$$n_1 = an. \tag{22}$$

Using (10)–(12), we obtain

$$n - kan \leq n_0 \leq n - kan + ms, \tag{23}$$

$$(k-1)an - sm \leq n_2 \leq (k-1)an. \tag{24}$$

It follows from (10), (21), and (22) that

$$a \geq \binom{t-k}{m-k} \bigg/ \binom{t}{m} = \prod_{i=0}^{k-1} \frac{m-i}{t-i} > \frac{1}{k^k} \prod_{i=0}^{k-1} \left(1 - \frac{i}{m}\right)$$
$$> \frac{1}{k^k}\left(1 - \frac{k}{m}\right)^{k/2} > \frac{1}{k^k}\left(1 - \frac{k^2}{2m}\right). \tag{25}$$

Let n satisfy the conditions of Lemma 6. Then, combining (22), (10), Lemma 5, and (19), we obtain

$$a \leq \prod_{i=0}^{k-1} \frac{m-i}{mk-i} + \frac{s}{n} = \frac{1}{k^k} \prod_{i=0}^{k-1} \frac{1 - i/m}{1 - i/(km)} + \frac{s}{n}$$
$$< \frac{1}{k^k} \frac{1 - (k-1)/m}{1 - (k-1)/(km)} + \frac{k^2}{n} \leq \frac{1}{k^k} \frac{1 - (k-1)/m + k^{k+2}/n}{1 - (k-1)/(km)}$$
$$\leq \frac{1}{k^k} - \frac{(k-1)/m - k^{k+2}/n - (k-1)/(km)}{k^k(1 - (k-1)/(km))} \leq \frac{1}{k^k}. \tag{26}$$

If n satisfies the conditions of Lemma 6, then, from (22)–(24), Lemma 5, and (19), we obtain

$$\sqrt{\frac{2\pi n_0(n_1+n_2)}{ns^2}} \frac{1}{e^{s/(12n)+1/(12n_1)+1/(12n_2)}} \geq \sqrt{\frac{2\pi(n-kan)(kan-sm)}{nk^4 e}}$$
$$\geq \sqrt{\frac{(1-k/k^k)(kn/k^k - k^2 m)}{k^4}} > 1. \tag{27}$$

The function x^{bx} is monotonically increasing for $b > 0$, $x > 1/e$. Therefore, it follows from (20), (22)–(24), and (27) that

$$\omega_{\mathcal{P}_0} \geq \frac{(n-kan)^{(n-kan)/s}(na)^{na/s}((k-1)an)^{(k-1)an/s}(kan-ms)^{kan-ms}}{n^{n/s}(na)^{na}((k-1)an)^{(k-1)an}}$$
$$= \left(\frac{(1-ka)^{(1-ak)}a^{ak}(k-1)^{a(k-1)}k^{aks}}{(k-1)^{as(k-1)}}\right)^{n/s}\frac{(1-sm/(kan))^{kan-sm}}{n^{ms}(ka)^{ms}}. \quad (28)$$

If n satisfies the conditions of Lemma 6, then it follows from (13), (19), and (25) that

$$sm/(kan) < \frac{k^2k^3/\varepsilon_k}{kk^{-k}(1-k^2/(2m))3k^{k+5}/\varepsilon_k} \leq \frac{1}{3k(1-k^2/(2k^3))} < \frac{1}{4}$$

for $k \geq 2$. Therefore, for $k \geq 2$ we have

$$(1-sm/(kan))^{kan-sm}/(ka)^{ms} = \exp\Big(-(kan-sm)\sum_{i=1}^{\infty}\frac{1}{i}\left(\frac{sm}{kan}\right)^i\Big)/(ka)^{ms}$$
$$\geq \exp\Big(-(kan-sm)\Big(\frac{sm}{kan}+\left(\frac{sm}{kan}\right)^2\Big)\Big)/(ka^{ms})$$
$$> \frac{e^{-sm}}{(ka)^{ms}} \geq \left(\frac{k^{k-1}}{e}\right)^{ms} > \left(\frac{1}{2}\right)^{ms}.$$

From this fact and (28) we obtain

$$\omega_{\mathcal{P}_0} \geq (2n)^{-ms}\exp\{(n/s)f(n,k,a)\}, \quad (29)$$

where

$$f(n,k,a) = (1-ak)\log(1-ak) + ak\log a + a(k-1)\log(k-1)$$
$$+as(k\log k-(k-1)\log(k-1)).$$

From the equality $s = s_k^0 + \varepsilon_k$ and from (14) we obtain

$$f(n,k,a) = (1-ak)\log(1-ak) + ak\log a$$
$$+a(k^2\log k-(k^k-k)\log(1-k^{-k+1})) + \varepsilon_k(k\log k-(k-1)\log(k-1)). \quad (30)$$

From (19), (25), and (26) we have

$$
\begin{aligned}
&(1-ak)\log(1-ak) - a(k^k-k)\log(1-k^{-k+1})\\
&= (1-ak)\log\frac{1-ak}{1-k^{-k+1}} + (1-ak^k)\log(1-k^{-k+1})\\
&\geq (1-ak^k)\log(1-k^{-k+1}) \geq -(1-ak^k)2k^{-k+1}\\
&\geq -(k^2/m)2^{-k+1} \geq -(k^2/k^3)2^{-k+1} \geq -\varepsilon_k/4 \qquad (31)
\end{aligned}
$$

for $k \geq 2$. By analogy with (15), we obtain from (25) that

$$
\begin{aligned}
&ak\log a + ak^2\log k\\
&= ak\log(ak^k) \geq \frac{1}{k^{k-1}}\left(1-\frac{k^2}{2m}\right)\log\left(1-\frac{k^2}{2m}\right) \geq -\frac{k^2}{2mk^{k-1}} \geq -\frac{\varepsilon_k}{8}. \qquad (32)
\end{aligned}
$$

The function $x\log x - (x-1)\log(x-1)$ is monotonically increasing for $x > 1$. Therefore, for $k \geq 2$ we have

$$k\log k - (k-1)\log(k-1) \geq 2\log 2. \qquad (33)$$

Using (30)–(33), we obtain

$$f(n,k,a) \geq \varepsilon_k.$$

It follows from (29) and Lemma 5 that

$$\omega_{\mathcal{P}_0} \geq \frac{\exp(n\varepsilon_k/k^2)}{(2n)^{mk^2}}.$$

Lemma 6 is proved.

§ 5. Proof of the Main Result

Let $\lambda_{k,s}(f) = Bk(f)/Bs(f)$, $\lambda_{k,s}(n) = \max\lambda_{k,s}(g)$, where the maximum is taken over all Boolean functions g of n variables.

Theorem 1. *For the Boolean function F_{n,s_k} the relation*

$$Bk(F_{n,s_k}) \succeq \exp(\alpha_k n)$$

holds, where $\alpha_k = \varepsilon_k^2/(2k^5)$.

PROOF. It follows from (5) that

$$Bk(F_{n,s_k}) \geq UBk(F_{n,s_k})(2kN)^{-1}, \tag{34}$$

where N is the number of variables of the Boolean function F_{n,s_k} (this number is given by (1)). By Lemma 5, we have $s \leq k^2$. Hence,

$$N \leq n^{k^2}. \tag{35}$$

Let $\mathcal{P}_0$ be a uniform k-program $\mathcal{P}_0$ of size $UBk(F_{n,s_k})$ computing the Boolean function $F_{n,s}$. Then the number of sequences of $2m_k$ nodes in $\mathcal{P}_0$ does not exceed $UBk(F_{n,s_k})^{2m_k}$. It follows from this and Lemma 6 that

$$UBk(F_{n,s_k})^{2m_k} \geq \frac{\exp(n\varepsilon_k/k^2)}{(2n)^{mk^2}}. \tag{36}$$

From (34)–(36), we have

$$Bk(F_{n,s_k}) \geq \frac{e^{n\varepsilon_k/(k^2 m_k)}}{(2n)^{k^2} 2kn^{k^2}} \geq \frac{e^{n\varepsilon_k^2/(k^5)}}{(2n)^{2k^2}}. \tag{37}$$

Consequently,

$$Bk(F_{n,s_k}) \succeq \exp\,(\alpha_k n),$$

where $\alpha_k = \varepsilon_k^2/(2k^5)$. Theorem 1 is proved.

Theorem 2. *For any k, $k \geq 2$, there exists a positive constant β_k such that*

$$\lambda_{k,s_k}(N) \succeq \exp\,(\beta_k N^{1/k^2}).$$

PROOF. In § 3 it was shown (see (3)) that $Bs_k(F_{n,s_k}) \leq sN$. From this fact, (35), and Theorem 1 it follows that

$$\lambda_{k,s_k}(N) \geq Bk(F_{n,s_k})/Bs_k(F_{n,s_k}) \succeq \exp\,(\alpha_k N^{1/k^2})/(k^2 N) \succeq \exp\,(\beta_k N^{1/k^2}),$$

where $\beta_k = \varepsilon_k^2/(3k^5)$. Theorem 2 is proved.

Let $s_k^1 = \lceil s_k^0 \rceil + 1$. It is possible to prove

Theorem 3. *The relation*

$$\lambda_{k,s_k^1}(N) \succeq \exp\,(N^{1/k^2}/(2k^5))$$

holds.

The table below gives the values of s_k^0, s_k, and s_k^1 for small k.

k	2	3	4	5	6	7	8	9	10
s_k^0	3	5,93	10,16	15,86	23,10	31,92	42,29	54,36	67,82
s_k	4	6	11	16	24	32	43	55	68
s_k^1	4	7	12	17	25	33	44	56	69

Acknowledgments. The author is grateful to Professor A. D. Korshunov who helped in editing of the present paper and to the reviewer for a number of opportune remarks.

References

1. I. Wegener (1987) *The Complexity of Boolean Functions*, John Wiley and Sons, Ltd. Chichester; B. G. Teubner, Stuttgart.

2. V. A. Kuz'min (1976) An estimate of the complexity of the realization of functions of the algebra of logic by the simplest forms of binary programs (in Russian), *Metody Diskret. Anal.* **29**, 11–39.

3. A. A. Razborov (1991) Lower bounds for deterministic and nondeterministic branching programs, *Fundamentals of Computation Theory* (Gosen, 1991), Lecture Notes in Comput. Sci. Vol. 529, Springer-Verlag, Berlin, pp. 47–60.

4. D. Sieling and I. Wegener (1995) New lower bounds and hierarchy results from restricted branching programs, *Graph-theoretical Concepts in Computer Science*, Lecture Notes in Comput. Sci. Vol. 903, Springer-Verlag, Berlin etc., pp. 359–370.

5. A. B. Borodin, A. A. Razborov, and R. Smolensky (1993) On lower bounds for read-k-times branching programs, *Comput. Complexity* **3**, No. 1, 1–18.

6. E. A. Okol'nishnikova (1991) Lower bounds on complexity for the realization of characteristic functions of binary codes by binary programs (in Russian), *Metody Diskret. Anal.* **51**, 61–83 (see also: Siberian Adv. Math. **3** (1993), No. 1, 152–166).

7. S. Žak (1984) An exponential lower bound for one-time only branching programs, *Mathematical Foundations of Computer Science*, Lecture Notes in Comput. Sci. Vol. 199, Springer-Verlag, Berlin etc., pp. 562–566.

8. P. E. Dunne (1985) Lower bounds on the complexity of 1-time only branching programs (preliminary version), *Fundamentals of Computation Theory*, Lecture Notes in Comput. Sci. Vol. 199, Springer-Verlag, Berlin etc., pp. 90–95.

9. L. Babai, P. Hajnal, and E. Szemerédi, and G. Turán (1987) A lower bound for read-once-only branching programs. *J. Comput. System Sci.* **35**, No. 2, 153–162.

10. E. A. Okol′nishnikova (1988) On a relation for the complexity of Boolean functions, VIII *All-Union Conference on Problems of Theoretical Cybernetics. Abstract of Reports* (in Russian), Gor′kov. University, Gor′kiĭ, p. 63.

11. E. A. Okol′nishnikova (1979) Circuits of functional gates with minimally sufficient conjunctions (in Russian), *Metody Diskret. Anal.* **33**, 53–67.

12. E. A. Okol′nishnikova (1981) The effect of one type of constraint on the complexity of circuits made of functional elements (in Russian), *Metody Diskret. Anal.* **36**, 46–58.

13. V. A. Emelichev, O. I. Mel′nikov, V. I. Sarvanov, and R. I. Tyshkevich (1990) *Lectures on Graph Theory* (in Russian), Nauka, Moscow.

Sobolev Institute of Mathematics
Universitetskiĭ pr., 4
Novosibirsk 630090
RUSSIA

TRANSLATED BY THE AUTHOR

SOME OPTIMIZATION PROBLEMS OF SCHEDULING THE TRANSMISSION OF MESSAGES IN A LOCAL COMMUNICATION NETWORK

A. V. Pyatkin

The problem of constructing an optimal schedule of information transmission in a local communication network is reduced to the problem of finding a *mixed* coloring of the edges in a directed multigraph with the least possible number of colors. Two algorithms for the latter problem are presented. The term "mixed coloring" means that some of the edges can be splitted into two parts and be colored in two colors so that the number of the color of the first part does not exceed the number of the color of the second part.

1. Semantical Formulation of the Problem

A local network is given consisting of the host computer and n buses connecting it to peripheral objects. These objects can transmit information to each other. If two objects are on the same bus, then their communication takes place without the host computer. In this paper such communication is not considered. The objects that are on different buses may communicate via the host computer only. The total amount of information to be transmitted by all objects of the ith bus to the objects of the jth bus is denoted by d_{ij}. (From the above it follows that $d_{ii} = 0$ for each i.) We assume that the capacities of all buses are the same and they are equal to c information units per time unit. Then the time t_{ij} of transmission d_{ij} information units from the ith bus to the jth one is defined according to the formula

$$t_{ij} = \frac{d_{ij}}{c}.$$

We also assume that all t_{ij} are nonnegative integers. The communication of buses can be organized in the following ways:

(1) the host computer connects the ith bus with the jth one directly. In this case the information from the ith object is transmitted to the jth one (or, conversely from the jth to the ith) directly, i.e., without storing in the host computer.

(2) the information that the ith bus must transmit to the jth bus may be first received from the ith bus and stored in the host computer and then be transmitted via the jth bus.

A. D. Korshunov (ed.), *Operations Research and Discrete Analysis*, 227–232.

Restriction. Each bus in every moment of time can communicate with at most one object (by an object we mean a peripheral device or the host computer). This restriction does not concern the host computer which can receive or transmit information to several buses simultaneously (in the case of the second way of communication).

The whole process of communication is a session, i.e., the amount of the information "claimed" before the communication is to be transmitted (in other words, d_{ij} do not change during the communication). It is required to construct a schedule of information transmission between the buses such that the total period of the session is least possible.

2. Constructing Mathematical Models

Suppose that the second way of communication is forbidden; i.e., the buses may communicate only directly. In this case, independently of whether the object on the ith bus must transmit information to the object on the jth bus (or conversely), it is necessary to connect the ith and the jth buses. Thus, it is possible to introduce the concept of "communication demand of the ith and the jth buses"

$$p_{ij} \stackrel{\text{def}}{=} t_{ij} + t_{ji}.$$

Let us consider a n-vertex undirected multigraph $G = (V, E)$ with vertex set V and edge set E such that to ith bus there corresponds the ith vertex and the number of edges connecting the ith and the jth vertices is equal to p_{ij}.

In this case, the construction of a communication schedule of the buses means that with each edge $e \in E$ there must be assigned some time $f(e) \in \{1, \dots, T\}$ when the communication demand is realized (T is the total duration of the session); moreover, for all edges e_1 and e_2 incident to the same vertex the inequality $f(e_1) \neq f(e_2)$ must hold. It is required to minimize T.

It is clear that if the word "time" above is replaced with the word "color," then we obtain the classical problem of the edge coloring of a multigraph (the desired T is the chromatic index of the multigraph G). This problem is NP-complete [1]. Only some approximate algorithms for solving it are known, in particular, the Vizing algorithm [2–4].

Let us return to the original problem when both the first and the second ways of message transmission are possible. The second way of information transmission from the object on the ith bus to the object on the jth bus, is, as a rule, performed when at a fixed time the jth bus is busy and the ith one is free. Then the information on the ith bus is transmitted to the host computer and, later, when the jth bus becomes free, it is transmitted to the object on the jth bus. Thus, not only the "communication demand" but also the *direction* of the information transmission becomes essential.

Therefore, consider a *directed* multigraph $G = (V, E)$ with n vertices in which the number of edges leading from the ith vertex to the jth vertex is equal to $t_{ij}, 1 \leq i, j \leq n$. We turn from the scheduling problem to the problem of constructing an edge coloring as follows: let $e = (i, j) \in E$ be an edge of the graph and $f(e)$ is its color. Let us assume that $f(e) = a$ if at time a the buses i and j are connected directly, and $f(e) = (a, b)$ if at time a the information from the ith object is transmitted to the host computer and at time b it is transmitted from the host computer to the jth object. The notation $f(e) = (a, b)$ must be understood as follows: the edge e is splitted into two parts and the part that incident to the vertex i is colored with color a and the other part is colored with color b. We call this a *mixed* coloring.

Here, it is necessary to account the following. At time b, in order to transmit information from the host computer to the object on the jth bus, it is necessary to have loaded it to the host computer via the ith bus. Thus, the following admissibility condition must hold: if $e \in E$ and $f(e) = (a, b)$, then $a < b$.

We call a coloring satisfying this condition an *admissible* coloring.

Denote by s the maximal degree of the graph, and by p, the maximal number of edges in the multiedge, i.e.,

$$s = \max_i \sum_{j=1}^{n} (t_{ij} + t_{ji}),$$

$$p = \max_{i,j} (t_{ij} + t_{ji}).$$

It is obvious that every correct coloring of edges (i.e., such coloring when each vertex is at most one edge of each color incident to this vertex) requires at least s colors. The Vizing algorithm makes it possible to color correctly the edges of a multigraph with at most $s + p$ colors without using mixed colorings of edges. It is easy to construct a correct mixed coloring of the edges of a multigraph with s colors, although, the coloring may fail to be admissible. In fact, if we split each edge into two parts, then we correctly color the edges of the resulting bichromatic graph. According to König theorem [5], it requires at most s colors.

In the present paper we establish the following

Theorem. *In each directed multigraph G with degree s there exists a correct admissible mixed coloring of edges with s colors.*

Search for a constructive proof of this theorem was performed in two essentially different directions:

(1) first, a partial correct mixed coloring of the edges of a graph is constructed; next, the number of the colored edges is increased in consequence of recoloring the already-colored edges;

(2) first, a inadmissible correct mixed coloring of the edges of a graph with s colors is constructed; next, by means of some transformation not increasing the number of colors the coloring becomes admissible.

The proposed algorithms **A1** and **A2** prove the theorem.

3. Algorithm A1

Let G be a directed multigraph with degree s whose edges are partially colored with at most s colors. For an arbitrary vertex A of G denote by $M(A)$ the set of colors in the collection $\{1, \ldots, s\}$ which were not used in coloring the edges that are incident to the vertex A. Let $f(AB)$ be the color of an edge leading from A to B. We call all uncolored edges the *chosen* edges. The algorithm is proposed that permits us to color the chosen edges while preserving the correctness and the admissibility of the coloring, i.e., the algorithm that constructs a correct admissible coloring of the edges of a multigraph G with s colors.

The step of the algorithm consists in analyzing one of the chosen edges and executing one of the procedures described below.

Let AB be a chosen edge. It is clear that $M(A) \neq \varnothing$ and $M(B) \neq \varnothing$.

I. If $M(A) \cap M(B) \neq \varnothing$, then the edge AB is colored with the color a, where $a \in (M(A) \cap M(B))$. In addition, the correctness of the coloring is preserved.

II. If $M(A) \cap M(B) = \varnothing$, then the minimal color a in $M(A)$ and the maximal color b in $M(B)$ are found. If $a < b$ then we let $f(AB) = (a, b)$, and the so-obtained coloring of the edges remains admissible.

III. When $a > b$, a color-alternating (a, b)-chain with its source at the vertex B is constructed (this chain is not empty, since $a \notin M(B)$). Suppose that, while constructing the chain, no mixed-colored edge occurs. In this case the chain is finite, since it crosses itself, i.e., its length does not exceed the number of the vertices of the graph. If the chain terminates at a vertex different from A, then the edges of the color a in the chain being considered are recolored with the color b, and the edges of the color b are recolored with the color a that preserves the correctness of the coloring, and it is assumed that $f(AB) = a$. Since there are no edges with color a incident to the vertex A (by the definition of $M(A)$) and the edge with the color a incident to the vertex B is recolored with the color b, it follows that the correctness of the coloring in recoloring the edge AB is preserved.

IV. If a chain terminates at the vertex A, then the recoloring $a \leftrightarrow b$ is performed and it is assumed that $f(AB) = (b, a)$.

V. Suppose that, while constructing the chain in the case where $a > b$, a mixed-colored edge e occurs. Then the recoloring $a \leftrightarrow b$ of all edges of the chain with the exception of the edge e is performed, it is assumed that $f(AB) = a$, and the edge e is extracted (i.e., the edge e becomes uncolored). The algorithm stops, when there remain no chosen edges.

REMARK. It is easy to check that, while coloring and recoloring the edges, the correctness and the admissibility of a coloring of the graph is preserved. Show that algorithm **A1** is finite; i.e., it contains no infinite loops. Procedures **I** and **III** decrease the number of uncolored edges by 1; Procedures **II** and **IV** also decrease the number of uncolored edges by 1, moreover, they increase the number of mixed edges by 1. In the beginning there are not mixed edges; they can appear only as a result of executing **II** or **IV**. Procedure **V** decrease by 1 the number of mixed edges preserving the same the number of chosen edges. Since Procedure **V** is performed only in the case when in the graph there is at least one mixed edge, not later than in m steps it will be necessary to perform one of the Procedures **I–IV** causing the number of uncolored edges to decrease. Thus, the algorithm is finite. In the worst case it terminates in $2N - 1$ steps, where N is the number of edges of the multigraph. Moreover, at most $O(n^4p^2)$ elementary operations are used.

4. Algorithm A2

Assume that a directed multigraph G with n vertices has a correct, but inadmissible, coloring with s colors. Let us choose all edges whose coloring is inadmissible.

The step of the algorithm consists in analyzing some chosen edge and executing one of the following procedures.

I. Let AB be a chosen edge and $f(AB) = (a, b)$. Since the edge AB is inadmissible, $a > b$. A color-alternating (a, b)-chain with the source at the vertex B is constructed. Suppose that this chain contains no mixed edges. Then this chain can contain at most n links; i.e., the chain is finite (in particular, it can be empty). If the chain terminates not at the vertex A, then the recoloring $a \leftrightarrow b$ is performed and it is assumed that $f\ (AB) = a$.

II. If the chain terminates at A, then the recoloring $a \leftrightarrow b$ is performed and it is assumed that $f(AB) = (b, a)$. (The condition $a > b$ is used.)

III. Suppose that, while constructing the chain, a mixed edge e occurs. Then $f(e) = (a, c)$ or $f(e) = (b, c)$. Consider the case $f(e) = (a, c)$. Then $c \neq a$, since the edge e is mixed. The recoloring $a \leftrightarrow b$ is performed and it is assumed that $f(AB) = a$, $f(e) = (b, c)$. In addition, the edge e may transform its epithet from admissible into inadmissible (or conversely). In this case, the edge must be added to the "list" of chosen edges.

The case $f(e) = (b, c)$, $c \neq b$ is considered in a similar way.

The algorithm stops when there remain no chosen edges.

Show that algorithm **A2** is finite. Let k be the number of chosen edges, and let N be the total number of mixed edges before the start of the algorithm. We may assume that $k \leq N/2$ and $N \leq n(n-1)p/2$.

Procedure **I** turns one inadmissible edge into a usual one, Procedure **II** transforms one inadmissible edge into an admissible one, and Procedure **III** transforms one mixed edge into a usual one. Thus, the number of steps of the algorithm does not exceed $k + N$, i.e., it does not exceed $3n(n-1)p/4$. In addition, at most $O(n^4p^2)$ elementary operations are used.

References

1. I. Holyer (1981) The NP-completeness of edge-coloring, *SIAM J. Comput.* **10**, No. 4, 718–720.
2. V. G. Vizing (1964) On an estimate for the chromatic class of a p-graph (in Russian), *Diskret. Anal.* **3**, 25–30.
3. V. G. Vizing (1965) Critical graphs of a given chromatic class (in Russian), *Diskret. Anal.* **5**, 9–17.
4. V. G. Vizing (1965) The chromatic class of a multigraph (in Russian), *Kibernetika* (Kiev) No. 3, 29–39.
5. A. A. Zykov (1987) *Fundamentals of Graph Theory* (in Russian), Nauka, Moscow.

Novosibirsk State University
ul. Pirogova, 2
Novosibirsk 630090
RUSSIA

TRANSLATED BY N. V. SHILOV

THE MAXIMUM-WEIGHT TRAVELING SALESMAN PROBLEM IN FINITE-DIMENSIONAL REAL SPACES[†)]

A. I. Serdyukov

The paper is devoted to further developing the approximation algorithms and characterizing their performance for certain classes of the maximum-weight traveling salesman problem. A polynomial algorithm is proposed that solves the problem with any prescribed accuracy for sufficiently large n in finite-dimensional real spaces with a Minkowski metric. The problem is shown to be polynomially solvable in two-dimensional polyhedral spaces whose unit ball is a triangle.

Introduce the following notation:

- B is a bounded closed convex set in $\mathbb{R}^k$ containing 0 in its interior;
- $f : \mathbb{R}^k \to \mathbb{R}^+$ is the Minkowski functional of the set B; i.e., $f(x) = \inf\{\lambda \in \mathbb{R} \mid x/\lambda \in B, \lambda > 0\}$;
- $\mu(x,y) = f(y-x)$ is the distance from x to y for an ordered pair $(x,y) \in \mathbb{R}^k \times \mathbb{R}^k$;
- S_n is the symmetric permutation group of degree n;
- $S_n(l) = \{s \in S_n \mid \text{the number of cycles in } s \text{ does not exceed } l\}$, $1 \le l \le n$;
- Y_n is the set of all sequences of n distinct points in $\mathbb{R}^k$;
- $M_n = M_n(X_n) = (\mu_{ij})_{n\times n}$ is the matrix of pairwise distances between the elements of a sequence $X_n = \{x_1, x_2, \dots, x_n\} \in Y_n$;
- $\mathcal{M}_n = \{M_n = M_n(X_n) \mid X_n \in Y_n\}$;
- $h(\overline{S}_n, M_n) = \max\limits_{s\in\overline{S}_n} \sum\limits_{i=1}^{n} \mu_{is(i)}$, where $\overline{S}_n \subseteq S_n$ and $M_n \in \mathcal{M}_n$;
- $\arg h(\overline{S}_n, M_n) = \{s \in \overline{S}_n \mid h(s, M_n) = h(\overline{S}_n, M_n)\}$, where $\overline{S}_n \subseteq S_n$ and $M_n \in \mathcal{M}_n$.

By the *maximum-weight traveling salesman problem* (MTSP) in $\mathbb{R}^k$ we mean the problem of finding a permutation

$$s = s(M_n) \in \arg h(S_n(1), M_n),$$

†) This work was supported by the Russian Foundation for Basic Research (Grant 93–01–00417).

A. D. Korshunov (ed.), Operations Research and Discrete Analysis, 233–239.

where $M_n \in \mathcal{M}_n$.

At present, we are not aware of any polynomial algorithms for solving the MTSP in $\mathbb{R}^k$ for $k = 2$ even for simple B (a circle, a square, etc.). In the present paper, we consider the approximation approach to solving the MTSP in $\mathbb{R}^k$ and provide estimates for certain performance characteristics (running time, error) of the suggested algorithms.

Let $\mathcal{A}$ be an algorithm for finding a permutation $s_{\mathcal{A}}(M_n) \in S_n(1)$, $n \in N$, $M_n \in \mathcal{M}_n$. For each n, we define the *error* of $\mathcal{A}$ as

$$\varepsilon_{\mathcal{A}}(n) = 1 - \inf\{h(s_{\mathcal{A}}(M_n), M_n)/h(S_n(1), M_n) \mid M_n \in \mathcal{M}_n\}.$$

Observe that the entries of the matrix $M_n \in \mathcal{M}_n$ satisfy the triangle inequality. Therefore, [1] implies

Theorem 1. *There exists an algorithm for solving the* MTSP *in* $\mathbb{R}^k$ *with running time* $O(n^3)$ *and error* $\varepsilon(n) \leq 1/4$.

This algorithm is designed for solving the MTSP on a class of matrices that is broader than $\mathcal{M}_n$ and does not take into account the specifics of the problem in question (the connection of the matrix M_n with the algebraic structure of $\mathbb{R}^k$ or the dependence of M_n on the set B).

Consider some typical spaces.

A pair $(\mathbb{R}^k, \mu)$ is called

— the *Euclidean space* if

$$B = \left\{x = (x^1, \ldots, x^k) \in \mathbb{R}^k \mid \left(\sum_{i=1}^{k}(x^i)^2\right)^{1/2} \leq 1\right\};$$

— a *Minkowski space* if B is a set centrally symmetric with respect to 0 (f is the norm);

— a *polyhedral space* if B is a polyhedron.

1. Euclidean spaces. We have the following

Theorem 2 [2]. *There exists an algorithm for solving the* MTSP *in* $\mathbb{R}^k$ *with running time* $O(n^3)$ *and error*

$$\varepsilon(n) \leq C_k/n^{\frac{2}{k+1}},$$

where C_k *depends on the dimension of the space only.*

2. Minkowski spaces. We have the following

Theorem 3 [3]. *If* $n \to \infty$, *then*

$$\inf\{h(S_n(1), M_n)/h(S_n, M_n) \mid M_n \in \mathcal{M}_n\} \to 1.$$

Theorem 4 [4]. *In* $\mathbb{R}^2$, *for each even* n,

$$h(S_n, M_n) = h(T_n, M_n),$$

where $T_n = \{s \in S_n \mid$ *each cycle in* s *contains exactly two elements*$\}$.

An approximation algorithm for solving the MTSP is presented in Subsection 4.

3. Polyhedral spaces. In this subsection, we develop the approach that was suggested in [3] for approximate solving the MTSP in Minkowski spaces which are also polyhedral.

Let B contain exactly m hyperfaces $\Gamma_1, \Gamma_2, \dots, \Gamma_m$. For each i, $1 \le i \le m$, denote by K_i the minimal cone with vertex the origin 0 which includes Γ_i. The next lemma follows from the convexity of B and the definition of f.

Lemma 1. *Let* $\tilde{\pi}$ *be a hyperplane parallel to some hyperface* Γ_i, $1 \le i \le m$. *Then, for any three points* $x \in \mathbb{R}^k$, $y \in \tilde{\pi}$, *and* $z \in \tilde{\pi}$,

$$\mu(x, z) \ge \mu(x, y) \text{ if } (y - x) \in K_i,$$
$$\mu(z, x) \ge \mu(y, x) \text{ if } (x - y) \in K_i.$$

Moreover, we have

Lemma 2. *For any four points* x, $\overline{x}$, y, *and* $\overline{y}$ *in* $\mathbb{R}^k$ *such that* $(\overline{x} - x) \in K_i$ *and* $(\overline{y} - y) \in K_i$ *for some* i, $1 \le i \le m$,

$$\mu(x, \overline{x}) + \mu(y, \overline{y}) \le \mu(x, \overline{y}) + \mu(y, \overline{x}).$$

PROOF. Without loss of generality, we assume that the points x, $\overline{x}$, y, and $\overline{y}$ belong to the cone K_i and $f(x) \le f(y)$. Three cases are possible:

CASE 1: $f(x) \le f(\overline{x}) \le f(y) \le f(\overline{y})$.

CASE 2: $f(x) \le f(y) \le f(\overline{x}) \le f(\overline{y})$.

CASE 3: $f(x) \le f(y) \le f(\overline{y}) \le f(\overline{x})$.

In Case 1, draw the hyperplanes through $\overline{x}$ and y in parallel to Γ_i. Let x_1 and y_1 be the points of intersection of the interval $[x, \overline{y}]$ with these hyperplanes. Then, by Lemma 1,

$$\mu(x, \overline{x}) \le \mu(x, x_1), \quad \mu(y, \overline{y}) \le \mu(y_1, \overline{y})$$

so that

$$\mu(x,\overline{x}) + \mu(y,\overline{y}) \le \mu(x,\overline{y}) \le \mu(x,\overline{y}) + \mu(y,\overline{x}),$$

as desired.

In Cases 2 and 3, draw the hyperplane through y in parallel to Γ_i. Let x_1 and y_1 be the points of intersections of the intervals $[x,\overline{x}]$ and $[x,\overline{y}]$ with this hyperplane. Then, by Lemma 1,

$$\begin{aligned}\mu(x,\overline{x}) + \mu(y,\overline{y}) &= \mu(x,x_1) + \mu(x_1,\overline{x}) + \mu(y,\overline{y})\\ &\le \mu(x,y_1) + \mu(y,\overline{x}) + \mu(y_1,\overline{y}) = \mu(x,\overline{y}) + \mu(y,\overline{x}).\end{aligned}$$

Lemma 2 is proved.

Lemma 3. *For each $n \in N$ and $M_n \in \mathcal{M}_n$,*

$$\arg h(S_n(\lfloor m/2 \rfloor), M_n) \subset \arg h(S_n, M_n).$$

There exists an algorithm that, in $O(n^3)$ time, finds a permutation on the left-hand side of the above inclusion.

PROOF. Denote by M_n the matrix of the pairwise distances between the points of the set $X_n = \{x_1, x_2, \dots, x_n\}$ in $\mathbb{R}^k$. It is known (see, for example, [5]) that there exist algorithms finding a permutation in $s \in \arg h(S_n, M_n)$ in $O(n^3)$ time. Clearly, s contains no loops since all the points in X_n are distinct.

Let $s = (c_1, c_2, \dots, c_t)$ be the representation of the permutation s in the form of cycles c_i, $1 \le i \le t$. Inspecting the pairs of elements belonging to different cycles of s, find the elements $\overline{n} \in c_v$ and $\widetilde{n} \in c_w$, $v \ne w$, $1 \le v, w \le t$, such that

$$\mu_{\overline{n}s(\overline{n})} + \mu_{\widetilde{n}s(\widetilde{n})} \le \mu_{\overline{n}s(\widetilde{n})} + \mu_{\widetilde{n}s(\overline{n})}.$$

Construct the permutation $\overline{s} \in \arg h(S_n, M_n) \cap \arg h(S_n(t-1), M_n)$ as follows:

$$\overline{s}(i) = \begin{cases} s(i) & \text{for } i \notin \{\overline{n}, \widetilde{n}\},\\ s(\widetilde{n}) & \text{for } i = \overline{n},\\ s(\overline{n}) & \text{for } i = \widetilde{n}.\end{cases} \tag{1}$$

Take $\overline{s}$ instead of s and repeat the above process.

Let $\widehat{s} = (\widehat{c}_1, \widehat{c}_2, \dots, \widehat{c}_q) \in \arg h(S_n, M_n) \cap \arg h(S_n(q), M_n)$ be a terminate permutation resulting from the repetition of the process. We now show that $q \le \lfloor m/2 \rfloor$. Assume the contrary. Observe that, for each cycle in $\widehat{s}$, the set of points $\{x_{\hat{s}(j)} - x_j\}$ intersects at least two cones in the set $\{K_1, \dots, K_m\}$. By assumption, there exist two elements $\overline{n}$ and $\widetilde{n}$ in different cycles of $\widehat{s}$ such that $x_{\overline{n}}$, $x_{s(\overline{n})}$, $x_{\tilde{n}}$, and $x_{s(\tilde{n})}$ satisfy the requirements of Lemma 2. But this contradicts the fact that $\widehat{s}$ is terminate permutation. Note that the constructing of $\widehat{s}$ is done in $O(n^3)$ time. The permutation $\widehat{s}$ satisfies all the requirements of the lemma. Lemma 3 is proved.

Lemma 3 implies

Theorem 5. *For* $m = 3$, *the* MTSP *is solvable in* $\mathbb{R}^2$ *in* $O(n^3)$ *time.*

Since the elements of the matrix M_n satisfy the triangle inequality, we have

Lemma 4 [6]. *Let* $s \in S_n(p)$ *and* $p \le n$. *Then there exists an algorithm finding in* $O(n \log n)$ *time a permutation* $s_0 \in S_n(1)$ *such that*

$$h(s_0, M_n) \ge (1 - 1/n)^{p-1} h(s, M_n).$$

Lemmas 3 and 4 imply

Theorem 6. *There exists an algorithm solving the* MTSP *in* $O(n^3)$ *time with the error*

$$\varepsilon(n) \le (\lfloor m/2 \rfloor - 1)/n .$$

4. An approximation algorithm for the MTSP in $\mathbb{R}^k$**.** Let $\delta > 0$. Following [3] (where the case of Minkowski spaces is considered), it is easy to show that there exists a polyhedron B_1 with a finite number of hyperfaces which circumscribes B_1 and satisfies $(1 - \delta) f(x) \le f_1(x)$ for each $x \in \mathbb{R}^k$, where f_1 is the Minkowski functional for B_1. Let B_1 have the least number $m = m(\delta)$ of hyperfaces among such polyhedra.

Lemma 5. *There exists an algorithm finding in* $O(n^3)$ *time a permutation* $s' \in S_n(m)$ *such that*

$$h(s', M_n) \ge (1 - \delta) h(S_n, M_n).$$

PROOF. Let $X_n = \{x_1, x_2, \dots, x_n\}$ be a set of points in $\mathbb{R}^k$, and let M_n be the matrix of the pairwise distances between the points. Using the algorithm for solving the assignment problem [5], find a permutation $s = (c_1, c_2, \dots, c_t) \in \arg h(S_n, M_n)$ in $O(n^3)$ time.

In each cycle c_i, choose a unforbidden element n_i, $1 \le i \le t$ (initially, all the elements are unforbidden). Among $\{n_i\}$, choose a pair $\{\overline{n}, \widetilde{n}\}$ such that

$$\mu_{\overline{n}s(\widetilde{n})} + \mu_{\widetilde{n}s(\overline{n})} \ge (1 - \delta)(\mu_{\overline{n}s(\overline{n})} + \mu_{\widetilde{n}s(\widetilde{n})}).$$

Construct the permutation $\overline{s} \in S_n(t - 1)$ of the form (1). For this permutation, implement the process that is similar to as described above with the only difference being that the elements $\overline{n}$ and $\widetilde{n}$ become forbidden at the subsequent steps. Continue this process of reconstructing the current permutation as long as possible.

Let $\widehat{s} = (\widehat{c}_1, \widehat{c}_2, \dots, \widehat{c}_q)$ be the resulting terminate permutation. Clearly, $h(\widehat{s}, M_n) \ge (1 - \delta) h(S_n, M_n)$ because, at each step, a pair of elements in $\{\mu_{is(i)}\}$ was replaced with new elements of the matrix M_n. We show that $q \le m$. Suppose the contrary. Then there exist elements $\overline{n}$ and $\widetilde{n}$ among $n_1, \dots, n_q$ such that

$(x_{s(\overline{n})} - x_{\overline{n}})$ and $(x_{s(\tilde{n})} - x_{\tilde{n}})$ belong to the same cone of the cones generated by the hyperfaces of the polyhedron B_1. By Lemma 2,

$$\mu_1(x_{\overline{n}}, x_{s(\overline{n})}) + \mu_1(x_{\tilde{n}}, x_{s(\tilde{n})}) \leq \mu_1(x_{\overline{n}}, x_{s(\tilde{n})}) + \mu_1(x_{\tilde{n}}, x_{s(\overline{n})}),$$

where $\mu_1(x, y) = f_1(y - x)$, $x \in \mathbb{R}^k$, $y \in \mathbb{R}^k$. Consequently,

$$(1 - \delta)(\mu_{\overline{n}s(\overline{n})} + \mu_{\tilde{n}s(\tilde{n})}) \leq \mu_{\overline{n}s(\tilde{n})} + \mu_{\tilde{n}s(\overline{n})}.$$

But this means that the $\widehat{s}$ is not a terminate permutation, a contradiction. This algorithm runs in $O(n^3)$ time. We may now set $s' = \widehat{s}$. Lemma 5 is proved.

Theorem 7. *For each $\delta > 0$, there exists an algorithm solving the* MTSP *in $O(n^3)$ time with error*

$$\varepsilon(n) = \delta + O(1/n).$$

PROOF. Using Lemmas 4 and 5, find a permutation $s_0 \in S_n(1)$ such that

$$\begin{aligned} &\inf\{h(s_0, M_n)/h(S_n(1), M_n) | M_n \in \mathcal{M}_n\} \\ &\geq \inf\{h(s_0, M_n)/h(S_n, M_n) | M_n \in \mathcal{M}_n\} \\ &\geq (1 - 1/n)^{q-1}(1 - \delta) \geq (1 - q/n)(1 - \delta). \end{aligned}$$

Consequently, $\varepsilon(n) \leq 1 - (1 - q/n)(1 - \delta) = \delta + O(1/n)$. The proof is complete.

References

1. A. V. Kostochka and A. I. Serdyukov (1985) Polynomial algorithms with the estimates 3/4 and 5/6 for the maximum-weight traveling salesman problem (in Russian), *Upravlyaemye Sistemy* **26**, 55–59.
2. A. I. Serdyukov (1986) An asymptotically exact algorithm for the maximum-weight traveling salesman problem in Euclidean spaces (in Russian), *Upravlyaemye Sistemy* **27**, 79–87.
3. A. I. Serdyukov (1991) Asymptotic properties of optimal solutions of extremal permutation problems in finite-dimensional normed spaces (in Russian), *Metody Diskret. Anal.* **51**, 105–111.
4. A. I. Serdyukov (1990) Extremal problems on permutations for special classes of matrices (in Russian), *Upravlyaemye Sistemy* **30**, 57–60.
5. E. A. Dinits and M. A. Kronrod (1969) An algorithm for solving the assignment problem (in Russian), *Dokl. Akad. Nauk SSSR* **189**, No. 1, 23–25.

6. A. I. Serdyukov (1991) Polynomial algorithms with estimates for the performance accuracy for a class of the maximum-weight traveling salesman problems (in Russian), in: *Combinatorial-Algebraic Methods* in *Discrete Optimization*, Nizhniĭ Novgorod University, Nizhniĭ Novgorod.

Sobolev Institute of Mathematics
Universitetskiĭ pr., 4
Novosibirsk 630090
RUSSIA

TRANSLATED BY S. I. SUSLOV

NONSTRICT VECTOR SUMMATION IN THE PLANE AND ITS APPLICATIONS TO SCHEDULING PROBLEMS[†)]

S. V. Sevast'yanov

Given a norm in the plane, we consider finite families of vectors with zero sum and each vector having at most unit length. The following question is addressed: Given such family of vectors in the plane, when does there exist a permutation of the vectors such that for any two successive partial sums of the vectors specified by the permutation at least one of them belongs to a given domain G? (Such a summation is called "nonstrict.") It is shown that under certain relations between the domain G and the unit ball of the norm there exists a permutation of the vectors which provides nonstrict summation of the vectors within G, and this permutation can be found in polynomial time. Application of this theorem to four scheduling problems with n jobs, m machines, and the minimum makespan criterion enables us: to extend the subclass of instances of the open-shop problem ($m = 3$) solvable in polynomial time; to construct polynomial-time approximation algorithms with currently best performance guarantees for the Akers–Friedman problem ($m = 3$), for its special case with two routes ($m = 3$), and for the flow-shop problem with four machines ($m = 4$).

Introduction

In the present paper, we proceed with investigating the application of the nonstrict vector summation (NVS) to scheduling problems which was initiated in [1]. Recall that in [1] the following three scheduling problems with n jobs and m machines were considered. In the first problem (known as the *flow-shop* problem) all jobs pass through machines in the same order $(1, 2, \dots, m)$. In the second problem (known as the *assembly line* problem) for each job but one specific of all its operations can be processed independently in time; the specific ("assembly") operation can be done only after all other operations of the job are finished.

Finally, in the third problem stated for three machines, the set of jobs is divided into two groups such that the jobs of the first group pass through machines in the order $(1, 2, 3)$, whereas for the jobs of the second group this order is $(2, 1, 3)$. (This problem is denoted here by R213.) It was shown that the above problems reduces to three different NVS-problems in $m - 1$-dimensional space so that, given an approximate solution of the NVS-problem, we can find an approximate solution of the

†) The present research was supported by the Russian Foundation for Basic Research (Grant 93–01–00489) and the International Science Foundation (Grant NQC000).

A. D. Korshunov (ed.), Operations Research and Discrete Analysis, 241–272.

corresponding scheduling problem. In the case of plane, a unified polynomial-time approximation algorithm is presented for the three NVS-problems with performance guarantees. The algorithm is based on the fact (proved in [1]) that, given an unbounded closed convex domain $G \subset \mathbb{R}^2$, any s-family of vectors can be nonstrictly summed within G provided that each O-chord of G has length (in norm s) at least 1.

Application of the above algorithms to the three scheduling problems mentioned above with three machines gives us a polynomial-time approximation algorithm to compute schedules with upper bounds on the deviation from the optimum guaranteed; two of the three bounds proven best possible.

In the present paper, we follow a similar scheme of presenting results. First, in § 2 sufficient conditions are established guaranteeing that, given an s-family of vectors in $\mathbb{R}^2$, a permutation of vectors can be found that provides their nonstrict summation within a domain $G \subset \mathbb{R}^2$. Application of this result to three NVS-problems in the plane gives a polynomial-time approximation algorithm to their solution with performance guarantees.

Next, in § 3 and § 4 we consider four scheduling problems with n jobs, m machines, and the minimum makespan criterion: the open-shop (OS(m)), the flow-shop (FS(m)) for $m = 3$, 4, the Akers–Friedman problem (AF(m)) for $m = 3$, and finally, its special case for $m = 3$ with two routes. (The latter is called the *cyclically closed route problem* and is denoted by R231.) A connection between these scheduling problems and NVS-problems in $(m-1)$- or $(m-2)$-dimensional space is established.

Finally, application of the algorithms of § 2 to solution of the NVS-problems to four scheduling problems enables us to enlarge the subclass of instances of the OS(3) problem solvable in polynomial time and to construct scheduling algorithms with better performance guarantees for the FS(4), AF(3), and R231 problems. Another consequence of the results obtained is a linear-time algorithm for the FS(3) problem with the best possible bound.

§ 1. Preliminaries

We start with stating four scheduling problems with the minimum makespan criterion and introduce some notions and notation. Later we formulate the results that are obtained for these problems and compare them with the known results. It is convenient to formulate our scheduling problems as special cases of the following general problem.

G-Problem. *Let* $\mathcal{M} = \{1, 2, \dots, m\}$ *be a set of machines, and let* $\mathcal{N} = \{1, 2, \dots, n\}$ *be a set of jobs. Each job* j *consists of* m *operations* $\{o_{j1}, \dots, o_{jm}\} = O^j$*, an operation* o_{ji} *being processed by machine* i *and requires* t_{ji} *time units,*

$1 \le i \le m$. *For each job $j \in \mathcal{N}$, a precedence relation on the set O^j is defined by a directed graph $G_j = (O^j, U_j)$, where the inclusion $(o_{ji}, o_{jk}) \in U_j$ means that operation o_{jk} cannot start until o_{ji} is finished. (We will write this also as $o_{ji} \to o_{jk}$.) The problem is to find a schedule $S = \{s_{ji} \ge 0 \mid j \in \mathcal{N},\, i \in \mathcal{M}\}$ (i.e., to assign to each operation o_{ji} a starting time s_{ji} of processing) which minimizes the maximum finish time over all operations:*

$$T(S) \doteq \max_{j,i}(s_{ji} + t_{ji})$$

(the last quantity is customarily referred to as "schedule length" or "makespan") and meets the following requirements:

- *No preemption in processing each operation is allowed, which means that once started at time s_{ji} operation o_{ji} has to be processed during the time interval $I_{ji} \doteq (s_{ji}, s_{ji} + t_{ji})$ until completion;*
- *at most one operation of each job can be processed at each time, i.e.,*

$$I_{ji} \cap I_{jk} = \varnothing \quad \text{for any } j \in \mathcal{N};\ i,\, k \in \mathcal{M},\ i \neq k;$$

- *each machine can process at most one operation at each time, i.e.,*

$$I_{ji} \cap I_{ki} = \varnothing \quad \text{for any } i \in \mathcal{M};\ j,\, k \in \mathcal{N},\ j \neq k. \tag{1}$$

In the framework of this general G-problem, the OS-, FS-, AF-, and R231-problems differ from one another only by sets U_j specifying different partial orders on the sets O^j.

In the OS-problem, we have $U_j = \varnothing$ for every $j \in \mathcal{N}$, which means that no precedence relation is prescribed on the set of operation of each job.

In the FS-problem, the graph G_j is a chain $o_{j1} \to o_{j2} \to \cdots \to o_{jm}$, $j \in \mathcal{N}$. Thus, for each job $j \in \mathcal{N}$, a machine passage route is given and all those routes are identical, namely $(1, 2, \ldots, m)$.

In the AF-problem, for every job $j \in \mathcal{N}$ we also prescribe a machine passage route $\left(i_1^j, \ldots, i_m^j\right)$ such that each machine occurs in the route exactly once, and different jobs may have different routes. (There may be at most $m!$ different routes simultaneously.)

Finally, the R231-problem is a special case of the AF-problem when there are only three machines and only two different routes: $(1,2,3)$ for the jobs in a set $\mathcal{N}_1$ and $(2,3,1)$ for the remaining jobs $j \in \mathcal{N}_2$ ($\mathcal{N} = \mathcal{N}_1 \cup \mathcal{N}_2$).

Put $M_i = \sum_{j=1}^n t_{ji}$, $M = \max_i M_i$, and $K = \max_{j,i} t_{ji}$, and let S_{opt} denote an optimal schedule. In [2], an interesting property of optimal schedules of the OS-problem was established. (We will refer to it as to the *Existence of an*

Idleless Optimal Schedule, or briefly, the EIOS-*property*.) At the same time, a fairly wide subclass of instances of the OS-problem solvable in polynomial time was described. Both the EIOS property and the subclass can be described in terms of the parameters M and K. Namely, it was shown that if

$$M \geq \eta(m)K \tag{2}$$

holds for a certain function $\eta(m)$ of the number of machines m, then

$$T(S_{\text{opt}}) = M, \tag{3}$$

i.e., in any optimal schedule the most loaded machine works without any idle time during the whole makespan, and the optimal schedule can be constructed in polynomial time.

It is natural to address the question of finding the least function $\eta(m) = \eta^*(m)$ such that (2) implies (3). It follows from the Gonzalez–Sahni algorithm [3] that $\eta^*(2) = 2$. In [4–6], upper bounds for $\eta^*(m)$ for arbitrary m were found improving the bound in [2]. In [6], a lower bound for $\eta^*(m)$ was also established. This bounds for $\eta^*(m)$ for $m \geq 3$ can be written as follows:

$$2m - 2 \leq \eta^*(m) \leq \min\left\{m^2 - 1 + \frac{1}{m-1}, \frac{16}{3}m\log_2 m + \frac{61}{9}m - 7.4\right\}.$$

In particular, for $m = 3$ we obtain

$$4 \leq \eta^*(3) \leq 8.5. \tag{4}$$

Using the NVS technique in § 2, we improve the upper bound for $\eta^*(3)$ to

$$\eta^*(3) \leq 7.$$

Another important property of the job-shop-type and open-shop-type problems is the so-called *optima localization* property. By this property, the optimum of any instance of problem X belongs to an interval $I_X = [M,\ M + \mu_X(m)K]$, where $\mu_X(m)$ is a certain function of the number of machines. Denote by $\mu_X^*(m)$ the exact upper bound for the set of values of $(T(S_{\text{opt}}) - M)/K$ over all instances of problem X with m machines. The similar function localizing the permutation optima of the FS-problem is denoted by $\tilde{\mu}(m)$. The minimal interval I_X specified by the function μ_X^* is denoted by I_X^*. In [6] and [1], the following relations were established:

$$\begin{gathered} m - 1 + \left\lfloor \frac{m-1}{2} \right\rfloor \leq \tilde{\mu}(m) \leq m^2 - 3m + 3 + \frac{1}{m-2}, \\ \tilde{\mu}(2) = \mu_{\text{FS}}^*(2) = 1, \quad \tilde{\mu}(3) = \mu_{\text{FS}}^*(3) = 3. \end{gathered} \tag{5}$$

For $m = 4$, from (5) we obtain

$$4 \leq \tilde{\mu}(4) \leq 7.5.$$

In § 4 we describe a reduction of the FS(m)-problem to an NVS-problem in $(m-2)$-dimensional space. This reduction enables us both to improve the upper bound for $\tilde{\mu}(4)$ ($\tilde{\mu}(4) \leq 6$) and to construct an approximation algorithm for the FS(3)-problem which computes a permutation schedule S_π with the best possible bound ($T(S_\pi) \leq M + 3K$) in linear time. Finally, a reduction of the AF(3)- and R231-problems to NVS-problems in the plane and application of the corollaries of Theorem 1 in § 2 enables us to derive bounds

$$\mu^*_{R231} \leq 5, \quad \mu^*_{\mathrm{AF}}(3) \leq 5.5.$$

(The latter improves the bound $\mu^*_{\mathrm{AF}}(3) \leq 6$ obtained in [7].) Since the best known lower bound for both functions is $\mu^*_{\mathrm{X}}(3) \geq 3$ ($X \in \{\text{R231, AF}\}$), the problem of finding the exact values of $\mu^*_{\mathrm{AF}}(3)$ and μ^*_{R231} remains open.

Note that finding an improved upper bound for the function μ^*_{X} for each X-problem, X $\in$ {FS, AF, R231}, considered in § 4 is accompanied by constructing an efficient approximation algorithm which computes a schedule with length in the interval I_{X}. This provides an upper bound on the deviation of the length of the schedule found from the optimum independent of the number of jobs.

§ 2. Nonstrict Vector Summation Within a Given Domain in the Plane

We first introduce some notion and definitions.

Let $\mathbb{R}^m$ be the m-dimensional vector space over the field of real numbers, and let $\Sigma(X)$ be the sum of vectors of a given finite family $X \subset \mathbb{R}^m$.

A permutation π of the numbers $1, 2, \ldots, n$ will be simply referred to as *permutation* π.

DEFINITION 1. Let s be a norm in $\mathbb{R}^m$. A finite family of vectors $X = \{x_1, \ldots, x_n\} \subset \mathbb{R}^m$ is called an *s-family* if

$$\Sigma(X) = 0 \tag{6}$$

and the norm of each vector is at most 1.

DEFINITION 2. Let G be a domain in $\mathbb{R}^m$, and let $X = \{x_1, \ldots, x_n\}$ be a family of vectors with zero sum. We say that a permutation $\pi = (\pi_1, \ldots, \pi_n)$ provides a *nonstrict summation of vectors in X within domain G* if, for any $k = 1, \ldots, n$, the relation $x_\pi^{k-1} \notin G$ implies $x_\pi^k \in G$, where $x_\pi^k \doteq \sum_{j=1}^k x_{\pi_j}$.

We say that a permutation π provides a *nonstrict summation of vectors* in $X \subset \mathbb{R}^m$ within a family of domains $\mathcal{P} = \{G_i \mid i = 1, \dots, \nu\}$ and denote this by $(X, \pi) \in S_m\mathcal{P}$ if π provides a nonstrict summation within each domain $G_i \in \mathcal{P}$, $i = 1, \dots, \nu$.

For every family of domains $G_i \subset \mathbb{R}^m$ the following remark is true.

REMARK 1. If a permutation π provides a nonstrict summation of vectors in X within the intersection of domains $G = \bigcap_{i=1}^k G_i$, then it provides a nonstrict summation of these vectors within the family of domains $\{G_i \mid i = 1, \dots, k\}$.

For any vector $a = (a(1), \dots, a(m)) \in \mathbb{R}^m$, $a \neq 0$, and a number $\beta \in R$ we define a half-space $P(a, \beta) = \{x \in \mathbb{R}^m \mid (a, x) \leq \beta\}$, where

$$(a, x) = \sum_{i=1}^m a(i)x(i).$$

Let e_i denote the ith basis vector in $\mathbb{R}^m$. We define some norm $\hat{s}$ in $\mathbb{R}^m$ by its unit ball $\widehat{H}_m$ with center the origin by the formula

$$\widehat{H}_m = \{x \in \mathbb{R}^m \mid |x(i_1)| \leq 1,\ |x(i_1) - x(i_2)| \leq 1 \text{ for all } i_1,\ i_2 = 1, \dots, m\}.$$

The ball $\widehat{H}_m$ can also be defined as follows.

Lemma 1. $\widehat{H}_m = \operatorname{conv}(B \cup (-B)) \doteq B'$, *where* $B = [0,1]^m$.

PROOF. It is clear that $B \subseteq \widehat{H}_m$. As $\widehat{H}_m$ is symmetric, $-B \subseteq \widehat{H}_m$. Since $\widehat{H}_m$ is convex, $B' = \operatorname{conv}(B \cup (-B)) \subseteq \widehat{H}_m$. We prove the reverse inclusion.

Let $z \in \widehat{H}_m$. We show that $z \in B'$. Put $z_{\min} = \min_i z(i)$, $z_{\max} = \max_i z(i)$, and $\bar{e} = (1, 1, \dots, 1) \in \mathbb{R}^m$. Then we have $-1 \leq z_{\min} \leq z_{\max} \leq 1$ and $z_{\max} - z_{\min} \leq 1$.

If $z_{\min} \geq 0$ or $z_{\max} \leq 0$, then $z \in B$ or $z \in -B$, respectively, which in both cases implies $z \in B'$. Suppose that $z_{\min} < 0 < z_{\max}$. Define the vectors $z' = z - z_{\max}\bar{e}$ and $z'' = z - z_{\min}\bar{e}$. Then, for every $i = 1, \dots, m$, we have $z'(i) = z(i) - z_{\max} \in [-1, 0]$ and $z''(i) = z(i) - z_{\min} \in [0, 1]$. Therefore, $z' \in -B$ and $z'' \in B$. Thus,

$$B' = \operatorname{conv}(B \cup (-B)) \ni \frac{-z_{\min}}{z_{\max} - z_{\min}} z' + \frac{z_{\max}}{z_{\max} - z_{\min}} z''$$

$$= \left(\frac{-z_{\min}}{z_{\max} - z_{\min}} + \frac{z_{\max}}{z_{\max} - z_{\min}} \right) z + \left(\frac{z_{\max} z_{\min}}{z_{\max} - z_{\min}} - \frac{z_{\max} z_{\min}}{z_{\max} - z_{\min}} \right) \bar{e} = z.$$

Lemma 1 is proved.

Now we formulate the four NVSi-problems, $i = 1, \dots, 4$, on finding a nonstrict summation of a given $\hat{s}$-family of vectors $X \subset \mathbb{R}^m$ within optimal families of half-spaces in $\mathbb{R}^m$. Note that the same indices $i = 1, 2, 3$ were used in [1] for other NVSi-problems. This however will not cause a confusion, since we do not refer to those problems in this paper.

NVS1(m)-Problem. *Given an $\hat{s}$-family $X = \{x_1, \dots, x_n\} \subset \mathbb{R}^m$, find a permutation $\pi = (\pi_1, \dots, \pi_n)$ and a collection of real numbers $b_{m+1} = (\beta_1, \dots, \beta_{m+1})$ minimizing*

$$\theta_1(b_{m+1}) = \sum_{i=1}^{m+1} \beta_i$$

subject to $(X, \pi) \in S_m\mathcal{P}_1(m, b_{m+1})$ for $\mathcal{P}_1(m, b_{m+1}) = \{P(e_1 - e_2, \beta_1), \dots, P(e_{m-1} - e_m, \beta_{m-1}), P(e_m, \beta_m), P(-e_1, \beta_{m+1})\}$.

NVS2(m)-Problem. *Given an $\hat{s}$-family $X = \{x_1, \dots, x_n\} \subset \mathbb{R}^m$, find a permutation $\pi = (\pi_1, \dots, \pi_n)$ and a family of real numbers $b_m = (\beta_1, \dots, \beta_m)$ minimizing $\theta_2(b_m) = \sum_{i=1}^m \beta_i$ subject to $(X, \pi) \in S_m\mathcal{P}_2(m, b_m)$ for $\mathcal{P}_2(m, b_m) = \{P(e_1 - e_2, \beta_1), \dots, P(e_{m-1} - e_m, \beta_{m-1}), P(e_m, \beta_m)\}$.*

Put $N_6 = \{1, \dots, 6\}$. Introduce a cyclic order $1 \to 2 \to \cdots \to 6 \to 1$ on the set N_6, and for each $i \in N_6$ denote its predecessor and its successor in this order by $p(i)$ and $s(i)$, respectively. Denote by I the set of nonneighboring pairs of items in N_6, i.e.,

$$I \doteq \{(i,j) \mid i, j \in N_6;\ i \neq j;\ i \neq s(j);\ i \neq p(j)\}.$$

NVS3-Problem. *Given an $\hat{s}$-family $X = \{x_1, \dots, x_n\} \subset \mathbb{R}^2$ and vectors $a_1 = e_1$, $a_2 = e_2$, $a_3 = e_2 - e_1$, $a_4 = -e_1$, $a_5 = -e_2$, and $a_6 = e_1 - e_2$, find a permutation $\pi = (\pi_1, \dots, \pi_n)$ and a collection of real numbers $b_6 = (\beta_1, \dots, \beta_6)$ minimizing*

$$\theta_3(b_6) = \max_{(i,j) \in I} (\beta_i + \beta_j)$$

subject to $(X, \pi) \in S_2\mathcal{P}_3(2, b_6)$ for $\mathcal{P}_3(2, b_6) = \{P(a_1, \beta_1), \dots, P(a_6, \beta_6)\}$.

NVS4-Problem. *Given an $\hat{s}$-family $X = \{x_1, \dots, x_n\} \subset \mathbb{R}^2$, find a permutation $\pi = (\pi_1, \dots, \pi_n)$ and a collection of real numbers $b_3 = (\beta_1, \beta_2, \beta_3)$ minimizing*

$$\theta_4(b_3) = \max\{\beta_1 + \beta_2,\ \beta_2 + \beta_3\}$$

subject to $(X, \pi) \in S_2\mathcal{P}_1(2, b_3)$.

We prove a lemma which provides a reduction from the NVS2-problem in m-dimensional space to the NVS1-problem in $(m-1)$-dimensional space.

Lemma 2. *Suppose that Algorithm $\mathcal{A}$, given all integers $n \geq 2$, $m \geq 2$ and an $\hat{s}$-family of vectors $X = \{x_1, \dots, x_n\} \subset \mathbb{R}^{m-1}$, solves the* NVS1$(m-1)$-*problem in $\mathcal{T}_{\mathcal{A}}(n)$ time with the objective function $\theta_1(b_m) \leq \theta_1^*(m)$. Then the*

NVS2(m)-problem can be solved in $O(\mathcal{T}_{\mathcal{A}}(n)+n)$ time with the objective function $\theta_2(b_m) \le \theta_1^*(m)$.

PROOF. Suppose we have Algorithm $\mathcal{A}$ claimed in the theorem, and let $X = \{x_1, \dots, x_n\}$ be an $\hat{s}$-family in $\mathbb{R}^m$. The desired algorithm $\mathcal{A}'$ for solution the NVS2(m)-problem consists of three stages.

Algorithm $\mathcal{A}'$

1. For each vector $x_j \in X \subset \widehat{H}_m$ define its projection $x'_j = (0, x_j(2), \dots, x_j(m))$ onto the hyperplane $\Gamma = \{x \in \mathbb{R}^m \,|\, x(1) = 0\}$, $j = 1, \dots, n$. Put $X' \doteq \{x'_1, \dots, x'_n\}$. It follows from the definition of $\hat{s}$ that $X' \subset \widehat{H}_{m-1}$, i.e., all vectors in X' have at most unit length in the norm $\hat{s}$ defined in the space $\Gamma = \mathbb{R}^{m-1}$ of coordinates $2, \dots, m$. Since $\sum_{j=1}^n x'_j = 0$, the family X' is an $\hat{s}$-family in $\mathbb{R}^{m-1}$.

2. Find an approximate solution to the NVS1($m-1$)-problem in the space Γ with input X', i.e., find a permutation $\pi^* = (\pi_1^*, \dots, \pi_n^*)$ which provides a non-strict summation of vectors X' within a family $\mathcal{P}' = \{P'(e_2 - e_3, \beta_2), P'(e_3 - e_4, \beta_3), \dots P'(e_{m-1} - e_m, \beta_{m-1}), P'(e_m, \beta_m), P'(-e_2, \beta_1)\} \doteq \{P'(a_j, \beta_j) \,|\, i = 1, \dots, m\}$ of half-spaces in $\mathbb{R}^{m-1}$:

$$(X', \pi^*) \in S_{m-1}\mathcal{P}' \tag{7}$$

such that

$$\theta_1(b_m) = \sum_{i=1}^{m} \beta_i \le \theta^*. \tag{8}$$

3. Let $y'_\nu = \sum_{j=1}^{\nu} x'_{\pi_j^*}$ be the νth node of the summation trajectory of vectors $x_j \in X$ according to the permutation π^*, and let $y_\nu = \sum_{j=1}^{\nu} x_{\pi_j^*}$ be an analogous node for summation of the vectors $x_j \in X$. Find the number k such that $y_k(1) = \max_{\nu=1,\dots,n} y_\nu(1)$, and define the desired permutation $\pi = (\pi_1, \dots, \pi_n)$ as a cyclic shift of π^* by k items:

$$\pi_j = \begin{cases} \pi_{j+k}^* & \text{for } j+k \le n, \\ \pi_{j+k-n}^* & \text{for } j+k > n. \end{cases}$$

Algorithm $\mathcal{A}'$ is described.

Assign to each half-space $P'(a, \beta) \subset \Gamma$ for $a \in \Gamma$ a half-space $P(a, \beta) \subset \mathbb{R}^m$, and let $\mathcal{P}$ be the family of half-spaces $P \subset \mathbb{R}^m$ assigned to half-spaces $P' \in \mathcal{P}'$. Put $y'_\nu = \sum_{j=1}^{\nu} x'_{\pi_j^*}$. Since each vector x'_j differs from x_j by a vector collinear to the basis vector e_1, we have $y_\nu - y'_\nu = \alpha_\nu e_1$ for some $\alpha_\nu \in \mathbb{R}$. Hence, for every

$a \in \Gamma$, we obtain $(y_\nu, a) = (y'_\nu, a) + \alpha_\nu(e_1, a) = (y'_\nu, a)$. If $y'_\nu \in P'(a_i, \beta_i) \in \mathcal{P}'$, then $y_\nu \in P(a_i, \beta_i) \in \mathcal{P}$. Thus, (7) implies

$$(X, \pi^*) \in S_m \mathcal{P}, \tag{9}$$

i.e., we have a nonstrict summation of the vectors of X within the family $\mathcal{P}$.

Denote $z_\nu = \sum_{j=1}^{\nu} x_{\pi_j}$. Due to $\sum_{j=1}^{n} x_j = 0$, we have

$$z_\nu = \begin{cases} y_{k+\nu} - y_k & \text{for} \quad k + \nu \leq n, \\ y_{k+\nu-n} - y_k & \text{for} \quad k + \nu > n. \end{cases}$$

Hence, $z_\nu(1) \leq 0$, i.e.,

$$z_\nu \in P(e_1, 0), \quad \nu = 1, \dots, n. \tag{10}$$

With respect to the family X, the cyclic shift of the permutation π^* by k items is equivalent to the translation of the origin to the kth node of the summation trajectory of the vectors of X (i.e., to the point y_k), the resulting summation trajectory equivalent to the original up to the parallel shift by vector $y \doteq -y_k$. Hence, this trajectory provides a nonstrict summation of vectors in X with respect to a family of half-spaces $\mathcal{P} + y \doteq \{P + y \mid P \in \mathcal{P}\}$, i.e., $(X, \pi) \in S_m\{\mathcal{P} + y\}$.

We show that such shift of all half-spaces in $\mathcal{P}$ by the same vector y does not change the value of the objective function θ_1. Indeed,

$$x \in P(a_i, \beta_i) + y \iff (a_i, x - y) \leq \beta_i$$

$$\iff (a_i, x) \leq \beta_i + (a_i, y) \doteq \beta'_i \iff x \in P(a_i, \beta'_i).$$

Thus, the permutation π provides a nonstrict summation of the vectors of X within the family of half-spaces $\{P(a_i, \beta'_i) \mid i = 1, \dots, m\}$, where

$$\sum \beta'_i = \sum \beta_i + \left(\sum_{i=1}^{m} a_i, y \right) = \sum \beta_i + ((e_2 - e_3) + (e_3 - e_4)$$

$$+ \dots + (e_{m-1} - e_m) + e_m - e_2,\ y) = \sum \beta_i.$$

Finally, it follows from (10) that if $z_\nu \in P(-e_2, \beta'_1)$, then

$$z_\nu \in P(e_1, 0) \cap P(-e_2, \beta'_1) \subset P(e_1 - e_2, \beta'_1).$$

This means that permutation π provides a nonstrict summation of vectors in X within the half-space $P(e_1 - e_2, \beta'_1)$, as well. Replacing the half-space $P(-e_2, \beta'_1)$

in the family $\mathcal{P} + y$ by the half-space $P(e_1 - e_2, \beta_1')$, we obtain the family of half-spaces for the NVS2(m)-problem with the following bound on the objective function

$$\theta_2(b_m) = \sum_{i=1}^{m} \beta_i' = \sum_{i=1}^{m} \beta_i \le \theta^*.$$

Evidently, the running time of Algorithm $\mathcal{A}'$ meets the upper bound of the lemma. Lemma 2 is proved.

In the rest of this section, we consider the NVSi-problems in the case of the two-dimensions. Each vector $x \in \mathbb{R}^2$, $x \neq 0$, specifies the straight line $l(x) = \{tx \mid t \in R\}$, the ray $Q(x) = \{tx \mid t \in \mathbb{R}^+\}$ which has the endpoint at the origin and passes through the point x, and the *left* and the *right* closed half-spaces $L(x)$ and $R(x)$, i.e., the half-spaces stretched to the left and to the right of the straight line $l(x)$ when it runs in the direction of the ray $Q(x)$.

For x, $y \in \mathbb{R}^2$, let $[x, y]$ denote the closed straight line segment connecting the points x and y, i.e., $[x, y] = \{\lambda x + (1 - \lambda)y \mid \lambda \in [0, 1]\}$.

DEFINITION 3. We say that three vectors $y_1, y_2, y_3 \in \mathbb{R}^2$ form a *normal triple* if there exist numbers $\lambda_i \geq 0$, $i = 1, 2, 3$, such that $\sum \lambda_i y_i = 0$ and $\sum \lambda_i > 0$.

It is easy to see that three vectors in the plane form a normal triple if and only if they are not in the same open half-space.

Let s be a norm in $\mathbb{R}^2$, and let H be its unit ball with center the origin, i.e., $H = \{x \in \mathbb{R}^2 \mid \|x\| \leq 1\}$. The value of the norm $\|x\|$ is called the *length* of x. In addition, for each vector x we define its *angular coordinate* $u(x) \in [0, 2\pi)$. We assume for definiteness that $u(x)$ increases in rotating the vector x clockwise (similar to the current agreement for complex numbers). We also assume that $u(0) = 0$.

For every two vectors a and b such that $u(a) \neq u(b)$, $a \neq 0$, $b \neq 0$, the *sector* of the plane from the ray $Q(a)$ counterclockwise to the ray $Q(b)$ is denoted by $C(a, b)$. More formally:

$$C(a,b) = \begin{cases} L(a) \cap R(b) & \text{if } \ b \in L(a), \\ L(a) \cup R(b) & \text{if } \ b \in R(a). \end{cases}$$

In the case $u(a) = u(b)$ we set $C(a, b) = C(b, a) = Q(a) = Q(b)$.

Denote by $C^0(a, b)$ the open sector $C(a, b) \setminus (Q(a) \cup Q(b))$ and let $\angle C(a, b)$ be the angular measure of the sector $C(a, b)$, i.e.,

$$\angle C(a,b) = \begin{cases} u(b) - u(a) & \text{if } \ u(b) \geq u(a), \\ u(b) - u(a) + 2\pi & \text{if } \ u(b) < u(a). \end{cases}$$

Theorem 1. *Let s be a norm in the plane, and let H be its unit ball. Suppose that G is a convex domain in $\mathbb{R}^2$ such that*

$$\text{if } x_1 \notin G,\ x_2 \notin G, \text{ and } [x_1, x_2] \cap H \neq \varnothing, \text{ then } \|x_2 - x_1\| > 1. \tag{11}$$

Then, for any s-family $X = \{x_1, \dots, x_n\} \subset \mathbb{R}^2$, there exists a permutation $\pi = (\pi_1, \dots, \pi_n)$ providing a nonstrict summation of vectors in X within G, and π can be found in $O(n \log n)$ time.

PROOF. It follows from (11) and the convexity of G that $H \subset G$ and each chord of G intersecting H has length at least 1. Conversely, if $H \subset G$ and each chord of G intersecting H has length at least 1, we obtain (11), provided G is closed. However, in the proof of Theorem 1 we will use (11) without the assumption that G is closed.

We now describe Algorithm $\mathcal{A}^*$ for computing the desired permutation π.

Algorithm $\mathcal{A}^*$

STAGE 1 (numbering the vectors and constructing a search tree).

We enumerate the vectors $x_i \in X$ in increasing order of their angular coordinates $u(x_i)$. If these coordinates are equal, we enumerate the vectors in increasing order of their length. When both coordinates are equal, we enumerate the vectors in an arbitrary order. Thus, all zero vectors receive the smallest numbers (from 1 to k_0, where $k_0 \geq 0$). This procedure runs in $O(n \log n)$ time.

Then, according to Tarjan [8], we construct a *balanced binary tree* (T_0) with the set of (internal) vertices $I_0 = \{1, \dots, n\}$ by keys $u(x_i)$, $i \in I_0$. Recall that such search tree has the following useful property: for each internal vertex i all vertices $\{j\}$ in its left subtree have keys $u(x_j) \leq u(x_i)$, whereas all vertices in its right subtree have keys $u(x_j) \geq u(x_i)$. The fact that the tree is "balanced" means that the distance from the root of the tree to any of the external vertices is $O(\log n)$.

While constructing such tree, we define the *range* $r(i)$ of each vertex i as the maximal integer ν such that i is a multiple of 2^ν. Thus, ranges will vary from 0 to $\lfloor \log_2 n \rfloor$, the maximal range $r = \lfloor \log_2 n \rfloor$ being attained at the unique vertex $i_0 = 2^r$ which is defined as the root of the tree T_0. For each vertex i not equal to the root i_0, we define its parent $p(i)$ choosing from the two vertices $i + 2^{r(i)}$ and $i - 2^{r(i)}$ as follows.

if $i + 2^{r(i)} > n$, then $p(i) := i - 2^{r(i)}$;
else if $r(i + 2^{r(i)}) = r(i) + 1$, then $p(i) := i + 2^{r(i)}$;
else $p(i) := i - 2^{r(i)}$.

Since $r(p(i)) \geq r(i) + 1$ holds in each case, the depth of the tree is at most the maximum rank, i.e., the value $\lfloor \log_2 n \rfloor$. It can also be easily checked that the so-defined tree is a binary tree, i.e., each of its vertices has at most two children.

STAGE 2 (computing the permutation π).

The stage consists of steps $k = 1, \ldots, n$. After step k is finished, we have the first k values $\pi_1, \ldots, \pi_k$ of the permutation π and the set of indices I_k of those vectors in X which are not delegated to the total $\sigma_k \doteq \sum_{i=1}^{k} x_{\pi_i}$. We have also a search tree T_k with the set of vertices I_k, the depth of the tree being at most r.

For $k = 1, \ldots, k_0$, we set $\pi_k = k$ which means that zero vectors go first. Removing the indices $\{1, \ldots, k_0\}$ in the set I_0, and deleting the corresponding vertices from the tree T_0, we obtain the set I_{k_0} and the tree T_{k_0}. (Removing any vertex from a search tree with n vertices requires time $O(\log n)$ [8, p. 46].)

Suppose that k steps are already finished; $k_0 \leq k \leq n - 2$.

STEP $k + 1$. Find indices p_k, $s_k \in I_k$ such that the angular coordinates of x_{p_k}, x_{s_k} are closest to the value $u' = u(-\sigma_k)$ (from below and from above, respectively) over all angular coordinates of vectors $x_i \in X_k$. Formally, let $I_k^-(u') \doteq \{i \in I_k \mid u(x_i) \leq u'\}$ and $I_k^+(u') \doteq \{i \in I_k \mid u(x_i) \geq u'\}$. Put

$$p_k = \begin{cases} \max\{j \in I_k^-(u')\} & \text{if } I_k^-(u') \neq \varnothing, \\ \max\{j \in I_k\} & \text{otherwise;} \end{cases}$$

$$s_k = \begin{cases} \min\{j \in I_k^+(u') \setminus \{p_k\}\} & \text{if } I_k^+(u') \setminus \{p_k\} \neq \varnothing, \\ \min\{j \in I_k\} & \text{otherwise.} \end{cases}$$

(Again, we will write p, s, σ instead of p_k, s_k, σ_k when this leads to no ambiguity.) It is not hard to describe an algorithm for searching vertices p_k and s_k in the search tree T_k that runs in $O(\log n)$ time.

For π_{k+1} we will always choose among the indices $\{p, s\}$. (The details of the choice rule will be described later.) Removing (in $O(\log n)$ time) the vertex π_{k+1} from T_k results in the tree T_{k+1}.

This completes the draft description of Algorithm $\mathcal{A}^*$.

Theorem 1 follows from three lemmas to be proven below.

Lemma 3. *The triple of vectors $\langle \sigma, x_p, x_s \rangle$ is normal at each step $k = 1, \ldots, n$ of Algorithm $\mathcal{A}^*$.*

PROOF. By the definition of x_p and x_s, at each step k we have

$$-\sigma_k \in C(x_p, x_s), \tag{12}$$

$$X_k \subset C(x_s, x_p). \tag{13}$$

This implies $\angle C(x_s, x_p) \geq \pi$, which is equivalent to

$$\angle C(x_p, x_s) \leq \pi. \tag{14}$$

(The contrary would contradict the property $\Sigma(X) = 0$.)

Now the normality of the triple $\langle \sigma, x_p, x_s \rangle$ follows from (12) and (14). Lemma 3 is proved.

Lemma 4. *Let λ_σ, λ_p, and λ_s be the coefficients witnessing the normality of the triple $\langle \sigma, x_p, x_s \rangle$, i.e.,*

$$\lambda_\sigma \sigma + \lambda_p x_p + \lambda_s x_s = 0;$$
$$\lambda_i \geq 0 \quad \text{for} \quad i \in \{\sigma, p, s\}; \quad \sum \lambda_i > 0.$$

Further, let

$$\sigma + x_p \notin H \quad \text{and} \quad \sigma + x_s \notin H. \tag{15}$$

Then

$$\lambda_\sigma < \max\{\lambda_p, \lambda_s\}. \tag{16}$$

Proof. Let $\lambda_\sigma \geq \max\{\lambda_p, \lambda_s\}$. Then $\lambda_\sigma > 0$.

If $\lambda_p \geq \lambda_s$ then

$$\sigma + x_p = \left(1 - \frac{\lambda_p}{\lambda_\sigma}\right) x_p + \frac{\lambda_s}{\lambda_\sigma}(-x_s) = \gamma_p x_p + \gamma_s(-x_s),$$

where $\gamma_p \geq 0$, $\gamma_s \geq 0$, $\gamma_p + \gamma_s \leq 1$, $x_p \in H$, $-x_s \in H$ (since H is symmetric). Hence, $\sigma + x_p \in \operatorname{conv}\{x_p, -x_s, 0\} \subset H$, which contradicts (15).

If $\lambda_p \leq \lambda_s$, we have

$$\sigma + x_s = \frac{\lambda_p}{\lambda_\sigma}(-x_p) + \left(1 - \frac{\lambda_s}{\lambda_\sigma}\right) x_s \in H,$$

which contradicts (15). Lemma 4 is proved.

We now are able to describe the performance of Algorithm $\mathcal{A}^*$ at step $k+1$ in detail.

Step $k+1$ (detailed description).

- If $\sigma + x_p \in H$, then $\pi_{k+1} := p$.
- If $\sigma + x_p \notin H$ and $\sigma + x_s \in H$, then $\pi_{k+1} := s$.
- If $\sigma + x_p \notin H$ and $\sigma + x_s \notin H$, then find nonnegative coefficients λ_σ, λ_p, and λ_s witnessing the normality of the triple $\langle \sigma, x_p, x_s \rangle$. By Lemma 4, we

have (16). Therefore, only the following three cases of relations between λ_σ, λ_p, and λ_s are possible.

CASE 1 ($\lambda_s \leq \lambda_\sigma < \lambda_p$): $\pi_{k+1} := p$.

CASE 2 ($\lambda_p \leq \lambda_\sigma < \lambda_s$): $\pi_{k+1} := s$.

CASE 3 ($\lambda_\sigma < \min\{\lambda_p, \lambda_s\}$):
(A) if $\sigma + x_p \in G$, then $\pi_{k+1} := p$;
(B) if $\sigma + x_p \notin G$, $\sigma + x_s \in G$, then $\pi_{k+1} := s$;
(C) if $\sigma + x_p \notin G$, $\sigma + x_s \notin G$, then $\pi_{k+1} := p$, $\pi_{k+2} := s$.
Step $k + 1$ is described completely.

As is easily seen, in Case 3C we have a description for steps $k + 1$ and $k + 2$.

Lemma 5. *Suppose that the relations*

$$\sigma_t \in G, \tag{17}$$

$$(\sigma_t + Q(-x')) \cap H \neq \varnothing \tag{18}$$

hold at step $t = k$ ($k_0 \leq k \leq n - 2$), *where* x' *is either* x_{p_t} *or* x_{s_t}. *Then* (17) *and* (18) *hold either at step* $t = k + 1$ *or at step* $t = k + 2$.

PROOF. Note that, in view of $\Sigma(X) = 0$, the equality in (14) holds only if σ_k and all vectors in X_k are collinear. Since the proof of Lemma 5 is trivial in this case (at each step $t = k + 1, \ldots, n$ the vector $x_{\pi_t} = x_p$ has the opposite direction to the direction of the vector σ_t), we assume that

$$\angle C(x_p, x_s) < \pi. \tag{19}$$

If $\sigma + x_p \in H$ or $\sigma + x_s \in H$, then we put $\pi_{k+1} := p$ or $\pi_{k+1} := s$, respectively, and (17) and (18) hold at step $t = k + 1$. Now assume that (15) holds. Assume also for definiteness that (18) holds at step $t = k$ for $x' = x_s$, i.e.,

$$(\sigma_t + Q(-x_s)) \cap H \neq \varnothing. \tag{20}$$

(In case $x' = x_p$ the proof of the lemma is similar.) By (15), according to the description of the algorithm we find the coefficients λ_σ, λ_p, and λ_s witnessing the normality of the triple $\langle \sigma, x_p, x_s \rangle$. Note that (19) implies $\lambda_\sigma > 0$. Furthermore, $\lambda_p \neq 0$. (Otherwise, the vector x_s would have the opposite direction to the vector σ.) This together with (20) implies $\sigma \in H$, $\sigma + x_s \in H$, which contradicts (15). In case $\lambda_s = 0$ the vector x_p has the opposite direction to vector σ, and (17) and (18) hold for $\sigma_{k+1} = \sigma + x_p$ at step $t = k + 1$. Therefore, we assume that

$$\lambda_i > 0, \quad i \in \{\sigma, p, s\}. \tag{21}$$

Consider the three possible cases of relations between λ_σ, λ_p, and λ_s.

CASE 1 ($\lambda_s \le \lambda_\sigma < \lambda_p$).

We have $\pi_{k+1} = p$, $\sigma_{k+1} = \sigma + x_p$. By (19) and (21), we obtain

$$-\sigma_{k+1} = -\sigma - x_p = \left(\frac{\lambda_p}{\lambda_\sigma} - 1\right) x_p + \frac{\lambda_s}{\lambda_\sigma} x_s \in C^0(x_p, x_s). \tag{22}$$

Hence, $x_{s_{k+1}} = x_s$. Furthermore, we have

$$y_1 \doteq -\frac{\lambda_s}{\lambda_\sigma} x_s \in H, \tag{23}$$

$$\sigma_{k+1} = \sigma + x_p = \sigma - \frac{\lambda_\sigma}{\lambda_p}\sigma - \frac{\lambda_s}{\lambda_p} x_s = \left(1 - \frac{\lambda_\sigma}{\lambda_p}\right)\sigma + \frac{\lambda_\sigma}{\lambda_p} y_1. \tag{24}$$

The relations (23) and (24) imply (17) for $t = k + 1$.

We prove (18). It follows from (20) that there exists a point $z = \sigma - \gamma x_s \in H$, where $\gamma \ge 0$. Putting $z_1 = \lambda_s z/(\lambda_\sigma \gamma + \lambda_s)$, we obtain $z_1 \in H$ and

$$z_1 = \frac{\lambda_s}{\lambda_\sigma \gamma + \lambda_s}(\sigma - \gamma x_s) = \frac{\lambda_s}{\lambda_\sigma \gamma + \lambda_s}\sigma + \frac{\lambda_\sigma \gamma}{\lambda_\sigma \gamma + \lambda_s} y_1$$

$$= (1 - \lambda')\sigma + \lambda' y_1 \in [y_1, \sigma], \tag{25}$$

where $\lambda' \doteq \lambda_\sigma \gamma/(\lambda_\sigma \gamma + \lambda_s)$. If $\lambda_\sigma/\lambda_p \ge \lambda'$, then (24) and (23) imply $\sigma_{k+1} \in [y_1, z_1] \subset H$, which contradicts (15). Thus, $\lambda_\sigma/\lambda_p < \lambda' = \lambda_\sigma\gamma/(\lambda_\sigma\gamma + \lambda_s)$. Hence, $\gamma - (\lambda_\sigma/\lambda_p)\gamma - \lambda_s/\lambda_p \doteq \alpha > 0$. Therefore,

$$\sigma_{k+1} - \alpha x_{s_{k+1}} = \sigma_{k+1} - \alpha x_s = \sigma - \frac{\lambda_\sigma}{\lambda_p}\sigma - \frac{\lambda_s}{\lambda_p} x_s - \left(\gamma - \frac{\lambda_\sigma}{\lambda_p}\gamma - \frac{\lambda_s}{\lambda_p}\right) x_s$$

$$= \left(1 - \frac{\lambda_\sigma}{\lambda_p}\right)(\sigma - \gamma x_s) = \left(1 - \frac{\lambda_\sigma}{\lambda_p}\right) z \in H \cap (\sigma_{k+1} + Q(-x_{s_{k+1}})),$$

which proves (18) for $t = k + 1$.

CASE 2 ($\lambda_p \le \lambda_\sigma < \lambda_s$).

Note that this case is not symmetric to Case 1 in view of (20).

Put $\pi_{k+1} = s$ and $\sigma_{k+1} = \sigma + x_s$. Let $z = \sigma - \gamma x_s$ ($\gamma \ge 0$) be a point in H whose existence follows from (20). Then

$$\sigma_{k+1} = \sigma + x_s \in \left[\sigma - \gamma x_s, \sigma + \frac{\lambda_s}{\lambda_\sigma} x_s\right] = \left[\sigma - \gamma x_s, -\frac{\lambda_p}{\lambda_\sigma} x_p\right] \subset H,$$

which implies (17) and (18) for $t = k+1$.

CASE 3 ($\lambda_\sigma < \min\{\lambda_p, \lambda_s\}$).

SUBCASE A ($\sigma + x_p \in G$).

We have $\pi_{k+1} = p$, $\sigma_{k+1} = \sigma + x_p \in G$, i.e. (17) holds for $t = k+1$. To prove (18), define the vectors y_1, z, and z_1 and the numbers α, γ, and λ' as in Case 1. Then using (22), we have $-\sigma_{k+1} \in C^0(x_p, x_s)$; hence $x_{s_{k+1}} = x_s$ and $x_{p_{k+1}} \in C(-x_s, x_p)$, i.e.,

$$-x_{p_{k+1}} \in C(x_s, -x_p). \tag{26}$$

If $\lambda_\sigma/\lambda_p \geq \lambda'$, then from (24) and (25) it follows that

$$z_1 \in [\sigma_{k+1}, \sigma] \subset \sigma_{k+1} + Q(-x_p).$$

Hence,

$$(\sigma_{k+1} + Q(-x_p)) \cap H \neq \varnothing. \tag{27}$$

Moreover, $\alpha \leq 0$ and

$$\sigma_{k+1} - \alpha x_s = \left(1 - \frac{\lambda_\sigma}{\lambda_p}\right) z \in H \cap (\sigma_{k+1} + Q(x_s)). \tag{28}$$

From (26)–(28) with regard to (19), we obtain

$$(\sigma_{k+1} + Q(-x_{p_{k+1}})) \cap H \neq \varnothing,$$

i.e. (18) holds for $t = k+1$.

If $\lambda_\sigma/\lambda_p < \lambda'$, then $\alpha > 0$ and

$$\sigma_{k+1} - \alpha x_{s_{k+1}} = \sigma_{k+1} - \alpha x_s = \left(1 - \frac{\lambda_\sigma}{\lambda_p}\right) z \in H \cap (\sigma_{k+1} + Q(-x_{s_{k+1}})),$$

i.e. (18) holds for $t = k+1$.

SUBCASE B ($\sigma + x_s \in G$).

We have $\pi_{k+1} = s$, $\sigma_{k+1} = \sigma + x_s \in G$, i.e., (17) holds for $t = k+1$. We prove (18).

Due to (19) and (21), we have

$$-\sigma_{k+1} = \frac{\lambda_p}{\lambda_\sigma} x_p + \left(\frac{\lambda_s}{\lambda_\sigma} - 1\right) x_s \in C^0(x_p, x_s).$$

It follows from this and (19) that

$$\angle C(-x_s, -\sigma_{k+1}) < \pi, \tag{29}$$
$$x_{s_{k+1}} \in C(x_s, -x_p) \subset C(x_s, \sigma_{k+1}),$$

i.e.,

$$-x_{s_{k+1}} \in C(-x_s, -\sigma_{k+1}). \tag{30}$$

Using $(\sigma_{k+1} + Q(-\sigma_{k+1})) \cap H \neq \varnothing$ and (20), we have

$$(\sigma_{k+1} + Q(-x_s)) \cap H \supset (\sigma + Q(-x_s)) \cap H \neq \varnothing.$$

With respect to (30) and (29) we obtain $(\sigma_{k+1} + Q(-x_{s_{k+1}})) \cap H \neq \varnothing$, which proves (18) for $t = k + 1$.

SUBCASE C ($\sigma + x_p \notin G$, $\sigma + x_s \notin G$).

We have $\pi_{k+1} = p$, $\pi_{k+2} = s$, $\sigma_{k+2} = \sigma + x_p + x_s$, and $\sigma_{k+2} \in C^0(x_p, x_s)$. From this and (19) we obtain $x_{p_{k+2}} \in C(-x_s, x_p)$, $x_{s_{k+2}} \in C(x_s, -x_p)$, i.e.,

$$-x_{p_{k+2}} \in C(x_s, -x_p), \tag{31}$$
$$-x_{s_{k+2}} \in C(-x_s, x_p). \tag{32}$$

Having defined z_1, y_1, λ', and α, as in Case 1, observe that the case $\sigma_{k+1} \in [z_1, \sigma]$ is impossible. (Otherwise, this would imply $\sigma_{k+1} \in G$.) Therefore, $z_1 \in (\sigma_{k+1}, \sigma]$, which implies $\lambda' < \lambda_\sigma/\lambda_p$ and $\alpha < 0$.

Consider the cases $\alpha \geq -1$ and $\alpha < -1$ separately.

CASE 1: $\alpha \in [-1, 0)$. Since $\sigma_{k+1} - \alpha x_s \in [\sigma_{k+1}, \sigma_{k+2}]$ and $\sigma_{k+1} \notin G$, from (11) and $\|\sigma_{k+2} - \sigma_{k+1}\| = \|x_s\| \leq 1$ we obtain (17) for $t = k + 2$.

We show that (18) holds for $t = k + 2$ and $x' = x_{s_{k+2}}$. Indeed,

$$\sigma_{k+2} - (1+\alpha)x_s = \sigma_{k+1} - \alpha x_s$$
$$= \left(1 - \frac{\lambda_\sigma}{\lambda_p}\right) z \in (\sigma_{k+2} + Q(-x_s)) \cap H. \tag{33}$$

Furthermore,

$$\begin{aligned}
\sigma_{k+2} + \frac{\lambda_p(1+\alpha)}{\lambda_s + \lambda_\sigma\gamma} x_p &= \sigma + x_s + \frac{\lambda_s + \lambda_\sigma\gamma + \lambda_p + \lambda_p\alpha}{\lambda_s + \lambda_\sigma\gamma} x_p \\
&= \sigma + x_s - \frac{\lambda_p + \lambda_p\gamma}{\lambda_s + \lambda_\sigma\gamma}\frac{\lambda_\sigma}{\lambda_p}\sigma - \frac{\lambda_p + \lambda_p\gamma}{\lambda_s + \lambda_\sigma\gamma}\frac{\lambda_s}{\lambda_p} x_s \\
&= \frac{\lambda_s + \lambda_\sigma\gamma - \lambda_\sigma - \lambda_\sigma\gamma}{\lambda_s + \lambda_\sigma\gamma}\sigma + \frac{\lambda_s + \lambda_\sigma\gamma - \lambda_s - \lambda_s\gamma}{\lambda_s + \lambda_\sigma\gamma} x_s \\
&= \frac{\lambda_s - \lambda_\sigma}{\lambda_s + \lambda_\sigma\gamma}\sigma - \frac{\lambda_s - \lambda_\sigma}{\lambda_s + \lambda_\sigma\gamma}\gamma x_s = \frac{\lambda_s - \lambda_\sigma}{\lambda_s + \lambda_\sigma\gamma}(\sigma - \gamma x_s) \\
&= \frac{\lambda_s - \lambda_\sigma}{\lambda_s + \lambda_\sigma\gamma} z \doteq \gamma' z \doteq z_2,
\end{aligned} \tag{34}$$

where $\gamma' \in (0,1)$. Hence,

$$(\sigma_{k+2} + Q(x_p)) \cap H \neq \varnothing. \tag{35}$$

From (32), (33), (35), and (19) we obtain

$$(\sigma_{k+2} + Q(-x_{s_{k+2}})) \cap H \neq \varnothing,$$

i.e. (18) holds for $t = k+2$.

CASE 2: $\alpha < -1$. It follows from (34) that $z_2 = \sigma_{k+2} + \lambda_2 x_p \in H$, where $\lambda_2 = \lambda_p(1+\alpha)/(\lambda_s + \lambda_\sigma \gamma) \in (-1,0)$. Since $[\sigma + x_s, \sigma_{k+2}] \cap H \neq \varnothing$, $\|\sigma_{k+2} - \sigma - x_s\| = \|x_p\| \leq 1$, and $\sigma + x_s \notin G$ due to (11), we obtain $\sigma_{k+2} \in G$. So, (17) holds for $t = k+2$. We show that (18) holds for $t = k+2$ and $x' = x_{p_{k+2}}$.

Since $z_2 \in \sigma_{k+2} + Q(-x_p)$, we have

$$(\sigma_{k+2} + Q(-x_p)) \cap H \neq \varnothing. \tag{36}$$

Furthermore, it follows from (33) that

$$\sigma_{k+2} - (1+\alpha)x_s = \left(1 - \frac{\lambda_\sigma}{\lambda_p}\right) z \in (\sigma_{k+2} + Q(x_s)) \cap H. \tag{37}$$

From (31), (36), (37), and (19) we obtain

$$(\sigma_{k+2} + Q(-x_{p_{k+2}})) \cap H \neq \varnothing,$$

i.e. (18) holds for $t = k+2$. Lemma 5 is proved.

Now we can proceed with the proof of Theorem 1.

It is clear that the complexity of the first step of Algorithm $\mathcal{A}^*$ is $O(n \log n)$, since both numbering the vectors and constructing the search tree T_0 can be done in this time. The same running time is sufficient for carrying out Stage 2 of the algorithm, since at each step $k = 1, \ldots, n$ searching the vertices p_k and s_k and subsequent removing one of them from the search tree T_k can be done in $O(\log n)$ time. We assume here that verifying inclusions $x \in H$ and $x \in G$ that we have to do at each step of Stage 2 for the vectors $x = \sigma + x_p$ and $x = \sigma + x_s$ requires some constant time. This is evidently true in those special cases of Theorem 1 that will be considered in Corollaries 1–3. Moreover, the bound on the running time claimed in the theorem remains true if each checking requires $O(\log n)$ time.

To complete the proof of Theorem 1, it remains to show that the permutation π constructed by Algorithm $\mathcal{A}^*$ provides a nonstrict summation of vectors X within G. This is easily proved by induction on steps $t = 0, 1, \ldots, n$. The base of induction relies on the relations $\sigma_t = 0 \in H \subset G$ $(t = 0, 1, \ldots, k_0)$ which imply (17) and (18) for $t = 0, 1, \ldots, k_0$, whereas the induction step is provided by Lemma 5. Theorem 1 is proved.

We state some corollaries of Theorem 1 that enable us to find efficiently approximate solutions to the NVSi-problems, $i = 1, 3, 4$, in two-dimensional space, as well as, due to Lemma 2, an approximate solution to the NVS2-problem in tree-dimensional space.

Consider the intersection G_1 of half-spaces of the family $\{P(a_i, \beta_i)\}$ for the NVS1(m)-problem (NVS4-problem) for $m = 2$ and $\beta_1 = \beta_2 = \beta_3 = 1$. It is easily seen that G_1 is a triangle circumscribed around the unit ball $\widehat{H}_2$ (Fig. 1).

For the NVS3-problem, the intersection G_2 of half-spaces $P(a_i, \beta_i) \in \mathcal{P}_3(2, b_6)$ for $\beta_i = 5/4$, $i = 1, \ldots, 6$, is a ball of radius 5/4 (see Fig. 2). Since in the both cases we have $\widehat{H}_2 \subset G_i$ and the $\hat{s}$-norm of every chord of G_i intersecting $\widehat{H}_2$ is at least 1, condition (11) of Theorem 1 is satisfied. This implies that each $\hat{s}$-family $X \subset \mathbb{R}^2$ can be nonstrictly summed within each of the domains G_1 and G_2, and the desired permutations can be found efficiently. By Remark 1, this implies that the vectors in X can be nonstrictly summed within the corresponding families of half-spaces.

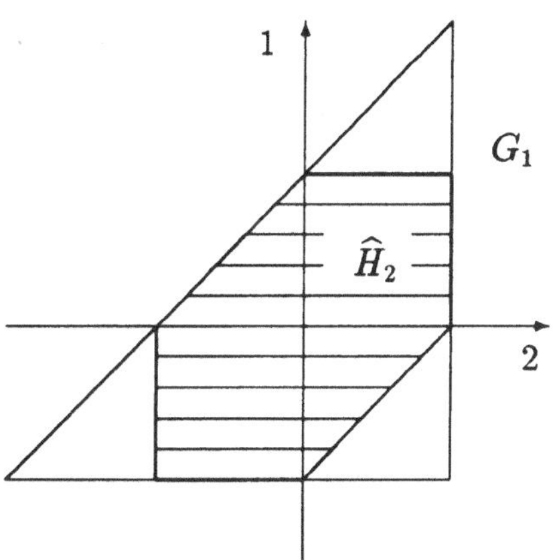

Fig. 1. Domain G_1.

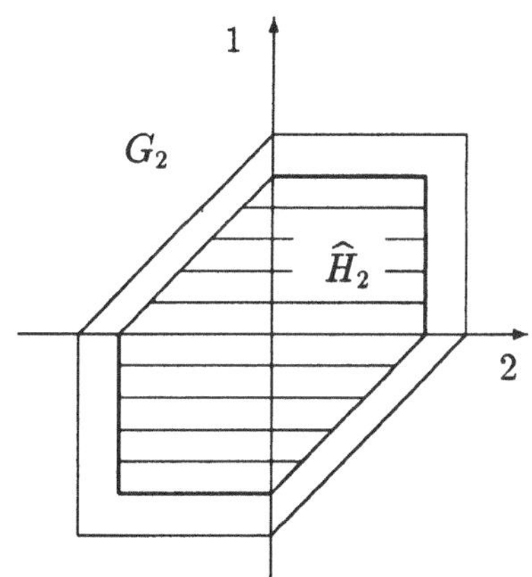

Fig. 2. Domain G_2.

Thus, we obtain the following corollaries of Theorem 1.

Corollary 1. *There is an $O(n \log n)$ algorithm providing an approximate solution of the* NVS1(2)*-problem with the objective function*

$$\theta_1(b_3) \leq 3.$$

Corollary 2. *There is an $O(n \log n)$ algorithm providing an approximate solution of the* NVS3*-problem with the objective function*

$$\theta_3(b_6) \le 2.5.$$

Corollary 3. *There is an $O(n \log n)$ algorithm providing an approximate solution of the* NVS4*-problem with the objective function*

$$\theta_4(b_3) \le 2.$$

§ 3. The Open-Shop Problem with Three Machines

In this section we establish a relation between the optimal solution of the OS-problem with m machines and an approximate solution for the NVS1$(m-1)$-problem. In the case $m = 3$ this relation and Corollary 1 enable us to specify a wide subclass of open-shop problems with three machines solvable in polynomial time.

Theorem 2. *Suppose that some Algorithm $\mathcal{A}$, given all integers n and m, and an $\hat{s}$-family $\{x_1, \dots, x_n\} \subset \mathbb{R}^m$, solves the* NVS1$(m)$*-problem in $\mathcal{T}_{\mathcal{A}}(m,n)$ time with the objective function*

$$\theta_1(b_{m+1}) \le \theta_1^*(m). \tag{38}$$

If every instance of the open shop problem with m machines and n jobs is such that

$$M \ge \left(\theta_1^*(m-1) + 2m - 2\right) K, \tag{39}$$

then, for this problem there exists an algorithm permitting to find an optimal schedule of length M in $O(\mathcal{T}_{\mathcal{A}}(m-1,n) + mn)$ time.

PROOF. Let $\pi = (\pi_1, \dots, \pi_n)$ be a permutation of numbers $1, 2, \dots, n$. Put

$$\Delta_i \doteq \max_{k=1,\dots,n} \left(\sum_{j=1}^{k} t_{\pi_j i} - \sum_{j=1}^{k-1} t_{\pi_j, i+1} \right), \quad i = 1, \dots, m, \tag{40}$$

where $t_{\pi_j, m+1} = t_{\pi_j 1}$. Define a family of vectors $\{d_1, \dots, d_n\} = D \subset \mathbb{R}^{m-1}$, where

$$d_j(i) = t_{ji} - t_{jm}, \quad i = 1, \dots, m-1. \tag{41}$$

Since

$$\sum_{j=1}^{k} t_{\pi_j i} - \sum_{j=1}^{k-1} t_{\pi_j,i+1} = \sum_{j=1}^{k-1}(t_{\pi_j i} - t_{\pi_j,i+1}) + t_{\pi_k i} = \sum_{j=1}^{k}(t_{\pi_j i} - t_{\pi_j,i+1}) + t_{\pi_k,i+1}$$
$$\leq K + \min\left\{\sum_{j=1}^{k-1}(t_{\pi_j i} - t_{\pi_j,i+1}), \sum_{j=1}^{k}(t_{\pi_j i} - t_{\pi_j,i+1})\right\} \tag{42}$$

and

$$t_{ji} - t_{j,i+1} = \begin{cases} d_j(i) - d_j(i+1) & \text{for} \quad i = 1, \ldots, m-2, \\ d_j(m-1) & \text{for} \quad i = m-1, \\ -d_j(1) & \text{for} \quad i = m, \end{cases} \tag{43}$$

using (40), (42), and (43), we obtain

$$\Delta_i \leq K + \max_{k=1,\ldots,n} \min\{d_\pi^{k-1}(i) - d_\pi^{k-1}(i+1), d_\pi^k(i) - d_\pi^k(i+1)\}, \qquad i = 1, \ldots, m-2, \tag{44}$$

$$\Delta_{m-1} \leq K + \max_{k=1,\ldots,n} \min\{d_\pi^{k-1}(m-1), d_\pi^k(m-1)\}, \tag{45}$$

$$\Delta_m \leq K + \max_{k=1,\ldots,n} \min\{-d_\pi^{k-1}(1), -d_\pi^k(1)\}, \tag{46}$$

where $d_\pi^k \doteq \sum_{j=1}^k d_{\pi_j}$.

We describe the claimed Algorithm $\mathcal{A}_1$ for constructing an optimal schedule for any instance of the open shop problem which satisfy (39).

Algorithm $\mathcal{A}_1$

STAGE 1. For the fixed values of M and K, we increase the durations of some operations so as to satisfy the equations

$$M_i = M, \quad i = 1, \ldots, m. \tag{47}$$

The step requires $O(mn)$ time. We assume that the equations (47) hold for the initial matrix (t_{ji}) so that add no change to the notation of all the variables dependent on $\{t_{ji}\}$.

It follows from (47), (41), and the definition of the norm $\hat{s}$ that if K is taken for a new time unit, then the family of vectors D becomes an $\hat{s}$-family in the space $\mathbb{R}^{m-1}$.

STEP 2. Applying Algorithm $\mathcal{A}$ to the family D, find a permutation $\pi = (\pi_1, \ldots, \pi_n)$ providing a nonstrict summation of vectors in D within a family of half-spaces $\mathcal{P}_1(m-1, b_m K)$, where $b_m = (\beta_1, \ldots, \beta_m)$ and

$$\sum_{i=1}^{m} \beta_i \leq \theta_1^*(m-1). \tag{48}$$

Then (44)–(46) imply the inequalities $\Delta_i \leq (1+\beta_i)K$. From this and (48) it follows that

$$\sum_{i=1}^{m} \Delta_i \leq (m + \theta_1^*(m-1))K. \tag{49}$$

The running time of Step 2 is the same as that of Algorithm $\mathcal{A}$, i.e., $O(\mathcal{T}_{\mathcal{A}}(m-1, n))$.

STEP 3 runs as the corresponding algorithm in [6]. We briefly describe it.

For each machine $i \in \mathcal{M}$, we construct a schedule S_i that starts at time 0, terminates at time M, and such that the operations of jobs $j \in \mathcal{N}$ are processed on this machine without any delay according to the permutation π. Obviously, the set of schedules $\{S_1, \ldots, S_m\}$ does not form a feasible schedule for the whole model. These schedules will be afterwards transformed so that each schedule S_i is shifted to the right by a certain amount, and then the part of the schedule that stretches to the right of M is cut off and moved to the vacant place between time zero and starting point of the schedule. Thus, it is more convenient to represent the schedules $\{S_i\}$ as rings $\{\mathcal{K}_i\}$ on a "drum" with circumference length equal to M. The ring $\mathcal{K}_1$ is assumed to be immovable, the starting point of S_1 being taken for the origin.

Let k^* be the first value of k for which the maximum in the right side of (40) is attained for $i = 1$. Thus,

$$\Delta_1 = \sum_{j=1}^{k^*} t_{\pi_j 1} - \sum_{j=1}^{k^*-1} t_{\pi_j 2}.$$

Perform $m-1$ successive rotations (we enumerate them $2, 3, \ldots, m$) of the rings $\mathcal{K}_2, \ldots, \mathcal{K}_m$ retaining the ring $\mathcal{K}_1$ immovable.

ROTATION 2. Rotate each ring $\mathcal{K}_2, \ldots, \mathcal{K}_m$ to the right by the amount Δ_1, retaining their interlocation. Due to the definition of Δ_1, the time $t^* \doteq \sum_{j=1}^{k^*} t_{\pi_j 1}$ coincides with finish times of operations of job π_{k^*} on machine 1 and job π_{k^*-1} on machine 2. The time t^* we taken as the *splitting point* of the schedule on the "drum."

ROTATION ν ($\nu = 3, \dots, m$). Rotate each of the rings $\mathcal{K}_\nu, \dots, \mathcal{K}_m$ to the right by the amount $\Delta_{\nu-1}$ and then by a minimal additional amount $\delta_{\nu-1} \geq 0$ such that time t^* coincides with the finish time of an operation on machine ν. Clearly,

$$\delta_\nu < K, \ \nu = 2, \dots, m-1. \tag{50}$$

The total rotation (over all $m-1$ rotations) of the ring $\mathcal{K}_m$ to the right with respect to $\mathcal{K}_1$ is equal to $\Delta' \doteq \sum_{i=1}^{m-1} \Delta_i + \sum_{i=2}^{m-1} \delta_i$ which is equivalent to its rotation to the left by $M - \Delta'$, or the same, to the rotation of ring $\mathcal{K}_1$ to the right with respect to $\mathcal{K}_m$ by the same amount. It is not hard to see that if

$$M - \Delta' \geq \Delta_m, \tag{51}$$

then the schedule S' obtained as a result of all $m-1$ rotations is feasible *on the drum.* This means that the time intervals for processing any two operations of the same job do not overlap in the schedule S'.

Now, to obtain a feasible schedule of length M, it suffices to split all rings of the drum at the point t^*. Since no time interval for processing an operation can be split under this procedure (by the choice of the amounts $\{\delta_\nu\}$, the point t^* coincides with the finish time of an operation on each machine), the resulting schedule is feasible. It is clear that the schedule is also feasible for the initial matrix of shorter processing times that was the input of the problem before Step 1.

This completes the description of Algorithm $\mathcal{A}_1$.

It follows from (49)–(51) that the schedule of length M constructed by the above algorithm is feasible, as soon as (39) holds. It is also clear that Algorithm $\mathcal{A}_1$ runs in time claimed in the theorem. Theorem 2 is proved.

Theorem 2 and Corollary 1 of Theorem 1 imply

Theorem 3. *Given an instance of the open shop problem with three machines and n jobs satisfy the condition*

$$M \geq 7K, \tag{52}$$

an optimal schedule has length M and can be found in $O(n \log n)$ time.

Note that if an instance does not satisfy (52) but satisfy (51), then Algorithm $\mathcal{A}_1$ still succeeds.

In all other cases, as it was shown in [4], application of a simple greedy algorithm instead of Algorithm $\mathcal{A}_1$ provides us (within linear time in n) an approximate schedule S of length $T(S) \leq M + 2K$.

Theorem 3 and relations (4) imply

Corollary 4. *The minimal function $\eta^*(m)$ that ensures the* EIOS*-property for the m-machine* OS*-problem satisfies the inequalities*

$$4 \leq \eta^*(3) \leq 7.$$

§ 4. Three Job-Shop-Type Problems

In this section we consider the FS-, AF(3)-, and R231-problems. In the case of the arbitrary number of machines m, the FS-problem reduces to the NVS2$(m-1)$-problem which in view of Lemma 2 provides its reducibility to the NVS1$(m-2)$-problem. Next, we show how the AF(3)-problem and R231-problem can be reduced to the NVS3- and NVS4-problems, respectively. By Corollaries 1–3 of Theorem 1, this reduction provides construction of polynomial-time approximation algorithms with the best known performance guarantees for the three scheduling problems under consideration, as well as enables us to find better bounds on the minimum optima localization intervals for these problems.

While solving each scheduling problem, we assume that $K = 1$ (which can be provided by defining K as a new time unit) and that (47) holds. (We can provide this by implementing Step 1 of Algorithm $\mathcal{A}_1$ of § 3.) Thus, the family of vectors $\{d_1, \dots, d_n\} = D$ defined by (41) becomes an $\hat{s}$-family of vectors in $\mathbb{R}^{m-1}$.

The Flow-Shop Problem

Theorem 4. *Suppose that some Algorithm $\mathcal{A}'$, given all positive integers n and m and every $\hat{s}$-family $\{x_1, \dots, x_n\} \subset \mathbb{R}^m$, solves the* NVS2$(m)$*-problem in time $T_{\mathcal{A}'}(m,n)$ with the objective function*

$$\theta_2(b_m) \leq \theta_2^*(m).$$

Then, for every instance of the FS*-problem with m machines and n jobs, a permutation schedule S_T can be constructed in $O(T_{\mathcal{A}'}(m-1,n) + mn)$ time with length*

$$T(S_\pi) \leq M + \big(m - 1 + \theta_2^*(m-1)\big) K.$$

PROOF. It was established in [1] that the length $T(S_n)$ of any permutation schedule S_π satisfy the inequality

$$T(S_\pi) \leq M + \sum_{i=1}^{m-1} \max_{k=1,\dots,n} \left(\sum_{j=1}^{k} t_{\pi_j i} - \sum_{j=1}^{k-1} t_{\pi_j, i+1} \right).$$

Using (40), (44), and (45), in terms of coordinates of vectors in $\{d_1, \dots, d_n\} = D$ this inequality can be written as

$$T(S_\pi) \le M + \sum_{i=1}^{m-1} \Delta_i \le M + (m-1)K$$

$$+ \sum_{i=1}^{m-2} \max_{k=1,\dots,n} \min\{d_\pi^{k-1}(i) - d_\pi^{k-1}(i+1), d_\pi^k(i) - d_\pi^k(i+1)\}$$

$$+ \max_{k=1,\dots,n} \min\{d_\pi^{k-1}(m-1), d_\pi^k(m-1)\}. \tag{53}$$

Applying Algorithm $\mathcal{A}'$ to the NVS2-problem with the family of vectors $D \subset \mathbb{R}^{m-1}$, we can find a permutation $\pi = (\pi_1, \dots, \pi_n)$ which provides a nonstrict summation of vectors in D within the family of half-spaces $\mathcal{P}_2(m-1, b_{m-1}K)$, where $b_{m-1} = (\beta_1, \dots, \beta_{m-1})$ and

$$\sum_{i=1}^{m-1} \beta_i \le \theta_2^*(m-1). \tag{54}$$

We can derive from (53) and (54) that the length $T(S_\pi)$ of the schedule S_π specified by the permutation π satisfy the inequality

$$T(S_\pi) \le M + (m-1)K + \sum_{i=1}^{m-1} \beta_i K \le M + \big(m - 1 + \theta_2^*(m-1)\big)K.$$

Theorem 4 is proved.

Theorem 4 and Lemma 2 immediately imply

Theorem 5. *Suppose that some Algorithm $\mathcal{A}$, given all positive integers n and m and every $\hat{s}$-family $\{x_1, \dots, x_n\} \subset \mathbb{R}^m$, solves the* NVS1($m$)*-problem in time $\mathcal{T}_\mathcal{A}(m, n)$ with the objective function*

$$\theta_1(b_{m+1}) \le \theta_1^*(m).$$

Then, for every instance of the FS*-problem with m machines and n job, a permutation schedule S_π can be constructed in $O(\mathcal{T}_\mathcal{A}(m-2, n) + mn)$ time with length*

$$T(S_\pi) \le M + \big(m - 1 + \theta_1^*(m-2)\big)K.$$

Using this result and Corollary 1 of Theorem 1, we obtain

Theorem 6. *Given an instance of the flow shop problem with four machines and n jobs, a permutation schedule S_π can be constructed in $O(n \log n)$ time with length*

$$T(S_\pi) \le M + 6K.$$

Corollary 5. *For any instance of the* FS(m)*-problem, its permutation optimum belongs to an interval $[M, M + \tilde{\mu}(m)K]$, where $\tilde{\mu}(m)$ for $m = 4$ satisfies the inequalities*

$$4 \le \tilde{\mu}(4) \le 6.$$

In the case of three machines, as it follows from Theorem 5, the flow shop problem reduces to the NVS1-problem in 1-dimensional space, i.e., to the problem of finding a nonstrict summation of $\hat{s}$-family of numbers (with absolute values at most 1 and with zero sum) within the family of half-spaces $\{P(-e_2, \beta_2), P(e_2, \beta_1)\}$ which gives a minimum to $\beta_1 + \beta_2$. (In other words, we wish to sum numbers nonstrictly within an interval of minimum length.) It is easily seen that we can sum such families of numbers within any interval of length 1 containing the origin, and this can be done in $O(n)$ time. Thus, we have a solution to the NVS1(1)-problem with bound $\beta_1 + \beta_2 \le 1$, which implies the following result.

Theorem 7. *Given an instance of the flow shop problem with three machines and n jobs, a permutation schedule S_π can be constructed in $O(n)$ time with length*

$$T(S_\pi) \le M + 3K.$$

This result improves both the result of [7] (an algorithm with running time $O(n \log n)$ and the bound on schedule length $M + 3K$), and the result of [9] (an algorithm with running time $O(n)$, which finds schedules with length at most $M + 4K$).

The Akers–Friedman Problem with Three Machines

Theorem 8. *Suppose that some Algorithm $\mathcal{A}$, given any positive integer n and every $\hat{s}$-family $\{x_1, \dots, x_n\} \subset \mathbb{R}^2$, solves the* NVS3*-problem in time $\mathcal{T}_{\mathcal{A}}(n)$ with the objective function*

$$\theta_3(b_6) \le \theta_3^*.$$

Then, for any instance of the Akers–Friedman problem with three machines and n jobs, a schedule can be constructed in $O(\mathcal{T}_{\mathcal{A}}(n) + n \log n)$ time with length

$$T(S) \le M + \left(3 + \theta_3^*\right)K.$$

PROOF. We now describe Algorithm $\mathcal{A}_3$ that solves the Akers–Friedman problem.

Algorithm $\mathcal{A}_3$

STAGE 1. Applying Algorithm $\mathcal{A}$, find a permutation $\pi = (\pi_1, \dots, \pi_n)$ providing a nonstrict summation of vectors in $D \subset \mathbb{R}^2$ within a family of half-spaces $\mathcal{P}_3(2, b_6)$ such that $\theta_3(b_6) \le \theta_3^*$. For simplicity of notation, we assume that the jobs are numbered according to the permutation π, where $\pi = (1, 2, \dots, n)$.

STAGE 2. Let o_i^j denote the ith operation of job j. Define some priority order $\succ_p$ on the set of operations:

(1) $o_1^j \succ_p o_3^k$, $o_2^j \succ_p o_3^k$ for all j, $k \in \mathcal{N}$, i.e., each first or second operation of any job is more preferable than any third operation;

(2) $o_i^j \succ_p o_\nu^k$ for every $\nu \in \{1, 2\}$ and all j and k such that $j < k$, i.e., the priority order on the set of first and second operations is specified by permutation π of job indices;

(3) $o_3^j \succ_p o_3^k$ for all j and k such that $j < k$.

The desired schedule S is now constructed by a greedy algorithm that

- loads a machine (does not allow it to be idle) if there are operations available for processing on that machine;
- if at some time a machine becomes available and there are several operations available for processing on that machine, the algorithm schedules the operation with highest priority $\succ_p$.

This completes the description of Algorithm $\mathcal{A}_3$.

For each machine i, $i = 1, 2, 3$, at any time we arrange a queue for the set of operations available for processing on machine i as a "heap" [10, p. 33], i.e., such binary tree T^i with vertices-operations in which, for the path from the root to any external vertex, the priority of vertices along the path is decreasing and the length of each such path is $O(\log n)$. Then both the insertion of a new operation to the tree and deleting the operation with the highest priority (which is evidently located at the root of the tree) can be implemented in $O(\log n)$ time. Thus, the total running time of the algorithm is $O(n \log n)$, as claimed in the theorem.

We establish the upper bound for the schedule length which is claimed in the theorem.

We can represent the schedule S_i for machine i as a sequence of *work intervals* $[s_\nu^i, f_\nu^i)$ and *idle intervals* $[f_\nu^i, s_{\nu+1}^i)$, $\nu = 1, 2, \dots$.

Consider the first operation $o_{ji} = o_1^j$ of a job j. Since o_{ji} is available for processing at time zero, it cannot let an operation with less priority be scheduled before it. Hence, for its finish time f_{ji} we have the bound as follows:

$$f_{ji} \le \sum_{q=1}^{j} t_{qi} \doteq T_i^j. \tag{55}$$

Consider the second operation $o_{ji} = o_2^j$ of a job j. Suppose that it is processed on machine i in time-interval $[s_{ji}, f_{ji}) \subseteq [s_\nu^i, f_\nu^i)$. Let $o_{j_1 i}, o_{j_2 i}, \ldots, o_{j_\nu i} = o_{ji}$ be the maximal sequence of operations that are executed on machine i one-by-one (with no delay) within time-interval $[s_\nu^i, f_{ji})$, and such that $o_{j_1 i} \succ_p o_{j_2 i} \succ_p \cdots \succ_p o_{ji}$. Then the operation $o_{j_1 i}$ is either the first operation executed on machine i in this interval, i.e., $s_{j_1 i} = s_\nu^i$, or it is preceded by a less priority operation. In both cases, the gap between the time of arrival of job j_1 at machine i and the time $s_{j_1 i}$ of starting the operation $o_{j_1 i}$ is less than $K = 1$.

From the definition of priority $\succ_p$ it follows that $o_{j_1 i} \in \{o_1^{j_1}, o_2^{j_1}\}$ and

$$j_1 < j_2 < \cdots < j_\nu = j. \tag{56}$$

Suppose that $o_{j_1 i} = o_2^{j_1}$ is the second operation of job j_1, whereas its first operation is processed by machine $i_1 \neq i$. Using (55), we have

$$s_{j_1 i} < f_{j_1 i_1} + 1 \leq T_{i_1}^{j_1} + 1 = T_i^{j_1 - 1} + 1 + \left(T_{i_1}^{j_1} - T_i^{j_1 - 1}\right).$$

Hence, by (56) we obtain an upper bound on the finish time of the operation o_{ji}:

$$f_{ji} = s_{j_1 i} + \sum_{k=1}^{\nu} t_{j_k i} < T_i^j + 1 + \max_k\left(T_{i_1}^k - T_i^{k-1}\right). \tag{57}$$

In this bound we have "forgotten" that it takes its origin from job j_1. So, we can say about machine i_1 only that $i_1 \neq i$.

Since the $\max_k$ in the right-hand side of (57) is nonnegative, the bound (57) is valid in the case $o_{j_1 i} = o_1^{j_1}$ as well. Therefore, the right-hand side of (57) is an upper bound on the finish time of the second operation of each job. This enables us to derive an upper bound on the time $\hat{t}_{ji_3}$ when the third operation o_{ji_3} of job j becomes available for processing. Assuming that the second operation of the job is processed on machine i_2, we obtain:

$$\begin{aligned} \hat{t}_{ji_3} &= f_{ji_2} < T_{i_2}^j + 1 + \max_k\left(T_{i_1}^k - T_{i_2}^{k-1}\right) \\ &= T_{i_3}^{j-1} + 1 + \left(T_{i_2}^j - T_{i_3}^{j-1}\right) + \max_k\left(T_{i_1}^k - T_{i_2}^{k-1}\right) \leq T_{i_3}^{j-1} + 1 + \beta, \end{aligned} \tag{58}$$

where

$$\beta = \max_{\{i', i'', i''' \mid i'' \neq i', i'' \neq i'''\}} \left\{\max_k\left(T_{i''}^k - T_{i'''}^{k-1}\right) + \max_k\left(T_{i'}^k - T_{i''}^{k-1}\right)\right\}.$$

From (58) and (57), it follows that the time $\hat{t}_{ji}$ when the second operation o_{ji} becomes available for processing can also be bounded by the same amount. All the more, it is true for any first operation o_{ji} (since $\hat{t}_{ji} = 0$). Thus, any operation o_{ji} becomes available for processing by at most time

$$\hat{t}_{ji} < T_i^{j-1} + 1 + \beta. \tag{59}$$

Let i be an arbitrary machine. We show that the duration D_i of its idle time is at most $1+\beta$. If the starting time t' of the last work interval of machine i is less than $1+\beta$, then $D_i < 1+\beta$. Let $t' \geq 1+\beta$ and let j^* be the maximal j such that

$$T_i^{j-1} + 1 + \beta \leq t'. \tag{60}$$

It can be deduced from (59) and (60) that all operations $o_{1i}, \ldots, o_{j^*i}$ become available for processing on machine i before time t'. Since time t' is immediately preceded by an idle time on machine i, all the operations that became available for processing on machine i strictly before time t' have to be finished before time t' as well due to the first rule of the greedy algorithm. Hence, $j^* < n$, and the duration D_i of the idle time of machine i before time t' (as well as that before the finish time C_i of the machine) satisfy relations

$$D_i \leq t' - T_i^{j^*} < 1 + \beta \tag{61}$$

by the definition of j^*.

Let us estimate the value of β via the values of β_i defined in the NVS3-problem. By the definition of the vectors $\{a_i \mid i \in \mathcal{N}_6\}$ and $\{d_j \mid j \in \mathcal{N}\}$, for every job $j \in \mathcal{N}$ and all two indices ν', $\nu'' \in \mathcal{M}$, $\nu' \neq \nu''$, we have $t_{j\nu'} - t_{j\nu''} = (a_i, d_j)$, where the correspondence between the indices i, ν', and ν'' is defined by the following table:

i	1	2	3	4	5	6	1
ν'	1	2	2	3	3	1	1
ν''	3	3	1	1	2	2	3

It can be seen from the table that if $t_{ji'} - t_{ji''} = (a_{i_1}, d_j)$ and $t_{ji''} - t_{ji'''} = (a_{i_2}, d_j)$, then i_1 and i_2 cannot either coincide or differ by 1 (more precisely, $i_1 \neq i_2$, $i_1 \neq s(i_2)$, $i_2 \neq s(i_1)$, due to the cyclic order defined on the set N_6 in section S 2). Otherwise, the second and the third rows of the table would contain two equal items i'' in the same column or in two neighboring columns (by a diagonal) which is impossible. (It is easily seen that each pair of the items (i', i') in the second row of the table is in a diagonal neighborhood with the pairs $(i'', i''), (i''', i''')$ from the third row, where i', i'', and i''' are different.) Thus, putting $d^k = \sum_{j=1}^{k} d_j$, we obtain

$$\beta = \max_{\{i',i'',i''' \mid i''\neq i',\, i''\neq i'''\}} \left(\max_{k\in\mathcal{N}} \min\left\{ \sum_{q=1}^{k-1} (t_{qi''} - t_{qi'''}) + t_{ki''}, \sum_{q=1}^{k} (t_{qi''} - t_{qi'''}) + t_{ki'''} \right\} + \max_{k\in\mathcal{N}} \min\left\{ \sum_{q=1}^{k-1} (t_{qi'} - t_{qi''}) + t_{ki'}, \sum_{q=1}^{k} (t_{qi'} - t_{qi''}) + t_{ki''} \right\} \right) \le \max_{(i_1,i_2)\in I} \left(\max_{k\in\mathcal{N}} \min\{ (a_{i_2}, d^{k-1}), (a_{i_2}, d^k) \} + \max_{k\in\mathcal{N}} \min\{ (a_{i_1}, d^{k-1}), (a_{i_1}, d^k) \} \right) + 2,$$

where $I = \{(i_1, i_2) \mid i_1, i_2 \in N_6, i_1 \neq i_2, i_1 \neq s(i_2), i_2 \neq s(i_1)\}$. Since the permutation $\pi = (1, 2, \ldots, n)$ provides a nonstrict summation of vectors in $D = \{d_1, \ldots, d_n\}$ within half-spaces of the family $\{P(a_1, \beta_1), \ldots, P(a_6, \beta_6)\}$, we obtain

$$\beta \le 2 + \max_{(i_1,i_2)\in I} (\beta_{i_1} + \beta_{i_2}) = 2 + \theta_3(b_6) \le 2 + \theta_3^*.$$

From (61) it follows that

$$C_i = M + D_i \le M + 1 + \beta \le M + 3 + \theta_3^*,$$

which provides the desired bound on the schedule length. Theorem 8 is proved.

Theorem 8 and Corollary 2 of Theorem 1 imply the following results.

Theorem 9. *Given an instance of the Akers–Friedman problem with three machines and n jobs, a schedule S can be constructed in $O(n \log n)$ time with length*

$$T(S) < M + 5.5K.$$

Corollary 6. *Given an instance of the Akers–Friedman problem, its optimum belongs to an interval $[M, M + \mu_{AF}(m)K]$, where the function $\mu_{AF}(m)$ for $m = 3$ satisfy the inequalities*

$$3 \le \mu_{AF}(3) \le 5.5.$$

The R231-Problem

Theorem 10. *Suppose that some Algorithm $\mathcal{A}$, given every positive integer n and every $\hat{s}$-family $\{x_1, \dots, x_n\} \subset \mathbb{R}^2$, solves the* NVS4*-problem in time $\mathcal{T}_{\mathcal{A}}(n)$ with the objective function*

$$\theta_4(b_3) \leq \theta_4^*.$$

Then, given an instance of the R231 *problem with n jobs, a schedule S can be constructed in $O(\mathcal{T}_{\mathcal{A}}(n) + n \log n)$ time with length*

$$T(S) \leq M + \big(3 + \theta_4^*\big)K.$$

PROOF. To determine the desired schedule S, we apply some Algorithm $\mathcal{A}_4$ that differs from Algorithm $\mathcal{A}_3$ by that, at Stage 1 we solve the NVS4-problem instead of the NVS3-problem.

Suppose that the NVS4-problem is solved and the corresponding permutation $\pi = (\pi_1, \dots, \pi_n)$ of vector indices provides a nonstrict summation of vectors in $D = \{d_1, \dots, d_n\} \subset \mathbb{R}^2$ within the family of half-spaces $\mathcal{P}_4(2, b_3) = \{P(e_1 - e_2, \beta_1), P(e_2, \beta_2), P(-e_1, \beta_3)\}$, where

$$\theta_4(b_3) = \max\{\beta_1 + \beta_2,\ \beta_2 + \beta_3\} \leq \theta_4^*. \tag{62}$$

Again, we assume for simplicity that $K = 1$ and the jobs are numbered according to the permutation π, or in other words that $\pi = (1, 2, \dots, n)$.

Repeating the argument in the proof of Theorem 8 for the Akers–Friedman problem, we estimate from above the time when the second and the third operations of job $j \in \mathcal{N}$ become available for processing. Firstly, each job $j \in \mathcal{N}_1$ arrives at machine 3 before time $T_3^{j-1} + 3 + \beta_1 + \beta_2$. Secondly, each job $j \in \mathcal{N}_2$ arrives at machine 3 before time $T_3^{j-1} + 1 + \beta_2$ and at machine 1 before time $T_1^{j-1} + 3 + \beta_2 + \beta_3$.

Hence, machine 3 completes its work before time $M + 3 + \beta_1 + \beta_2$, and machine 1 does so before time $M + 3 + \beta_2 + \beta_3$, which by (62) provides the desired bound for the length of schedule S. Theorem 10 is proved.

Theorem 10 and Corollary 3 of Theorem 1 imply the following results.

Theorem 11. *Given an instance of the* R231*-problem with n jobs, a schedule S can be constructed in $O(n \log n)$ time with length*

$$T(S) < M + 5K.$$

Corollary 7. *Given an instance of the* R231*-problem, its optimum belongs to an interval $[M, M + \mu_{\text{R231}} K]$, where the constant μ_{R231} satisfy the inequalities*

$$3 \leq \mu_{\text{R231}} \leq 5.$$

Acknowledgments. The author thanks È. Kh. Gimadi and N. I. Glebov for useful discussion of this article.

References

1. S. V. Sevast'yanov (1994) Nonstrict vector summation in scheduling problems (in Russian), *Sibirsk. Zh. Issled. Oper.* **1**, No. 2, 67–99.
2. T. Fiala (1983) An algorithm for the open-shop problem, *Math. Oper. Res.* **8**, No. 1, 100–109.
3. T. Gonzalez and S. Sahni (1976) Open shop scheduling to minimize finish time, *J. Assoc. Comput. Mach.* **23**, No. 4, 665–679.
4. I. Bárány and T. Fiala (1982) Nearly optimum solution of multimachine scheduling problems, *Szigma–Mat.-Közgaz-Dasági Folyóirat.* **15**, No. 3, 177–191.
5. S. V. Sevast'yanov (1992) A polynomially solvable case of the open-shop problem with an arbitrary number of machines (in Russian), *Kibernet. Sistem. Anal.* (Kiev), No. 6, 135–154.
6. S. V. Sevast'yanov (1995) Vector summation in Banach space and polynomial algorithms for flow shops and open shops, *Math. Oper. Res.* **20**, No. 1, 90–103.
7. S. V. Sevast'yanov (1982) Algorithms with estimates for the Johnson and Akers–Friedman problems in the case of three machines (in Russian), *Upravlyaemye Sistemy* **22**, 51–57.
8. R. E. Tarjan (1983) *Data Structures and Network Algorithms*, SIAM, Philadelphia, PA.
9. B. I. Dushin (1980) Note on the algorithm for Johnson's single-route problem (in Russian), *Kibernetika* (Kiev), No. 2, 129–131.
10. J. W. J. Williams (1964) Heapsort, *Comm. ACM* **7**, No. 6, 347–348.

Sobolev Institute of Mathematics
Universitetskiĭ pr., 4
Novosibirsk 630090
RUSSIA

TRANSLATED BY THE AUTHOR

BOUNDS FOR APPROXIMATE SOLUTION OF A SCHEDULING PROBLEM†)

P. I. Sharygin

We investigate a scheduling problem with two different criteria. When considering the criterion of the minimal length of a schedule, the value of the objective function of the problem under consideration is a lower bound for the objective function of the strip packing problem. We demonstrate by an example that there exist some lists of jobs in which the ratio of values of optimum functions can attain 5/4. For the case of the maximum total profit we show by examples that the values of the objective function can differ from the optimal values arbitrarily. We propose some exact lower bounds provided that the jobs parameters satisfy some additional constraints.

In the present paper, we consider a particular case of the scheduling problem (SP) with limited resources and due dates (for the general formulation of the SP see [1]). We assume satisfied the following conditions:

— there is a limited nonstorable resource of one kind whose intensity of supply is constant in the entire makespan;

— all jobs are independent,

— starting and finishing times of jobs are not appointed, the intensity of the resource consumed by a job is constant in the makespan of the job.

If we associate with every job a rectangle whose height and width indicate the intensity of resource consumption and the makespan of the job, respectively. This problem can be regarded as a problem of packing rectangles into a semi-infinite strip whose height indicates the intensity of the total resource supply. Here any rectangle may be fractioned by vertical cuts into lesser strips called *slices*. In every admissible packing the totality of horizontal projections of the slices forms an interval equal to the width of the rectangle. We also admit vertical permutation of slices.

In Section 1 the problem is considered with the criterion of the minimal makespan, i.e., the minimal length of a schedule (SP_{min}). From the formulation of the problem we can see that without fractioning rectangles into slices the SP_{min} is a problem of packing rectangles (PPR) into a strip of minimal length. Additional opportunities provided by the admissibility of vertical permutation of slices make it possible in some cases to transform an optimal solution for the PPR (which is indisputably a schedule) to an even shorter schedule for the SP_{min}. Hence, a solution

†) The research was supported by the Russian Foundation for Basic Research (Grant 93-01-00489).

A. D. Korshunov (ed.), Operations Research and Discrete Analysis, 273–283.

for the SP_{min} gives a lower bound for the problem of packing rectangles. Therefore, the relation between optimal values of the objective functions for the problem of packing rectangles and the SP_{min} is of particular importance to our study.

Denote by $S(L)$ the optimal value of the makespan in the SP_{min} with the list of jobs L, and by $P(L)$, the optimal value of the packing length in the PPR with a list of the rectangles L corresponding to the jobs in L. Let

$$\rho = \sup_L \frac{P(L)}{S(L)}.$$

From the proof of the upper bound for the $P(L)$ obtained in [2] with the help of the NFDH algorithm one can easily derive the bound $P(L) \leq 3S/B$ for every list L, where S is the total area of the rectangles and B is the height of the strip. This bound and the obvious inequality $S(L) \geq S/B$ imply the inequality $\rho \leq 3$. The author is not aware of any other bounds for $P(L)$ in terms of the total area of the rectangles.

In [1] there is an example of a list of jobs with $\rho \geq 6/5$. In this paper, we present a list of jobs which makes it possible to improve the lower bound up to 5/4.

In Section 2, the problem with the maximum total profit criterion (SP_{max}) was is considered when the profit gained by a job execution is calculated as the product of the resource consumption intensity and the duration of the job. It is shown that the value of the objective function obtained by means of the so-called list algorithms can differ considerably from the optimal one. Also, precise lower bounds are given for certain cases with additional constraints imposed on parameters of the jobs.

Note that these problems are NP-complete [3].

1. The Criterion of the Minimum Makespan of a Schedule

Let us formulate the problem. Consider a semi-infinite strip of height B and a list $L = \{(b_i, t_i)_{i=1}^n\}$ of rectangles (items). Let the height of item i be equal to $b_i \leq B$ and the width of item i be equal to t_i. Pack the items into the strip so as to minimize the length of the part of the strip in use. The items cannot rotate or overlap one another although fractioning items into slices and vertical permutations are permitted.

Proposition 1. *There exists a list L_0 such that*

$$P(L_0)/S(L_0) = 5/4.$$

PROOF. Let the strip be of height 12. Consider the list L_0 consisting of 9 items. Information on the items is given in the table.

Item	Height	Width	Time
1	3	2	0
2	2	1	2
3	5	1	3
4	2	2	1
5	8	1	0
6	5	1	1
7	6	2	2
8	1	3	0
9	1	3	1

The numbers, height, and width of the items and the starting times of the corresponding jobs in an optimal schedule are indicated in the columns.

As one can see in Fig. 1,*a*, below, the makespan of an optimal schedule is 4. We prove that it is impossible to find a packing of length 4. Consider a rectangle of height 12 and width 4 and the list of items mentioned above.

Divide the rectangle into the following disjoint rectangles:

- $\{S_i\}_{i=1}^4$ of height 12 and width 1, called *columns*,
- $\{R_i\}_{i=1}^{12}$ of height 1 and width 4, called *strips*.

Items $A_1, \dots, A_k$ will be called

- *linked horizontally* (we denote this by $A_1 h \dots h A_k$) if there exists a column whose intersection with each of the items has nonzero area,
- *linked vertically* (we denote this by $A_1 v \dots v A_k$) if there exists a strip whose area intersection with each of the items has nonzero area.

We use the following notation:

$$\overline{A_1 h \dots h A_k} \iff A_1 h \dots h A_k \text{ is not valid;}$$

$$\overline{A_1 v \dots v A_k} \iff A_1 v \dots v A_k \text{ is not valid.}$$

Note the following simple property of the linked items $A_1, \dots, A_k$: if $A_i h A_j$, then $A_1 h \dots h A_k$, $1 \le i < j \le k$. Indeed, let x_i and y_i be the numbers of the left and right columns linking item A_i. The intervals $[x_i, y_i]$ and $[x_j, y_j]$ intersect if and only if $(x_i \le y_j)\&(x_j \le y_i)$. If the intervals $[x_i, y_i]$, $i = 1, \dots, k$, intersect pairwise, then $x_i \leq y_j$, $1 \leq i, j \leq k$. Therefore, $x = \max_i x_i \leq \min_i y_j = y$, and the interval $[x, y]$ is nonempty. Vertically linked items have a similar property.

We introduce the following notation:

$$H(A_{i_1}, \dots, A_{i_k}) \Longleftrightarrow b_{i_1} + \dots + b_{i_k} > 12,$$

$$L(A_{i_1}, \dots, A_{i_k}) \Longleftrightarrow t_{i_1} + \dots + t_{i_k} > 4.$$

We show that without permuting slices of some items we cannot allocate all items inside the large rectangle. Assume this it is not true. Since linking of items is not admitted, for any two items A and B we have $(AhB)\&(AvB) =$ FALSE.

Consider the four possible cases.

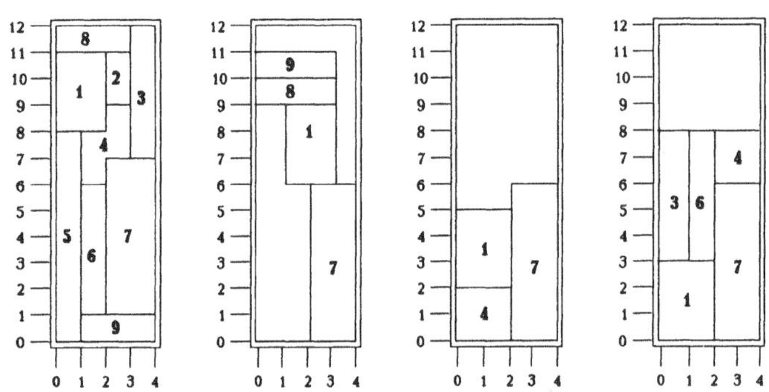

a *b* *c* *d*

Fig. 1

CASE 1: $1h7$ (Fig. 1,*b*). From $1h7$, $1h8$, $1h9$, $7h8$, $7h9$, and $8h9$ it follows that $1h7h8h9$. Therefore, in the column linking simultaneously items 1, 7, 8, and 9 an empty space of height 1 remains which can not be filled since other items of height 1 now be absent.

CASE 2: $\overline{(1h7)}\&\overline{(4h7)}$ (Fig. 1,*c*). From $H(1,4,5)$ and $H(5,7)$ it follows that there is no admissible position for item 5.

CASE 3: $\overline{(1h7)}\&\overline{(1h4)}$ (Fig. 1,*d*). From $H(3,4,7)$ and $H(6,4,7)$ we obtain $1h3$ and $1h6$. But from $H(4,7,5)$, $H(1,3,5)$, and $H(1,6,5)$ it follows that there is no admissible position for item 5.

CASE 4: $\overline{(1h7)}\&(1h4)\&(4h7)$ (see Fig. 2,*a–c*). From $H(1,4,5)$ and $H(5,7)$ it follows that $1h5$ and $4h5$. For definiteness, we assume that column 1 links items

1 and 5. Since $b_1 + b_5 = 11$, in column 1 there remains a lacuna of height 1 which can be filled only by items 8 or 9 (for definiteness, let it be item 8). Thus, column 1 is entirely filled. $H(1,3,6)$ implies $3h7$ or $6h7$ (for definiteness, we assume $3h7$). Hence, from $H(3,4,7)$ it follows that S_4h3h7, and the remaining lacuna of height 1 in column 4 can be filled by item 9 only. Thus, $S_4h3h7h9$. Finally, $H(4,6,7)$ implies $S_2h1h4h6h8h9$, i.e., column 2 is entirely filled. From $L(1,9)$ and $L(8,9)$ it follows that $5v9$, and from $L(8,9)$ and $L(8,7)$ it follows that $3v8$.

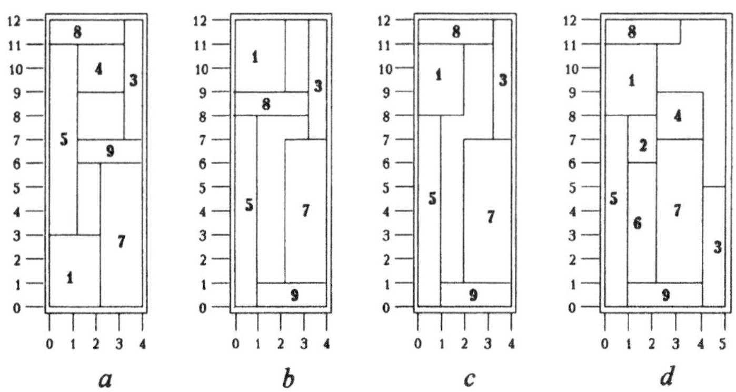

Fig. 2

There are three possible mutual allocations to within symmetry for items 1, 5, and 8 on the vertical.

CASE 1: item 5 is placed between items 1 and 8 (Fig. 2,a). From $b_3 + b_7 > b_8 + b_5$ it follows that $1v7$. Then item 9 is placed between items 3 and 7. In this case we have uniquely defined positions for items 1, 3, 5, 7, 8, and 9 which is in contradiction with the admissible placement of item 6.

CASE 2: item 8 is placed between items 1 and 5 (Fig. 2,b, up to the mutual permutation of items 7 and 9).

CASE 3: item 1 is placed between items 5 and 8 (Fig. 2,c, up to the mutual permutation of items 7 and 9).

In Cases 2 and 3 there exists a unique strip R_i such that R_iv3v5. Therefore, from $L(3,1,4)$, $L(3,8,4)$, $L(7,1,4)$, $L(7,5,4)$, $L(7,8,4)$, $L(9,1,4)$, $L(9,5,4)$, and $L(9,8,4)$ we arrive at a contradiction with the admissible placement of item 4.

Thus, in all the cases there is no packing of the chosen list into the given rectangle. The optimal packing is shown in Fig. 2,d. Proposition 1 is proved.

Further improvement on the lower bound for ρ by the indicated method is impossible since for every schedule of length 3 for jobs whose makespans are 1, 2 or 3 it is easy to pack the jobs into a strip of length 3.

2. The Criterion of the Maximum Total Profit

The problem SP_{max} can be formulated as follows. Given are a rectangle (container) of height B and width T and a list $L = \{(b_i, t_i)_{i=1}^{n}\}$ of rectangles (items). The height of item i is $b_i \leq B$ and its width is $t_i \leq T$. With item i we associate a number c_i called the cost of the item. The problem is to select and pack a subset L' of the items in the list L so that the following conditions hold:

- all items are placed inside the container,
- the sum of the numbers assigned to the items in L' has a maximal value over all such subsets.

The items cannot rotate or overlap one another although fractioning items into slices and vertical permutations are permitted.

REMARK. If an approximate method results in $L' = L$, then we have an optimal packing. Hence, below we suppose that at least one item from the list L does not belong to L'.

Consider the case in which the cost of an item equals its area, i.e. the product of its height and width. Then we must fill the container with items so as to maximize the total area of the items placed in the container.

By $\mathrm{Opt}(L)$ we denote the value of an optimal solution of the problem in the list L, and by $A(L)$, the value of a solution resulting from processing the list L by the algorithm A.

We present some examples which show that each listed algorithm (i.e., an algorithm processing a list in order in which it is given for processing) using preliminarily the sortings described below computes a solution with an arbitrarily large deviation from an optimal one, i.e., for any $q > 1$, there exists a list L_0 such that $\mathrm{Opt}(L_0)/A(L_0) = q$.

EXAMPLE 1. Sorting by nonincrease of the width of an item:

$$L_0 = \{(B/q, T),\ (B, 0.5T),\ (B, 0.5T)\},\ \mathrm{Opt}(L_0) = BT,\ A(L_0) = BT/q.$$

EXAMPLE 2. Sorting by nonincrease of the height of an item:

$$L_0 = \{(B, T/q),\ (0.5B, T),\ (0.5B, T)\},\ \mathrm{Opt}(L_0) = BT,\ A(L_0) = BT/q.$$

EXAMPLE 3. Sorting by nonincrease of the area of an item:

$$L_0 = \{(B/q,T),\ ((B,0.5T/q)_{i=1,\dots,2q})\},\ \mathrm{Opt}(L_0) = BT,\ A(L_0) = BT/q.$$

EXAMPLE 4. Sorting by nondecrease of the width, height, or area of an item:

$$L_0 = \{(B/\sqrt{q},T/\sqrt{q}),(B,T)\},\ \mathrm{Opt}(L_0) = BT,\ A(L_0) = BT/q.$$

Further, we consider the problem with some restrictions on the heights and widths of an item. We now prove some estimates for the following simple algorithm.

ALGORITHM A1

- We sort the list to be processed in order of the nonincrease of the width of an item and then begin to process the list.
- We place the currently processed item in the left upper corner of the container. Moving it along the upper frontier, we find the first position for the item in which it does not overlap the current packing.
- If we find such position, before the right frontier of the item intersects the right frontier of the container, we stop moving the item. The item is lowered and its slices are moved down until no gaps remain under the item. This gives a new current packing and we begin to process the next item in the list.
- If no such position is found, then we reject the item and begin to process the next item in the list.

In this manner we process the entire list.

The result of applying Algorithm A1 to the list L_0 in Section 1 for the container of height 12 and width 4 is shown in Fig. 3. Algorithm A1, after sorting, processes the list in the following order:

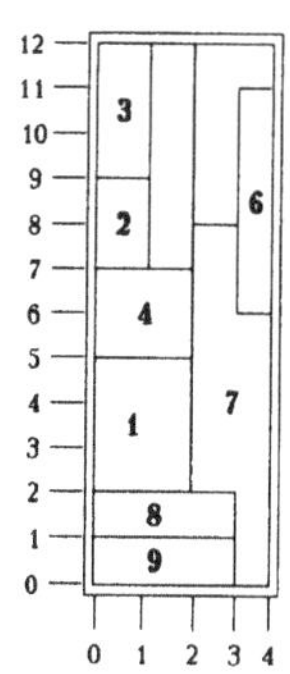

Fig. 3

$$(8, 9, 1, 4, 7, 2, 3, 5, 6).$$

Item 5 is not inserted in the packing. Let b and t be numbers such that the parameters of items in the processed list

$$L = \{(b_i, t_i)_{i=1,\dots,n}\}$$

satisfy the conditions

$$0 < b_i \le b, \quad 0 < t_i \le t.$$

The total area filled in the container when using Algorithm A1 with the list of jobs satisfying the conditions mentioned above is denoted by $S_{A1}(b,t)$.

We introduce some definitions and notation.

— The straight line parallel to the side frontiers of the container and lying at distance τ from the left frontier will be called the *moment* of τ.

— The straight line parallel to the upper and lower frontiers of the container and lying at distance β from the lower frontier will be called the *level* of β.

Denote by $\tau+0$ ($\tau-0$) a moment arbitrarily close to the moment of τ from the right (left) but different from it. The height of the packing in the moment of τ is denoted by $y(\tau)$.

If $\tau < 0$ and $\tau > T$, we put $y(\tau) = 0$.

For the moments of $\tau \in [0,T]$ other than the left and right frontiers of any item, $y(\tau) = \sum b_i$, where the sum is taken over all the items intersected by the moment of τ.

For the moments of $\tau \in [0,T]$ coinciding with either the left or right frontier of an item, we put $y(\tau) = \max\{y(\tau-0), y(\tau+0)\}$.

Proposition 2. *For every $b \in (0,B]$ and every $t \in (0,T]$ the inequality*

$$S_{A1}(b,t) > (B-b)\max\{(T-t), tT\lfloor T/t\rfloor/(T+t)\}$$

holds.

PROOF. Consider the packing obtained right after the first rejection of an item which could not be placed into the rectangle. The width of this item is denoted by θ.

We prove that the packing covers a rectangle of height $B-b$ and width $T-\theta$. Consequently, $S_{A1}(b,t) > (B-b)(T-\theta)$. Since the item being processed is lowered in such a way that no gaps remain under it for every moment of $\tau \in [0,T]$ the remaining empty space in the rectangle is entirely above the packing.

We prove that in every moment of $\tau \in [0, T-\theta]$ the height of the packing is greater than the level of $B-b$. To prove this fact we need the following

Lemma 1. *For arbitrary $\tau \in [0,\tau_0]$, $\tau_0 > T-\theta$, the height $y(\tau)$ of the packing satisfies the inequality $y(\tau) > B-b$.*

PROOF. Let the moment of τ_0 be the first when the condition $y_0 = y(\tau_0+0) \le B-b$ is true. The following cases are possible.

CASE 1: there exists a moment of $\tau > \tau_0$ such that $y(\tau) > B-b$.

CASE 2: for all $\tau \in (\tau_0, T]$, the height of the packing satisfies the inequality $y(\tau) \le B-b$.

In the case 1 we suppose that $\tau_1 > \tau_0$ is such that the moment of τ_1 is the first after that of τ_0 when the height of the packing is greater than $B - b$, i.e. $y(\tau_1 - 0) \leq B - b < y(\tau_1)$.

Then there exists an item p in the packing whose left frontier coincides with τ_1. Let us return to the step of the algorithm in which we packed p. Denote by $y'(\tau)$ the height of the packing at this step in the moment of τ. When packing the item p one can see that when moving p along the upper frontier of the container we must stop no later than at the moment of τ_0, since $y'(\tau) \leq y(\tau) \leq B - b \leq B - b_p$ for every moment of $\tau \in (\tau_0, \tau_1)$ and $y'(\tau) \leq y(\tau) - b_p \leq B - b_p$ for all $\tau \in [\tau_1, \tau_1 + t_p - 0]$. Hence, in the moment of τ_0 the item p does not overlap the packing and its movement to the moment of a later τ_1 contradicts the principles of Algorithm A1. This contradiction proves that Case 1 is impossible.

In Case 2, we have $\tau_0 > T - \theta$ because otherwise there exist a gap in the right upper corner of the container of height b and width θ sufficient for allocation of the item which had not been allocated. Lemma 1 is proved.

Therefore, $y(\tau) > B - b$ for any $\tau \in [0, T - \theta]$, i.e. the filled rectangle has an area greater than $(B - b)(T - t)$. Consequently,

$$S_{A1}(b, t) > (B - b)(T - t). \tag{1}$$

Assume that $\theta \leq tT/(T + t)$. Hence, we have

$$\begin{aligned}(B - b)(T - t) &\geq (B - b)\left(T - \frac{tT}{T + t}\right) = (B - b)T^2/(T + t)\\ &\geq (B - b)tT\lfloor T/t \rfloor/(T + t).\end{aligned}$$

Suppose that θ satisfies the inequalities

$$tT/(T + t) < \theta \leq t. \tag{2}$$

At every step of the algorithm the boundary between the empty space in the container and the packing is a staircase broken line, and the item p being processed when moving along the upper frontier of the container stops either at the moment of 0 or at the moment of some $\tau' > 0$ which forms a "descending step," i.e., $y'(\tau' - 0) > y'(\tau' + 0)$, where $y'(\tau)$ is the height of the packing at the current step of the algorithm. Hence, the left frontier of item p coincides with either the left frontier of the container or the right frontier of item q.

Consider items $\{p_i\}_{i=1}^{s}$ whose right frontiers lie strictly to the right of the moment of $T - \theta$ and whose left frontiers does not lie to the right of the moment of $T - \theta$. By Lemma 1 such items exist. The width of item p_i is denoted by t_{i_1}. Its left frontier coincides with either the left frontier of the container or the moment coinciding with the right frontier of some item whose width is denoted by t_{i2}. Also, the left frontier of the item coincides with either the left frontier of the container or the moment coinciding with the right frontier of some item whose width is denoted by t_{i3}. Continuing this process we reach an item whose left frontier coincides with the left frontier of the container. The width of this item is denoted by t_{ik}.

Thus, the right frontier of item p_i coincides with the moment that lies at distance $\tau_i = t_{i1} + \ldots + t_{ik}$ from the left frontier of the container. Taking into account (2) and the inequality $i_j \geq \theta$, $1 \leq j \leq k$, we obtain

$$T - t \leq T - \theta < \tau_i \leq kt, \quad ktT/(T+t) < k\theta \leq \tau_i \leq T.$$

Therefore, $T/t - 1 < k < T/t + 1$. Since k is integer, we have $k = \lfloor T/t \rfloor$ or $k = \lceil T/t \rceil$ and the value of τ_i is not less than $k\theta > tT\lfloor T/t \rfloor/(T+t)$. This relation holds for any i, $1 \leq i \leq s$.

By Lemma 1, the sum of the heights of items $\{p_i\}_{i=1}^{s}$ is greater than $B - b$. Therefore, up to the moment of $\tau' = tT\lfloor T/t \rfloor/(T+t)$ the height of the packing is greater than $B - b$. Hence,

$$S_{\mathrm{A1}}(b,t) > (B-b)tT\lfloor T/t \rfloor/(T+t).$$

Together with (1) this gives the desired result. Proposition 2 is proved.

REMARK. It is easy to see that the time complexity of Algorithm A1 is equal to $O(n^2)$.

Corollary 1. *If $T = kt$ and k is integer, then*

$$S_{\mathrm{A1}}(b,t) > (B-b)kT/(k+1).$$

Let the parameters of the problem depend on the number n of items, i.e., $b = b_n$, $B = B_n$, and $T = T_n$. Then the following proposition is valid.

Corollary 2. *If $b_n = o(B_n)$ and $t_n = o(T_n)$, then Algorithm* A1 *is asymptotically optimal.*

PROOF. The value delivered by an optimal algorithm does not exceed B_nT_n. Therefore,

$$1 \geq \frac{S_{\mathrm{A1}}(B_n,t_n)}{B_nT_n} \geq \frac{(B_n-b_n)(T_n-t_n)}{B_nT_n} = \left(1-\frac{b_n}{B_n}\right)\left(1-\frac{t_n}{T_n}\right) = 1-\varepsilon_n,$$

where

$$\varepsilon_n = \frac{b_n}{B_n} + \frac{t_n}{T_n} - \frac{b_n}{B_n}\frac{t_n}{T_n} \to 0 \text{ as } n \to \infty.$$

Corollary 2 is proved.

We provide the following example. Let $b = B$ and $t = T$. For sufficiently small $\varepsilon \in (0, B)$, we consider the list $L = \{(\varepsilon, T), (B, 0.5T), (B, 0.5T)\}$. Then the value $\varepsilon = \varepsilon T$ obtained by Algorithm A1 will be sufficiently close to the value $(B - b)\max\{(T - t), tT[T/t]/(T + t)\} = 0 \cdot 0.5T = 0$.

References

1. È. Kh. Gimadi (1988) Some mathematical models and methods for planning large-scale projects (in Russian), in: *Modeli i Metody Optimizatsii*, Trudy Inst. Math. Vol. 10, Nauka, Novosibirsk, pp. 89–115.
2. E. G. Coffman (jr.), M. R. Garey, D. S. Johnson, and R. E. Tarjan (1980) Performance bounds for level-oriented two-dimensional packing algorithms, *SIAM J. Comput.* **9**, No. 4, 808–826.
3. M. R. Garey and D. S. Johnson (1979) *Computers and Intractability*, Freeman, San Francisco.

Novosibirsk State University
ul. Pirogova, 2
Novosibirsk 630090
RUSSIA

TRANSLATED BY A. V. RYAZANOV

ON CERTAIN BOUNDS FOR THE WEIGHT OF *l*-BALANCED BOOLEAN FUNCTIONS[†)]

Yu. V. Tarannikov

The density of a Boolean function f of n arguments is the quantity $\rho(f) = W_f/2^n$, where W_f is the number of vectors of length n at which f equals 1. The function f is called l-balanced if $|W_{f_1} - W_{f_2}| \leq l$ for each of its subfunctions f_1 and f_2 of the same number of arguments. It is established that, for a fixed l and large n's, the densities of l-balanced Boolean functions of n arguments approach one of the following five numbers: 0, 1/3, 1/2, 2/3, or 1.

In the theory of control systems, finding classes of Boolean functions which are informative in some sense occupies an important place.

Boolean functions may be defined in different ways, for example, by tables, disjunctive normal forms, Zhegalkin polynomials, etc. In many cases, it is convenient to define a function by specifying the subset of vertices of the Boolean cubes at which the function equals 1 (or 0). The problem of classifying Boolean functions depending on the structure of this subset is of interest. For example, for monotone functions, symmetric functions, and some other Boolean functions, these subsets have a rather distinctive structure.

In this paper, as done in [1] for binary sequences, the classification of Boolean functions is studied according to their "degree of homogeneity," the parameter characterizing how uniformly the unit values of a function are distributed throughout its n-dimensional domain.

At present, several notions of "homogeneity" of the Boolean functions are under study and so far there is no clearly established terminology for classifying these notions. Often a function is called homogeneous if its "center of gravity" is located exactly at the middle of the Boolean cube. We point out, however, that the property defined this way may fail when passing from a function defined on the cube to a subfunction defined on a subcube. A notion of l-balancedness, proposed by E. P. Lipatov, is free of this drawback.

We define the *weight* of a Boolean function f to be the quantity W_f equal to the number of vectors at which the function f takes the value 1. We call the quantity $\rho(f) = W_f/2^n$ the *density* of a Boolean function f of n arguments.

†) This research was supported by the Russian Foundation for Basic Research (Grant 93–01–01527) and the Lyapunov French-Russian Center of Applied Mathematics and Informatics at Moscow State University.

A. D. Korshunov (ed.), Operations Research and Discrete Analysis, 285–299.

Let l be a nonnegative integer. A Boolean function $f(x_1, x_2, \ldots, x_n)$ is called *l-balanced* if $|W_{f_1} - W_{f_2}| \leq l$ for any of its subfunctions f_1 and f_2 of the same number of arguments.

In [2], all 1-balanced Boolean functions are described. Certain bounds for the weight of l-balanced Boolean functions appear in [3]. The purpose of the present article is to prove that, for large n, the densities of l-balanced functions approach one of the following five numbers: 0, 1/3, 1/2, 2/3, or 1. More precisely, this result is stated as follows.

Theorem. *For each positive integer l and each positive ε, there exists a positive integer N such that, for each positive integer n, $n \geq N$, and for each l-balanced Boolean function f of n arguments, only one of the following five inequalities holds:*

$$W_f \leq 2l;$$

$$|\rho(f) - 1/3| < \varepsilon;$$

$$|\rho(f) - 1/2| < \varepsilon;$$

$$|\rho(f) - 2/3| < \varepsilon;$$

$$W_f \geq 2^n - 2l.$$

The theorem is established at the end of the paper after a number of lemmas.

Lemma 1. *For each positive integer m, there exists an $N(m)$ such that, for $n > N(m)$ and $l < m/2$, there are no l-balanced Boolean functions of n arguments with weight m.*

PROOF. To a function $f\ (x_1, \ldots, x_n)$ with weight m, assign the matrix of size $m \times n$ whose rows are all the m vectors of length n at which the function f equals 1. For $n \geq 2^m + 1$, the matrix has two coinciding columns. Assume that these columns correspond to the variables x_i and x_j. Then the weight of the subfunctions $f|_{x_i=0,\ x_j=1}$ and $f|_{x_i=1,\ x_j=0}$ equals 0 and the weight of one of the subfunctions $f|_{x_i=0,\ x_j=0}$ and $f|_{x_i=1,\ x_j=1}$ is at least $l + 1$. Hence, the function f is not l-balanced. Thus, the assertion of Lemma 1 is true for $N(m) = 2^m$.

Lemma 2. *For each positive integer l, one can find a positive constant $c(l)$ such that, for each l-balanced Boolean function f with weight W_f, $W_f > 2l$, the inequality $\rho\ (f) \geq c\ (l)$ holds.*

PROOF. Let $f = f_0$ be a function satisfying the hypothesis of Lemma 2. If $W_f > 5l$, then decompose the function f with respect to any variable into two subfunctions of $n-1$ arguments and denote by f_1 the subfunction of less weight (if the weights are equal, select f_1 arbitrarily). The property of l-balancedness implies that $W_{f_1} > 2l$; in addition, it is evident that $\rho\ (f) \geq \rho\ (f_1)$. Further, if $W_{f_1} > 5l$, then decompose the function f_1 with respect to any variable into two subfunctions of $n-2$ arguments and denote by f_2 the subfunction with the lesser weight (if the weights are equal, select f_2 arbitrarily), and so on. At some (sth) step, $s \geq 0$, obtain a function f_s whose weight satisfies $2l+1 \leq W_{f_s} \leq 5l$ and whose density satisfies $\rho\ (f) \geq \rho\ (f_s)$.

By Lemma 1, for each m, $m > 2l$, there exists only a finite number of l-balanced functions with weight m. Therefore, the quantity $c(l) = \min \rho(g)$ is well-defined, where the minimum is taken over all l-balanced functions g such that $2l+1 \leq W_g \leq 5l$.

From the definition of $c(l)$, it follows that $c(l) > 0$ and $\rho\ (f_s) \geq c(l)$. Hence, $\rho\ (f) \geq c(l)$. Lemma 2 is proved.

A Boolean function f is called *symmetric* if each of its values do not change under a permutation of arguments. A symmetric Boolean function is completely defined by the characteristic sequence $\widetilde{\pi} = (\pi_0, \pi_1, \ldots, \pi_n)$, where π_i is the value of f at any vector containing i ones and $n-i$ zeros. We say that a symmetric Boolean function of n arguments has *period* T if $\pi_i = \pi_{i+T}$ for all i, $0 \leq i \leq n-T$. Let $0 \leq k_1 \leq k_2 \leq n$. We define the *restriction* $f[k_1, k_2]$ of a symmetric Boolean function f of n arguments to be the function of $(k_2 - k_1)$ arguments obtained by the substitution in f the constant 1 for some k_1 arguments and the constant 0 for some other $n - k_2$ arguments.

Denote by S_n the n-element set, and by $P_r(S_n)$ the set of all r-element subsets (r-subsets) of S_n. A decomposition $P_r(S_n) = A_1 \cup \cdots \cup A_t$ is called an *ordered t-decomposition.*

Ramsey's Theorem [4, p. 78]. *Let $r \geq 1$ and $q_i \geq r$ $(i = 1, \ldots, t)$. Then there exists the least positive integer $\mathcal{M} = \mathcal{M}(q_1, \ldots, q_t; r)$ such that, for each $n \geq \mathcal{M}$ and each ordered t-decomposition $P_r(S_n) = A_1 \cup \cdots \cup A_t$, there exist subsets A_i and $S' \subset S_n$ such that A_i contains all r-subsets of S', where $|S'| \geq q_i$ (such an S' is called a $(q_i,\ A_i)$-subset).*

Lemma 3. *To each positive integer n_1, there is a positive integer N such that each Boolean function f of N arguments has a symmetric subfunction of n_1 arguments.*

For an equivalent formulation of Lemma 3, see [5].

PROOF. With each r-subset $\mathcal{A} = \{i_1, i_2, \ldots, i_r\}$, $1 \leq i_1 < i_2 < \cdots < i_r \leq n$, of the set $S_n = \{1, 2, \ldots, n\}$, associate a binary vector $\widetilde{\alpha}(\mathcal{A}) = (\alpha_1, \alpha_2, \ldots, \alpha_n)$ such that

$$\alpha_i = \begin{cases} 0 & \text{if} \quad i \notin \mathcal{A}, \\ 1 & \text{if} \quad i \in \mathcal{A}. \end{cases}$$

Let f be an arbitrary function of n arguments. For each r, $r = 1, 2, \ldots$, consider the decomposition $P_r(S_n) = A_{r,0} \cup A_{r,1}$, $r = 1, 2, \ldots$, defined as follows:

$$\begin{cases} \mathcal{A} \in A_{r,0} & \text{if} \quad f(\widetilde{\alpha}(\mathcal{A})) = 0, \\ \mathcal{A} \in A_{r,1} & \text{if} \quad f(\widetilde{\alpha}(\mathcal{A})) = 1. \end{cases}$$

We prove that, for all positive integers k and q, $q \geq k$, there exists a positive integer $N(k,q)$ such that under the decompositions $P_r(S_n)$, $r = 1, 2, \ldots, k-1$, defined above, there exists (given some $i_1, i_2, \ldots, i_{k-1} \in \{0,1\}$) a q-subset of the set $S_{N(k,q)}$ such that all its r-subsets are contained in A_{r,i_r}, $r = 1, 2, \ldots, k-1$, respectively. We call this subset a $(q, A_{1,i_1}, A_{2,i_2}, \ldots, A_{k-1,i_{k-1}})$-subset (for $k = 1$, we have a q-subset).

Define the values of $N(k,q)$ consecutively. For $k = 1$, it suffices to put $N(k,q) = q$. Suppose $N(k-1, q')$ is defined, where $q' = \mathcal{M}(q, q; k-1)$. Then, by Ramsey's theorem (given some $i_{k-1} \in \{0,1\}$), there exists a $(q, A_{k-1,i_{k-1}})$-subset of the set $S_{q'}$. By the definition of $N(k-1, q')$, all the r-subsets of $S_{N(k-1,q')}$, $r = 1, 2, \ldots, k-2$, are contained in the sets A_{r,i_r}, respectively. Therefore, the $(q, A_{k-1,i_{k-1}})$-subset identified is a $(q, A_{1,i_1}, A_{2,i_2}, \ldots, A_{k-1,i_{k-1}})$-subset of the set $S_{N(k-1,q')}$. Thus, it suffices to put $N(k,q) = N(k-1, \mathcal{M}(q,q;k-1))$. Hence, the numbers $N(k,q)$ are found.

Put $N = N(n_1, n_1)$. As proved above, there exists a $(n_1, A_{1,i_1}, A_{2,i_2}, \ldots, A_{n_1-1,i_{n_1-1}})$-subset $\mathcal{A}$ of the set S_N such that all its r-subsets are contained in the sets A_{r,i_r}, $r = 1, 2, \ldots, n_1 - 1$, respectively.

In the function f of N arguments, substitute zeros for all the variables x_i with $i \notin \mathcal{A}$. The subfunction of n_1 arguments obtained is denoted by f'. At each vector of length n_1 which contains precisely r ones, $r \in \{1, 2, \ldots, n_1 - 1\}$, the function f' takes the value i_r. Hence, f' is a symmetric function. Lemma 3 is proved.

Lemma 4. *For all positive integer l and n_2, there exists a positive integer N such that each l-balanced Boolean function f of N arguments has a periodic, symmetric subfunction of n_2 arguments with period T not exceeding 2^{l+1}.*

PROOF. Put $n_1 = \max\{n_2 + 2, 2^{l+1} + l + 3\}$. By Lemma 3, there exists a positive integer N such that each Boolean function f of N arguments has a symmetric subfunction f_1 of n_1 arguments. Consider the characteristic sequence $\widetilde{\pi} = (\pi_0, \pi_1, \ldots, \pi_{n_1})$ of this subfunction. By the Dirichlet principle, among its first $2^{l+1} + l + 1$ elements, there exist two coinciding intervals of length $l + 1$: $(\pi_h, \pi_{h+1}, \ldots, \pi_{h+l})$ and $(\pi_{h+T}, \pi_{h+T+1}, \ldots, \pi_{h+T+l})$, where $T \leq 2^{l+1}$, $h \geq 0$, and $h + T + l \leq 2^{l+1} + l + 1$. The coincidence of the intervals means that $\pi_i = \pi_{T+i}$ for $i = h, h+1, \ldots, h+l$. We prove that $\pi_i = \pi_{T+i}$ holds for all i, $i = 1, 2, \ldots, n_1 - T - 1$.

Suppose that the equality $\pi_i = \pi_{T+i}$ is established for all positive integers i such that $h \leq i \leq j$, where j satisfies $h + l \leq j \leq n_1 - 2$. We establish the equality $\pi_{j+1} = \pi_{T+j+1}$. To this end, consider the functions $f' = f_1[j-l, j+2]$ and $f'' = f_1[T+j-l, T+j+2]$. These functions are subfunctions of the function f; therefore, they are l-balanced. For the weights of f' and f'', we have

$$W_{f'} = \sum_{i=0}^{l+2} \binom{l+2}{i} \pi_{j-l+i}, \quad W_{f''} = \sum_{i=0}^{l+2} \binom{l+2}{i} \pi_{T+j-l+i}$$

and

$$\begin{aligned} |W_{f'} - W_{f''}| &= \left| \sum_{i=0}^{l+2} \binom{l+2}{i} (\pi_{j-l+i} - \pi_{T+j-l+i}) \right| \\ &= \left| \binom{l+2}{l+1} (\pi_{j+1} - \pi_{T+j+1}) + \binom{l+2}{l+2} (\pi_{j+2} - \pi_{T+j+2}) \right| \\ &= |(l+2)(\pi_{j+1} - \pi_{T+j+1}) + (\pi_{j+2} - \pi_{T+j+2})| \\ &\geq (l+2)|\pi_{j+1} - \pi_{T+j+1}| - 1. \end{aligned}$$

Hence, if $\pi_{j+1} \neq \pi_{T+j+1}$, then $|W_{f'} - W_{f''}| \geq l+1$, which contradicts the l-balancedness of f. Therefore, $\pi_{j+1} = \pi_{T+j+1}$.

Similarly, suppose that the equality $\pi_i = \pi_{T+i}$ is established for all integers i such that $j \leq i \leq h+l$, where $2 \leq j \leq h$. We establish the equality $\pi_{j-1} = \pi_{T+j-1}$. To do this, consider the functions $f' = f_1[j-2, j+l]$ and $f'' = f_1[T+j-2, T+j+l]$. These functions are subfunctions of f; therefore, they are l-balanced. As above, we have

$$W_{f'} = \sum_{i=0}^{l+2} \binom{l+2}{i} \pi_{j-2+i}, \quad W_{f''} = \sum_{i=0}^{l+2} \binom{l+2}{i} \pi_{T+j-2+i}$$

and

$$\begin{aligned} |W_{f'} - W_{f''}| &= \left| \sum_{i=0}^{l+2} \binom{l+2}{i} (\pi_{j-2+i} - \pi_{T+j-2+i}) \right| \\ &= \left| \binom{l+2}{0} (\pi_{j-2} - \pi_{T+j-2}) + \binom{l+2}{1} (\pi_{j-1} - \pi_{T+j-1}) \right| \\ &= |(l+2)(\pi_{j-1} - \pi_{T+j-1}) + (\pi_{j-2} - \pi_{T+j-2})| \\ &\geq (l+2)|\pi_{j-1} - \pi_{T+j-1}| - 1. \end{aligned}$$

Hence, if $\pi_{j-1} \neq \pi_{T+j-1}$, then $|W_{f'} - W_{f''}| \geq l+1$, which contradicts the l-balancedness of f. Therefore, $\pi_{j-1} = \pi_{T+j-1}$.

Thus, the equality $\pi_i = \pi_{T+i}$ is established for all integers i satisfying $1 \le i \le n_1 - T - 1$. Consider the function $f_2 = f_1[1, n_2 + 1]$ of n_2 arguments. (The function f_2 is well-defined since $n_1 \ge n_2 + 2$.) The function f_2 is clearly a periodic, symmetric function with the period not exceeding 2^{l+1} and, by construction, is a subfunction of f, which completes the proof of Lemma 4.

REMARK 1. A periodic, symmetric (with period T) Boolean function of a given number of arguments is completely defined by the initial interval of the characteristic sequence (called the characteristic interval hereafter) which consists of its first T elements. Denote this interval by $\widetilde{\pi}[T] = (\pi_0, \pi_1, \dots, \pi_{T-1})$.

We also use the notation $\varepsilon_k = \cos(2\pi k/T) + i\sin(2\pi k/T)$ and write $r_k = |1 + \varepsilon_k|$.

REMARK 2. If a, b, c, and d are integers then, as known, $\varepsilon_{ab}^{cd} = \varepsilon_{abc}^{d} = \varepsilon_{a}^{bcd}$.

REMARK 3. For $0 \le k_1 < k_2 \le \lfloor (T-1)/2 \rfloor$, the inequality $r_{k_1} > r_{k_2}$ holds.

REMARK 4. For $k < T/3$, the inequality $r_k > 1$ holds.

REMARK 5. For even T, the equality $r_{T/2} = 0$ holds.

Lemma 5. *For integer n, T, and j satisfying $n > 0$ and $0 \le j < T$,*

$$\sum_{i=0}^{\lfloor (n-j)/T \rfloor} \binom{n}{j+iT} = \frac{2^n}{T} + \frac{2}{T} \sum_{k=1}^{\lfloor (T-1)/2 \rfloor} r_k^n \cos \frac{\pi k(n-2j)}{T}.$$

Similar formulas appear in [6] and [7].

PROOF. Denote

$$a_j(n) = \sum_{i=0}^{\lfloor (n-j)/T \rfloor} \binom{n}{j+iT},$$

$j = 0, 1, \dots, T-1$, and $\overrightarrow{a}(n) = (a_0(n), a_1(n), \dots, a_{T-1}(n))$. Then $\overrightarrow{a}^T(n) = A\overrightarrow{a}^T(n-1)$, where A is a matrix of order T of the following form:

$$A = \begin{pmatrix} 1 & 0 & 0 & \dots & 0 & 0 & 1 \\ 1 & 1 & 0 & \dots & 0 & 0 & 0 \\ 0 & 1 & 1 & \dots & 0 & 0 & 0 \\ \vdots & \vdots & \vdots & \ddots & \vdots & \vdots & \vdots \\ 0 & 0 & 0 & \dots & 1 & 0 & 0 \\ 0 & 0 & 0 & \dots & 1 & 1 & 0 \\ 0 & 0 & 0 & \dots & 0 & 1 & 1 \end{pmatrix}.$$

The matrix A has the eigenvalues $\lambda_k = 1 + \varepsilon_k$, $k = 0, 1, \dots, T-1$, and eigenvectors $\overrightarrow{c}_k = (c_k^0, c_k^1, \dots, c_k^{T-1})$, where $c_k^j = \varepsilon_{-kj}$, $0 \le k, j \le T-1$.

Considering Remark 5, for $n > 0$ and $0 \le j \le T-1$, we have

$$\begin{aligned} a_j(n) &= \sum_{k=0}^{T-1} \frac{(\overrightarrow{a}(0), \overrightarrow{\overline{c_k}})}{(\overrightarrow{c_k}, \overrightarrow{\overline{c_k}})} c_k^j \lambda_k^n \\ &= \sum_{k=0}^{T-1} \frac{1}{T} \varepsilon_{-kj}(1+\varepsilon_k)^n = \frac{1}{T} \sum_{k=0}^{T-1} r_k^n \varepsilon_{k(n-2j)/2} \\ &= \frac{2^n}{T} + \frac{1}{T} \sum_{k=1}^{\lfloor (T-1)/2 \rfloor} r_k^n (\varepsilon_{k(n-2j)/2} + \varepsilon_{(T-k)(n-2j)/2}) \\ &= \frac{2^n}{T} + \frac{2}{T} \sum_{k=1}^{\lfloor (T-1)/2 \rfloor} r_k^n \cos \frac{\pi k(n-2j)}{T}, \end{aligned}$$

which proves Lemma 5.

Lemma 6. *Let $F = \{f_n\}_{n=1}^{\infty}$ be an infinite sequence of periodic, symmetric Boolean functions, where f_n is a function of n arguments defined by the characteristic interval $\widetilde{\pi}[T] = (\pi_0, \pi_1, \ldots, \pi_{T-1})$. If all the functions in the sequence F are l-balanced, then, for each positive integer k, $k < T/3$, the equality $\sum_{j=0}^{T-1} \pi_j \varepsilon_{kj} = 0$ is true.*

Proof. Assume that the assertion of the lemma is false. Denote by K the minimal k such that $\sum_{j=0}^{T-1} \pi_j \varepsilon_{kj} \neq 0$; here $K < T/3$. Then $\sum_{j=0}^{T-1} \pi_j \varepsilon_{kj} = 0$ holds for all positive integers k less than K and $\sum_{j=0}^{T-1} \pi_j \varepsilon_{Kj} \neq 0$.

Consider a sequence of Boolean functions $F' = \{f'_n\}$, $n = 1, 2, \ldots$, such that $f'_n = f_{n'+T-1}$, where $n' = 2nT$. Put $f_n^m = f'_n[m, n'+m]$, $m = 0, 1, \ldots, T-1$. Clearly, all the functions f_n^m are l-balanced, symmetric, and periodic with period T.

For the weight of f_n^m, by Lemma 5, we have

$$\begin{aligned} W_{f_n^m} &= \sum_{i=0}^{n'} \binom{n'}{i} \pi_{i+m} = \sum_{j=0}^{T-1} \sum_{i=0}^{\lfloor (n'-j)/T \rfloor} \binom{n'}{j+iT} \pi_{j+m} \\ &= \sum_{j=0}^{T-1} \pi_{j+m} \sum_{i=0}^{\lfloor (n'-j)/T \rfloor} \binom{n'}{j+iT} \\ &= \sum_{j=0}^{T-1} \pi_{j+m} \left(\frac{2^{n'}}{T} + \frac{2}{T} \sum_{k=1}^{\lfloor (T-1)/2 \rfloor} r_k^{n'} \cos \frac{\pi k(n'-2j)}{T} \right) \end{aligned}$$

$$= \sum_{j=0}^{T-1} \pi_{j+m} \left(\frac{2^{n'}}{T} + \frac{2}{T} \sum_{k=1}^{\lfloor (T-1)/2 \rfloor} r_k^{n'} \cos \frac{2\pi kj}{T} \right)$$

$$= \frac{2^{n'}}{T} \sum_{j=0}^{T-1} \pi_{j+m} + \sum_{k=1}^{\lfloor (T-1)/2 \rfloor} r_k^{n'} \frac{2}{T} \sum_{j=0}^{T-1} \pi_{j+m} \cos \frac{2\pi kj}{T}$$

$$= \frac{2^{n'}}{T} \sum_{j=0}^{T-1} \pi_j + \sum_{k=1}^{\lfloor (T-1)/2 \rfloor} r_k^{n'} \frac{2}{T} \sum_{j=0}^{T-1} \pi_j \cos \frac{2\pi k(j-m)}{T}.$$

By assumption, $\sum_{j=0}^{T-1} \pi_j \varepsilon_{kj} = 0$ for all positive integers k less than K. It follows that $\sum_{j=0}^{T-1} \pi_j \varepsilon_{kj} \varepsilon_{-km} = 0$ for each m and, thus, $\sum_{j=0}^{T-1} \pi_j \cos \frac{2\pi k(j-m)}{T} = 0$. Further, $\sum_{j=0}^{T-1} \pi_j \varepsilon_{Kj} \neq 0$ and $K < T/3$ imply that the values of the three sums $\sum_{j=0}^{T-1} \pi_j \varepsilon_{Kj} \varepsilon_{-Km}$, $m = 0, 1, 2$, are pairwise distinct. However, the absolute values of these sums are equal; therefore, among the values of their real parts, $\sum_{j=0}^{T-1} \pi_j \cos \frac{2\pi K(j-m)}{T}$, $m = 0, 1, 2$, there exist at least two distinct values. Suppose the values of $\sum_{j=0}^{T-1} \pi_j \cos \frac{2\pi K(j-m_1)}{T}$ and $\sum_{j=0}^{T-1} \pi_j \cos \frac{2\pi K(j-m_2)}{T}$, $m_1, m_2 \in \{0, 1, 2\}$, are distinct. Then, by Remarks 3 and 4, we have

$$|W_{f_n^{m_1}} - W_{f_n^{m_2}}| = \left| \sum_{k=K}^{\lfloor (T-1)/2 \rfloor} r_k^{n'} \frac{2}{T} \sum_{j=0}^{T-1} \pi_j \cos \frac{2\pi k(j-m_1)}{T} \right.$$

$$\left. - \sum_{k=K}^{\lfloor (T-1)/2 \rfloor} r_k^{n'} \frac{2}{T} \sum_{j=0}^{T-1} \pi_j \cos \frac{2\pi k(j-m_2)}{T} \right|$$

$$\geq \left| r_K^{n'} \frac{2}{T} \left(\sum_{j=0}^{T-1} \pi_j \cos \frac{2\pi K(j-m_1)}{T} - \sum_{j=0}^{T-1} \pi_j \cos \frac{2\pi K(j-m_2)}{T} \right) \right|$$

$$- \left| \sum_{k=K+1}^{\lfloor (T-1)/2 \rfloor} r_k^{n'} \frac{2}{T} \sum_{j=0}^{T-1} \pi_j \cos \frac{2\pi k(j-m_1)}{T} \right.$$

$$\left. - \sum_{k=K+1}^{\lfloor (T-1)/2 \rfloor} r_k^{n'} \frac{2}{T} \sum_{j=0}^{T-1} \pi_j \cos \frac{2\pi k(j-m_2)}{T} \right|$$

$$\geq r_K^{n'} \frac{2}{T} \left| \sum_{j=0}^{T-1} \pi_j \cos \frac{2\pi K(j-m_1)}{T} - \sum_{j=0}^{T-1} \pi_j \cos \frac{2\pi K(j-m_2)}{T} \right|$$

$$-r_{K+1}^{n'}\frac{T-1}{2}\frac{2}{T}2T = C_1 r_K^{n'} - C_2 r_{K+1}^{n'},$$

where C_1 and C_2 are independent of n', and also $C_1 > 0$. Since $r_K > r_{K+1}$ and $r_K > 1$, we have $C_1 r_K^{n'} - C_2 r_{K+1}^{n'} \to \infty$ as $n' \to \infty$. Hence, all the functions in the sequence F' and, therefore, in the sequence F cannot be simultaneously l-balanced, which contradicts the hypothesis of Lemma 6.

REMARK 6. Each function in the sequence F is a subfunction of all subsequent functions in F since it is obtained from them by substituting the constant 0 for redundant variables. Therefore, if there is a function in F which is not l-balanced, then F contains only a finite number of l-balanced functions.

Lemma 7. *Let f be a periodic symmetric Boolean function with period $T = 6$ which is defined by the characteristic interval $\tilde{\pi}[6] = (\pi_0, \pi_1, \ldots, \pi_5)$, and let $\sum_{j=0}^{5} \pi_j \varepsilon_j = 0$. Then the function f has period $T' \leq 3$.*

PROOF. The negation of a periodic symmetric Boolean function is a periodic symmetric Boolean function with the same period. Therefore, it suffices to consider only the case of $\sum_{j=0}^{5} \pi_j \leq 3$. Note also that it suffices to find the period T' in the sequence $(\pi_0, \pi_1, \ldots, \pi_5)$.

1. Let $\sum_{j=0}^{5} \pi_j = 0$. Clearly then, all π_j's have the value 0 and f has the period 1.

2. Let $\sum_{j=0}^{5} \pi_j = 1$, i.e., only one of the π_j's is other than 0. Then $\sum_{j=0}^{5} \pi_j \varepsilon_j \neq 0$; therefore, this case is not possible.

3. Let $\sum_{j=0}^{5} \pi_j = 2$ and let $\pi_{m_1} = \pi_{m_2} = 1$, $0 \leq m_1 < m_2 \leq 5$. Then $\sum_{j=0}^{5} \pi_j \varepsilon_j = 0$ implies $\varepsilon_{m_1} = -\varepsilon_{m_2}$. From this it follows that $m_2 - m_1 = 3$ and, thus, $\pi_j = \pi_{j+3}$ holds for $j = 0, 1, 2$. Therefore, f has the period 3.

4. Let $\sum_{j=0}^{5} \pi_j = 3$. It is easy to see that now $\sum_{j=0}^{5} \pi_j \varepsilon_j = 0$ holds only in two cases: $\tilde{\pi}[6] = (0, 1, 0, 1, 0, 1)$ and $\tilde{\pi}[6] = (1, 0, 1, 0, 1, 0)$. Therefore, f has the period 2.

We have considered all the cases. Lemma 7 is proved.

An hth root of one ξ is called *primitive* if all the hth roots of one can be represented as ξ^i, $i \in \{0, 1, \ldots, h-1\}$.

Lemma 8 [8]. *Let an hth primitive root of one ξ be a root of a polynomial $F(x)$ with rational coefficients. Then all the hth primitive roots of one are roots of the polynomial $F(x)$.*

Lemma 9. *Let f be a periodic symmetric Boolean functions with period T defined by the characteristic interval $\tilde{\pi}[T] = (\pi_0, \pi_1, \ldots, \pi_{T-1})$. Assume also that $\sum_{j=0}^{T-1} \pi_j \varepsilon_{kj} = 0$ for each positive integer k, $k < T/3$. Then f has period $T' \leq 3$.*

PROOF. Consider the polynomial $\Pi(x) = \sum_{j=0}^{T-1} \pi_j x^j$; by construction, its degree does not exceed $T-1$. For $k < T/3$, by the hypothesis of Lemma 9, $\sum_{j=0}^{T-1} \pi_j \varepsilon_{kj} = 0$. By Remark 2, we have $\varepsilon_{kj} = \varepsilon_k^j$; therefore, ε_k is a root of $\Pi(x)$.

Each Tth root of unity is a primitive hth root of unity for some positive integer h which is a divisor of T. Let ε_j be an arbitrary primitive hth root of unity, $h > 3$. Then $\varepsilon_{T/h}$ is also a primitive hth root of unity. Since $T/h < T/3$, the number $\varepsilon_{T/h}$ is, by the above, a root of $\Pi(x)$; by Lemma 8, we obtain that ε_j is also a root of $\Pi(x)$. Hence, all the Tth roots of one, with the possible exception of ε_0, $\varepsilon_{T/2}$ (if T divides by 2), $\varepsilon_{T/3}$ and $\varepsilon_{2T/3}$ (if T divides by 3), are roots of the polynomial $\Pi(x)$.

Denote by $\Psi(x)$ the polynomial $\prod_{\substack{j=1 \\ j\neq T/3, T/2, 2T/3}}^{T-1} (x - \varepsilon_j)$. All the roots of $\Psi(x)$ are roots of $\Pi(x)$ and the field of polynomials with complex coefficients has no zero divisors; therefore, $\Pi(x)$ divides by $\Psi(x)$. Denote by $\Theta(x)$ the quotient: $\Pi(x) = \Psi(x)\Theta(x)$, $\deg \Theta(x) \leq T - 1 - \deg \Psi(x)$. Depending on whether T divides by 2 or 3, four cases are possible.

1. Let T be not divisible by either 2 or 3. Then

$$\Psi(x) = \prod_{j=1}^{T-1} (x - \varepsilon_j) = \frac{\prod_{j=0}^{T-1} (x - \varepsilon_j)}{x - \varepsilon_0} = \frac{x^T - 1}{x - 1} = \sum_{j=0}^{T-1} x^j.$$

From this it follows that $\Pi(x) = \sum_{j=0}^{T-1} x^j \Theta(x)$. From $\deg \Theta(x) \leq T - 1 - \deg \Psi(x) = 0$ we obtain that $\Theta(x)$ equals some constant a. Consequently, $\Pi(x) = a \sum_{j=0}^{T-1} x^j$ and, thus, the coefficients to all the degrees of polynomial $\Pi(x)$ are equal. Obviously then, $\pi_j = \pi_{j+1}$ holds for each nonnegative integer j. Therefore, f has period 1.

2. Let $T = 2t$ and T be not divisible by 3. Then

$$\Psi(x) = \prod_{\substack{j=1 \\ j\neq t}}^{T-1} (x - \varepsilon_j) = \frac{\prod_{j=0}^{T-1} (x - \varepsilon_j)}{(x - \varepsilon_0)(x - \varepsilon_t)} = \frac{x^{2t} - 1}{(x-1)(x+1)} = \sum_{j=0}^{t-1} x^{2j}.$$

From this we obtain $\Pi(x) = \sum_{j=0}^{t-1} x^{2j} \Theta(x)$. It follows from $\deg \Theta(x) \leq T - 1 - \deg \Psi(x) = 2t - 1 - (2t - 2) = 1$ that $\Theta(x)$ has the form $ax + b$. That, $\Pi(x) = (ax + b) \sum_{j=0}^{t-1} x^{2j}$ and, thus, the sequence of coefficients of $\Pi(x)$ has period 2. Therefore, f also has period 2.

3. Let $T = 3t$ and T be not divisible by 2. Then

$$\Psi(x) = \prod_{\substack{j=1 \\ j \neq t, 2t}}^{T-1} (x - \varepsilon_j) = \frac{\prod_{j=0}^{T-1} (x - \varepsilon_j)}{(x - \varepsilon_0)(x - \varepsilon_t)(x - \varepsilon_{2t})}$$

$$= \frac{x^{3t} - 1}{(x-1)(x^2 + x + 1)} = \sum_{j=0}^{t-1} x^{3j}.$$

From this we obtain $\Pi(x) = \sum_{j=0}^{t-1} x^{3j}\Theta(x)$. It follows from $\deg\Theta(x) \leq T - 1 - \deg\Psi(x) = 3t - 1 - (3t - 3) = 2$ that $\Theta(x)$ has the form $ax^2 + bx + c$. Hence, $\Pi(x) = (ax^2 + bx + c)\sum_{j=0}^{t-1} x^{3j}$ and the sequence of coefficients of $\Pi(x)$ has period 3. Therefore, f also has period 3.

4. Let $T = 6t$. Then

$$\Psi(x) = \prod_{\substack{j=1 \\ j \neq 2t, 3t, 4t}}^{T-1} (x - \varepsilon_j) = \frac{\prod_{j=0}^{T-1} (x - \varepsilon_j)}{(x - \varepsilon_0)(x - \varepsilon_{2t})(x - \varepsilon_{3t})(x - \varepsilon_{4t})}$$

$$= \frac{x^{6t} - 1}{(x-1)(x+1)(x^2 + x + 1)}$$

$$= \frac{x^6 - 1}{(x-1)(x+1)(x^2 + x + 1)} \sum_{j=0}^{t-1} x^{6j} = (x^2 - x + 1) \sum_{j=0}^{t-1} x^{6j}.$$

From this we obtain $\Pi(x) = (x^2 - x + 1)\sum_{j=0}^{t-1} x^{6j}\Theta(x)$. It follows from $\deg\Theta(x) \leq T - 1 - \deg\Psi(x) = 6t - 1 - (6t - 4) = 3$ that $\Theta(x)$ has the form $ax^3 + bx^2 + cx + d$. Hence, $\Pi(x) = \left((ax^3 + bx^2 + cx + d)(x^2 - x + 1)\right)\sum_{j=0}^{t-1} x^{6j}$ and the sequence of coefficients of $\Pi(x)$ has period 6. Therefore, f also has period 6. We show that 6 is not a minimal period. It follows from $\sum_{j=0}^{6t-1} \pi_j\varepsilon_{tj} = 0$ that $\sum_{j=0}^{6t-1} \pi_j\varepsilon_{tj} = \sum_{i=0}^{t-1}\sum_{j=0}^{5} \pi_{j+6i}\varepsilon_{t(j+6i)} = \sum_{i=0}^{t-1}\sum_{j=0}^{5} \pi_j\varepsilon_{tj} = t\sum_{j=0}^{5} \pi_j\varepsilon_j^t = 0$ and, thus, $\sum_{j=0}^{5} \pi_j\varepsilon_j^t = 0$. Since ε^t is the first root of the 6th roots of unity, the hypotheses of Lemma 7 are completely satisfied and thus f has period not greater than 3.

We have considered all the cases. Lemma 9 is proved.

Denote by $\mathcal{F}(l, T_{\max})$ the set consisting of all l-balanced periodic symmetric Boolean functions with period not exceeding $T_{\max}$.

Lemma 10. *For each positive integer l, the set $\mathcal{F}(l, 2^{l+1}) \setminus \mathcal{F}(l, 3)$ is finite.*

PROOF. Let $\widetilde{\pi}[T]$ be a characteristic interval of length T, where $T \leq 2^{l+1}$. Consider the sequence of periodic symmetric Boolean functions $F(\widetilde{\pi}[T]) = \{f_n\}$, $n = 1, 2, \dots$, defined by the characteristic interval $\widetilde{\pi}[T]$. If the sequence $F(\widetilde{\pi}[T])$ contains an infinite number of l-balanced functions, then, by Lemmas 6 and 9 and Remark 6, all the functions in the sequence $F(\widetilde{\pi}[T])$ are periodic symmetric Boolean functions with the period not exceeding 3. Thus, in any case, the sequence $F(\widetilde{\pi}[T])$ contains only a finite number of functions of the set $\mathcal{F}(l, 2^{l+1}) \setminus \mathcal{F}(l, 3)$.

There exists only a finite number of characteristic intervals of the length not exceeding 2^{l+1}. Therefore, the set $\mathcal{F}(l, 2^{l+1}) \setminus \mathcal{F}(l, 3)$ is covered by a finite number of sequences $F(\widetilde{\pi}[T])$. Hence, the set $\mathcal{F}(l,\ 2^{l+1}) \setminus \mathcal{F}(l,\ 3)$ is finite.

Lemma 11. *For any positive integer l and n', there exists a positive integer N such that each l-balanced Boolean function f of N arguments has a periodic symmetric subfunction f' of n' arguments with period not exceeding 3.*

PROOF. By Lemma 10, there exists a positive integer n_2, not less than n', for which there are no Boolean functions belonging to the set $\mathcal{F}(l, 2^{l+1}) \setminus \mathcal{F}(l, 3)$ and having at least n_2 arguments. By Lemma 4, there exists a positive integer N such that each l-balanced Boolean function f of N arguments has a periodic symmetric subfunction f' of n_2 arguments with period not exceeding 2^{l+1} and, consequently, not exceeding 3. Obviously, the function $f'[0, n']$ is a periodic symmetric subfunction of n' arguments with period not exceeding 3. Lemma 3 is proved.

Lemma 12. *Let f be a periodic symmetric Boolean function of n' arguments with period $T \leq 3$. Then $W_f \in \{0, \lfloor 2^{n'}/3 \rfloor, \lceil 2^{n'}/3 \rceil, 2^{n'}/2, \lfloor 2^{n'+1}/3 \rfloor, \lceil 2^{n'+1}/3 \rceil, 2^{n'}\}$.*

PROOF. 1. Let $T = 1$. Then $W_f = \sum_{i=0}^{n'} \binom{n'}{i} \pi_i = \pi_0 \sum_{i=0}^{n'} \binom{n'}{i} = \pi_0 2^{n'}$ and, hence, in this case $W_f \in \{0, 2^{n'}\}$.

2. Let $T = 2$. Then, by Lemma 5, we have

$$W_f = \sum_{i=0}^{n'} \binom{n'}{i} \pi_i = \pi_0 \sum_{i=0}^{\lfloor n'/2 \rfloor} \binom{n'}{2i} + \pi_1 \sum_{i=0}^{\lfloor (n'-1)/2 \rfloor} \binom{n'}{1+2i}$$
$$= \frac{2^{n'}}{2}\pi_0 + \frac{2^{n'}}{2}\pi_1 = (\pi_0 + \pi_1)\frac{2^{n'}}{2}$$

and, hence, in this case $W_f \in \{0, 2^{n'}/2, 2^{n'}\}$.

3. Let $T = 3$. By Lemma 5, taking it to account that $r_1 = 1$, we have

$$W_f = \sum_{i=0}^{n'} \binom{n'}{i} \pi_i = \pi_0 \sum_{i=0}^{\lfloor n'/3 \rfloor} \binom{n'}{3i} + \pi_1 \sum_{i=0}^{\lfloor (n'-1)/3 \rfloor} \binom{n'}{1+3i}$$

$$+\pi_2 \sum_{i=0}^{\lfloor (n'-2)/3 \rfloor} \binom{n'}{2+3i} = \left(\frac{2^{n'}}{3} + \frac{2}{3}\cos\frac{\pi n'}{3}\right)\pi_0$$

$$+ \left(\frac{2^{n'}}{3} + \frac{2}{3}\cos\frac{\pi(n'-2)}{3}\right)\pi_1$$

$$+ \left(\frac{2^{n'}}{3} + \frac{2}{3}\cos\frac{\pi(n'-4)}{3}\right)\pi_2 = \frac{2^{n'}}{3}(\pi_0 + \pi_1 + \pi_2)$$

$$+\frac{2}{3}\left(\pi_0 \cos\frac{\pi n'}{3} + \pi_1 \cos\frac{\pi(n'-2)}{3} + \pi_2 \cos\frac{\pi(n'-4)}{3}\right).$$

From

$$\cos\frac{\pi n'}{3} + \cos\frac{\pi(n'-2)}{3} + \cos\frac{\pi(n'-4)}{3} = 0,$$

we conclude that

$$\left|\pi_0 \cos\frac{\pi n'}{3} + \pi_1 \cos\frac{\pi(n'-2)}{3} + \pi_2 \cos\frac{\pi(n'-4)}{3}\right| \le 1.$$

From this we obtain the inequalities

$$\left\lfloor \frac{2^{n'}(\pi_0 + \pi_1 + \pi_2)}{3} \right\rfloor \le W_f \le \left\lceil \frac{2^{n'}(\pi_0 + \pi_1 + \pi_2)}{3} \right\rceil.$$

Hence, in this case $W_f \in \{0, \lfloor 2^{n'}/3 \rfloor, \lceil 2^{n'}/3 \rceil, \lfloor 2^{n'+1}/3 \rfloor, \lceil 2^{n'+1}/3 \rceil, 2^{n'}\}$.

We have considered all the cases. Lemma 12 is proved.

Lemma 13. *For each positive integer l and each positive ε, there exists a positive integer N such that, for each positive integer n, $n \ge N$, and each l-balanced Boolean function f of n arguments, one of the following five inequalities holds:*

$$\rho(f) < \varepsilon;$$
$$|\rho(f) - 1/3| < \varepsilon;$$
$$|\rho(f) - 1/2| < \varepsilon;$$
$$|\rho(f) - 2/3| < \varepsilon;$$
$$\rho(f) > 1 - \varepsilon.$$

PROOF. By Lemmas 11 and 12, for all positive integer l and n', there exists a positive integer N such that each l-balanced Boolean function f of N arguments has a periodic symmetric subfunction of n' arguments whose weight belongs to the set $\{0, \lfloor 2^{n'}/3 \rfloor, \lceil 2^{n'}/3 \rceil, 2^{n'}/2, \lfloor 2^{n'+1}/3 \rfloor, \lceil 2^{n'+1}/3 \rceil, 2^{n'}\}$. Then the property of l-balancedness implies that the weights of all the f's subfunctions f' of n arguments must simultaneously satisfy one of the following inequalities

$$\begin{aligned}
&W_{f'} < l+1;\\
&\left|W_{f'} - 2^{n'}/3\right| < l+1;\\
&\left|W_{f'} - 2^{n'-1}\right| < l+1;\\
&\left|W_{f'} - 2^{n'+1}/3\right| < l+1;\\
&W_{f'} > 2^{n'} - (l+1).
\end{aligned}$$

Put $n' = \lceil \log_2 \frac{l+1}{\varepsilon} \rceil$ and divide both sides of each of the five inequalities by $2^{n'}$. Since $\min_{f'} \rho(f') \leq \rho(f) \leq \max_{f'} \rho(f')$, Lemma 13 follows.

The theorem now is a direct corollary of Lemmas 2 and 13.

Corollary 1. *Let, for some l, $F = \{f_k\}$, $k = 1, 2, \ldots$, be a sequence of distinct l-balanced Boolean functions, the number of arguments on which the functions in this sequence depend being nondecreasing. Then the set of the cluster points of the weights of the functions in the sequence F is included in the set $\{0, 1/3, 1/2, 2/3, 1\}$.*

Corollary 2. *Let $F = \{f_k\}$, $k = 1, 2, \ldots$, be a sequence of Boolean functions, the number of arguments on which the functions in this sequence depend being nondecreasing and tending to infinity. Let the set of cluster points of the weights of the functions in the sequence F be disjoint from the set $\{1/3, 1/2, 2/3\}$, and let the weights of the functions in the sequence F and the weights of their negations tend to infinity. Then, for each positive integer l, there exists a positive integer $k(l)$ such that all functions in the sequence F, beginning with $f_{k(l)}$, are not l-balanced.*

Acknowledgements. The author expresses his sincere gratitude to O. B. Lupanov, O. M. Kasim-Zade, and M. I. Grinchuk for their interest in this research and valuable remarks.

References

1. E. P. Lipatov (1982) A classification of binary tuples and properties of homogeneity classes (in Russian), in: *Problemy Kibernetiki*. Vol. 39, Nauka, Moscow, pp. 67–84.

2. Yu. V. Tarannikov (1991) The class of 1-balanced functions and the complexity of its realization (in Russian), *Vestnik Moskov. Univ. Ser.* I *Mat. Mekh.* No. 2, 83–85.
3. Yu. V. Tarannikov (1995) On the number of unit values of l-balanced Boolean functions (in Russian), *Diskret. Anal. Issled. Oper.* **2**, No. 1, 80–81.
4. K. A. Rybnikov (1985) *Introduction to Combinatorial Analysis* (in Russian), Moscow University, Moscow.
5. J. Nešetřil (1984) Some nonstandard Ramsey-like applications, *Theoret. Comput. Sci.* **34**, No. 1–2, 3–15.
6. J. Riordan (1968) *Combinatorial Identities*, John Wiley and Sons Inc., New York, London, and Sydney.
7. G. P. Gavrilov and A. A. Sapozhenko (1992) *Problems and Exercises on Discrete Mathematics* (in Russian), Nauka, Moscow.
8. B. L. Van der Waerden (1971) *Algebra.* I, Springer-Verlag, Berlin etc.

Faculty of Mechanics and Mathematics
Moscow State University
Vorob'ëvy Gory
Moscow 119899
RUSSIA

TRANSLATED BY S. I. SUSLOV

SPECTRAL PROPERTIES OF PERFECT BINARY $(n, 3)$-CODES

A. Yu. Vasil'eva

For an arbitrary perfect binary $(n, 3)$-code, the distribution is considered of code vertices over the members of a partition of the n-dimensional unit cube. Emphasis is made on the case concerning the set of parallel faces of the same dimension. The face structure of the code complemention is investigated, and, in terms of the result, two characteristic properties of nonlinear codes are established.

In the paper we study the structure of perfect (close-packed) binary $(n, 3)$-codes, i.e., the codes of length n with distance 3. The structure of these codes is known symmetrical to a high degree. In particular, according to the property found by G. S. Shapiro and D. L. Slotnick [1], the weight spectra of all perfect $(n, 3)$-codes (to which the zero vertex of cube belongs) coincide. It is also known that the code vertices are uniformly distributed over faces of large dimension. In the case of an arbitrary q-ary code, P. Delsarte [2, 3] found the minimal size of the faces with this property. Later, A. K. Pulatov [4] independently found such size for q-ary ($q = 2, 3, \dots$) perfect $(n, 3)$-codes.

In § 1 a generalization of Shapiro–Slotnick Theorem is formulated and proved, and a particular case concerning the distribution of code vertices over faces is considered in § 2. From the result obtained a new simple proof of the Delsarte–Pulatov Theorem (§ 2) follows.

Furthermore, the generalization turns out useful for studying the problem of the structure of the set of faces of the code complemention. In § 3 it is demonstrated that the properties of the Hamming code [5] related with such faces are inherent to this code only. Thus, some criteria are stated for distinguishing between the sets of linear and nonlinear perfect codes.

§ 1. Generalization of the Shapiro–Slotnick Theorem

We start by introducing the following notation and concepts:

- $E^n = \{\mathbf{x} = (x_1, \dots, x_n) \mid x_i \in \{0, 1\}\}$ is the n-dimensional unit cube;
- the number of unit components of the vector $\mathbf{x}$ is denoted by $w(\mathbf{x})$ and called the *weight* of the vertex $\mathbf{x}$;
- $\rho(\mathbf{x}, \mathbf{y}) = \Sigma|x_i - y_i|$ is the distance between vertices $\mathbf{x}$ and $\mathbf{y}$;

A. D. Korshunov (ed.), Operations Research and Discrete Analysis, 301–309.

- $S_i^n(\mathbf{x}) = \{\mathbf{y} \in E^n \mid \rho(\mathbf{x},\mathbf{y}) = i\}$ is the n-dimensional sphere of radius i centered at a vertex $\mathbf{x} \in E^n$;
- an arbitrary perfect binary (further, the word "binary" is omitted) $(n, 3)$-code C is a subset of vertices of E^n at distance at least 3 from each other such that the set of balls of radius 1 with centers in C covers the cube.

Let $A_0, A_1, \dots, A_p$ be an arbitrary system of pairwise disjoint subsets in E^n such that $A_0 \cup \dots \cup A_p = E^n$. We call the system of sets $\{A_i\}$ *fibering*, the sets A_i, *layers*, and the set A_0, the *initial layer* in E^n if

$$A_i = \{\mathbf{x} \in E^n \mid \rho(\mathbf{x}, A_0) = i\} \quad \text{for each} \quad i,\ i \leq i \leq p,$$

where $\rho(\mathbf{x}, A_0) = \min_{y \in A_0} \rho(\mathbf{x}, \mathbf{y})$.

Denote by a_i the size of the layer A_i, and by v_i, the number of code vertices in the layer A_i, i.e., $a_i = |A_i|$ and $v_i = |A_i \cap C|$. The vector $(v_0, v_1, \dots, v_p)$ is called the *spectrum* of a code C with respect to the fibering $\{A_i\}$.

We call $\{A_i\}$ a *faithful* fibering if

(a) for each i, $0 \leq i \leq p$, for all vertices $\mathbf{x}, \mathbf{y} \in A_i$ the equality $|S_1^n(\mathbf{x}) \cap A_i| = |S_1^n(\mathbf{y}) \cap A_i| =: k_i$ holds;

(b) for each i, $0 \leq i \leq p-1$, for all vertices $\mathbf{x}$, $\mathbf{y} \in A_i$ the equality $|S_1^n(\mathbf{x}) \cap A_{i+1}| = |S_1^n(\mathbf{y}) \cap A_{i+1}| =: u_i$ holds.

In this terms, the Shapiro–Slotnick Theorem is reformulated as follows.

Let $\mathbf{x}$ be a vertex of a perfect $(n, 3)$-code C, and let $\{A_i\}$ is a faithful fibering of the cube E^n such that $A_0 = \{\mathbf{x}\}$. Then the spectrum of the code C with respect to the fibering $\{A_i\}$ is independent of the choice of the vertex $\mathbf{x}$ as well as of the choice of the code C.

This statement admits the following generalization for the case of an arbitrary faithful fibering.

Theorem 1. *The spectrum of each perfect $(n, 3)$-code with respect to a faithful fibering of the cube E^n is uniquely determined by the number of code vertices belonging to the initial layer.*

PROOF. Let C be a perfect $(n, 3)$-code, $\{A_i\}$ be a faithful fibering of the cube E^n, and $v_0 = \tilde{v}$. Fix an arbitrary i, $1 \leq i \leq p-1$.

Note that for each $\mathbf{x} \in A_i$ the inclusion $S_1^n(\mathbf{x}) \subset A_{i-1} \cup A_i \cup A_{i+1}$ is true. Moreover, from the definition of faithful fibering it is immediate that each vertex $\mathbf{x} \in A_i$ has a fixed number of adjacent vertices belonging to the layer A_{i-1}; this number is equal to $d_i = n - k_i - u_i$.

For each vertex $\mathbf{x} \in A_i$ one of the following conditions holds, and in each case the number of vertices may easily be counted.

1. $\mathbf{x} \in C$. The number of vertices is v_i.

2. The vertex $\mathbf{x}$ is not a code vertex and lies at distance 1 from some code vertex in the same layer. It is easy to check that the number of vertices satisfying this condition is $v_i k_i$.

3. The vertex $\mathbf{x}$ is not a code vertex and lies at distance 1 from some code vertex in the layer A_{i-1}. The number of such vertices is $v_{i-1}u_{i-1}$.

4. The vertex $\mathbf{x}$ is not a code vertex and lies at distance 1 from some code vertex in the layer A_{i+1}. The number of such vertices is $v_{i+1}d_{i+1}$.

Thus,

$$a_i = v_i + v_i k_i + v_{i-1}u_{i-1} + v_{i+1}d_{i+1} = u_{i-1}v_{i-1} + (k_i + 1)v_i + d_{i+1}v_{i+1}.$$

This equality holds for each i, $1 \le i \le p-1$.

For $i = 0$ and $i = p$, the relations for a_i have the following form:

$$a_0 = (k_0 + 1)v_0 + d_1 v_1, \quad a_p = u_{p-1}v_{p-1} + (k_p + 1)v_p.$$

As a result, we obtain the following system of equations in the unknowns $v_0, v_1, \dots, v_p$:

$$\begin{cases} a_0 = (k_0 + 1)v_0 + d_1 v_1; \\ a_i = u_{i-1}v_{i-1} + (k_i + 1)v_i + d_{i+1}v_{i+1}, \quad 1 \le i \le p-1; \\ a_p = u_{p-1}v_{p-1} + (k_p + 1)v_p. \end{cases} \tag{1}$$

Moreover, $v_0 = \tilde{v}$. Hence, the system (1) in fact is the system of $p+1$ equations in p unknowns. It has a solution, since there exists a perfect $(n,3)$-code C whose spectrum satisfies this system. It is easy to see that the solution is unique. Thus, the spectrum of each perfect $(n,3)$-code with respect to a faithful fibering is determined from the number of code vertices belonging to the initial layer. Theorem 1 is proved.

§ 2. The Face Spectrum of a Perfect $(n,3)$-Code

Fix an integer number k, $0 \le k \le n$, and let $m = n - k$. Let Γ be an arbitrary face of dimension k in E^n. Denote by $F(\Gamma)$ the family of all faces of dimension k which are parallel to Γ. Let $A_0 = \Gamma$ and assume that the set A_i consists of the faces in the family $F(\Gamma)$ which are at distance i from the face Γ, i.e.,

$$A_i = \cup\{\Gamma' \in F(\Gamma) \mid \rho(\Gamma, \Gamma') = i\}, \quad 0 \le i \le p,$$

where $\rho(A, B) := \min_{x \in A,\, y \in B} \rho(\mathbf{x}, \mathbf{y})$.

Lemma 1. *The system of sets $\{A_i\}$ is a faithful fibering.*

PROOF. From the definition of fibering, it follows immediately that $\{A_i\}_{i=0}^{p}$ is a fibering, where $p = m$.

Now, verify that the fibering $\{A_i\}$ is correct. Indeed, for each i and $\mathbf{x} \in A_i$ we have $k_i = |A_i \cap S_1^n(\mathbf{x})| = k$ and $u_i = m - i$. Lemma 1 is proved.

We call the above defined faithful fibering $\{A_i\}_{i=0}^{m}$ a *face* fibering; the face $\Gamma = A_0$, an *initial face*, and the spectrum $(v_0, \dots, v_p)$ of a perfect $(n,3)$-code with respect to such a fibering the *face spectrum* of a code.

Corollary. *The face spectrum of an arbitrary perfect $(n,3)$-code is determined from the number of vertices belonging to the initial face.*

Note that if we have a face fibering then $|A_i| = \binom{m}{i}2^k$ and $d_i = n - k_i - u_i = i$. Then from Lemma 1 and (1) we obtain the following system of linear equations

$$\begin{cases} 2^k = (k+1)v_0 + v_1; \\ 2^k\binom{m}{i} = (m-i+1)v_{i-1} + (k+1)v_i + (i+1)v_{i+1}, \quad 1 \le i \le m-1; \\ 2^k = v_{m-1} + (k+1)v_m. \end{cases} \tag{2}$$

Theorem 2. *Let C be a perfect $(n,3)$-code, $\{A_i\}$ is a face fibering of the cube E^n, and v_0 is the number of code vertices belonging to the initial layer. Then the face spectrum of the code C with respect to face fibering $\{A_i\}$ is given by the coefficients of the power series expansion of the function*

$$p(x) = \frac{2^k}{n+1}(1+x)^m + \left(v_0 - \frac{2^k}{n+1}\right)(1+x)^{(n-1)/2-k}(1-x)^{(n+1)/2}. \tag{3}$$

PROOF. Let

$$p(x) = \sum_{j=0}^{m} v_j x^j$$

be the generating function of (2). Introducing additionally v_{-1}, v_{m+1}, v_{m+2}, $\binom{m}{m+1}$ and letting the latter equal zero, rewrite (2) as follows:

$$(m-i+1)v_{i-1} + (k+1)v_i + (i+1)v_{i+1} = 2^k\binom{m}{i}, \quad 0 \le i \le m+1. \tag{4}$$

Then the function $p(x)$ may be written as

$$p(x) = \sum_{j=0}^{m+1} v_j x^j,$$

and its derivative may be written as

$$p'(x) = \sum_{j=0}^{m+1} j v_j x^{j-1} = \sum_{j=1}^{m} j v_j x^{j-1}.$$

Multiplying the ith equation of (4) by x^i and summing the equations over i, we obtain

$$m \sum_{i=0}^{m+1} v_{i-1} x^i - \sum_{i=0}^{m+1} (i-1) v_{i-1} x^i$$
$$+(k+1) \sum_{i=0}^{m+1} v_i x^i + \sum_{i=0}^{m+1} (i+1) v_{i+1} x^i = 2^k \sum_{i=0}^{m+1} \binom{m}{i} x^i.$$

Then

$$mxp(x) - x^2 p'(x) + (k+1)p(x) + p'(x) = 2^k (1+x)^m,$$
$$p'(x) = \frac{mx + k + 1}{x^2 - 1} p(x) + \frac{2^k (1+x)^m}{1 - x^2}. \tag{5}$$

Thus, the solution of (2) reduces to solution of the linear differential equation (5) with initial condition $p(0) = v_0$.

Solving the Cauchy problem (for example, integrating this equation by the method of varying arbitrary constants [6]), we find that the function $p(x)$ has the form shown above and the coefficients of its decomposition compose the face spectrum. Theorem 2 is proved.

The following theorem is a corollary of Theorem 2.

Theorem *(the Delsarte–Pulatov Theorem for perfect $(n,3)$-codes). Let C be a perfect $(n,3)$-code and $k \geq (n+1)/2$. Then in each k-dimensional face in E^n there are $2^k/(n+1)$ code vertices.*

PROOF. Let Γ be an arbitrary face of dimension k in E^n. Consider a face fibering $\{A_i\}_{i=0}^{n-k}$ such that $A_0 = \Gamma$. Then the numbers $v_i = |C \cap A_i|$ are the coefficients of the function $p(x)$ of (3). Since the number of layers is equal to $n-k$, it follows that $p(x)$ is a polynomial of degree at most $n-k$.

Suppose that $|C \cap \Gamma| \neq 2^k/(n+1)$, i.e., $v_0 - 2^k/(n+1) \neq 0$. Then from (3) it follows that $p(x)$ is not a polynomial, since $(1+x)^{(n-1)/2-k}$ is a linear fractional function expandible in an infinite series in powers of x. This is a contradiction. Hence, $|C \cap \Gamma| = 2^k/(n+1)$. The theorem is proved.

Thus, the face spectrum of an arbitrary perfect $(n,3)$-code is established in case $k \geq (n+1)/2$. Moreover, from (3) we may to obtain an explicit formula for the face spectrum in the case of smaller dimensions of the initial face.

Corollary of Theorem 2. *Let $k \leq (n-1)/2$ and the initial face of the face fibering has the dimension k. Then the face spectrum of every perfect $(n,3)$-code has the form*

$$v_i = \frac{2^k}{n+1}\binom{n-k}{i} + \left(v_0 - \frac{2^k}{n+1}\right)$$
$$\times \sum_{j+l=i}(-1)^l \binom{(n-1)/2-k}{j}\binom{(n+1)/2}{l}, \quad 1 \leq i \leq n-k. \tag{6}$$

§ 3. On the Structure of Complementions of Perfect $(n,3)$-Codes

Let $f(\mathbf{x})$ be a Boolean function over the cube E^n, and let Γ be a k-dimensional face. As usual, we call the face Γ *maximal* (inextensible) face of $f(\mathbf{x})$ if $f(\mathbf{x}) = 1$ for every vertex $\mathbf{x} \in \Gamma$ and in each $(k+1)$-dimensional face $\Gamma' \supset \Gamma$ there exists a vertex $\mathbf{y} \in \Gamma'$ such that $f(\mathbf{y}) = 0$. Let $C \subseteq E^n$ be a perfect $(n,3)$-code. Denote by D_C the complemention of the code C, and by $d_C(\mathbf{x})$, the characteristic function of D_C, i.e.,

$$D_C = E^n \setminus C, \quad d_C(\mathbf{x}) = \begin{cases} 1 & \mathbf{x} \in D_C, \\ 0 & \mathbf{x} \in C. \end{cases}$$

The proof of all claims of § 3 is based on partitioning E^n into face layers such that the initial face is some maximal face of the function $d_C(\mathbf{x})$.

Lemma 2. *Let $n > 3$, and let a k-dimensional face Γ be the maximal face of the function $d_C(\mathbf{x})$. Then $\log(n+1) \leq k \leq (n-1)/2$.*

Throughout the paper, log denotes the logarithm to the base 2.

PROOF. The inequality $k \leq (n-1)/2$ is immediate from the Delsarte–Pulatov Theorem.

We show that $k \geq \log(n+1)$. Consider the face fibering $\{A_i\}$ with the initial face Γ: $A_0 = \Gamma$. The faces parallel to Γ and the adjacent faces form the first layer A_1. The number of these faces is equal to $n-k$. Since Γ is the maximal face of the function $d_C(\mathbf{x})$, it follows that in each neighboring face there is at least one code vertex. Thus, $v_1 \geq n-k$. Moreover, from the first equation of (2) it follows that $v_1 = 2^k - (k+1)v_0 = 2^k$ (since $v_0 = 0$). Therefore, $n-k \leq 2^k$, hence it is easy to see that in this case $k \geq \log(n+1)$. Lemma 2 is proved.

Further we suppose that the zero vertex in E^n belongs to the code C.

Let Γ be a k-dimensional face. The number of code vertices in this face does not exceed $2^k/(k+1)$. We say that the code vertices of C are *close-packed* in the face Γ if Γ is covered by balls of radius 1 with centers in the set $C \cap \Gamma$. It is easy to see that only in this case $|\Gamma \cap C| = 2^k/(k+1)$.

Lemma 3. *Let C be a perfect $(n,3)$-code, $\mathbf{0} \in C$, Γ is the maximal face of the function $d_C(\mathbf{x})$, the dimension of the face Γ is equal to k, $m = n - k$; $\{A_i\}$ is a face fibering such that $A_0 = \Gamma$; $\{B_i\}$ is a face fibering such that $B_0 \in F(\Gamma)$ and $\mathbf{0} \in B_0$.*

Then the following statements are equivalent:

(a) $k = (n-1)/2$;

(b) *the even layers $\{A_i\}$ are entirely in the set D_C, and the odd layers consist of the faces $\Gamma' \in F(\Gamma)$ in which the vertex of the code C are close-packed;*

(c) *the odd layers $\{B_i\}$ are entirely in the set D_C, and the even layers consist of the faces $\Gamma' \in F(\Gamma)$ in which the vertex of the code C are close-packed;*

(d) $A_2 \cap C = \varnothing$.

PROOF. 1. We prove that (a)$\Rightarrow$(b). From (6) we find that the face spectrum $(v_0, \ldots, v_m)$ of the code C with respect to the fibering $\{A_i\}$ has the following form (since $v_0 = 0$):

$$v_{2r} = 0, \quad v_{2r-1} = \frac{2^k}{k+1}\binom{m}{2r-1}, \quad r = 1, 2, \ldots, (n+1)/4.$$

The number of faces in the family $F(\Gamma)$ forming the layer A_i is equal to $\binom{m}{i}$. Hence, in each face of the family $F(\Gamma)$ being in an odd layer there are $2^k/(k+1)$ code vertices, i.e., the code vertices are close-packed in this face. Moreover, in all even layers there are no code vertices.

2. It is easy to check that (b)$\Leftrightarrow$(c).

3. It is obvious that (b)$\Rightarrow$(d).

4. We prove that (d)$\Rightarrow$(a). It follows from the first equation of (2) that $v_1 = 2^k - (k+1)v_0 = 2^k$. By virtue of (d), $v_0 = v_2 = 0$. Hence, from the second equation of (2) we have $2^k m = (k+1)v_1 = (k+1)2^k$. Therefore, $m = k+1$, $k = (n-1)/2$. Lemma 3 is proved.

It is known [5] that the dimensions of all maximal faces of the function $d_H(\mathbf{x})$, where H is the Hamming code (a unique linear code up to isomorphism), are the same and equal to $(n-1)/2$. The converse statement is also valid.

Theorem 3. *Let C be a perfect $(n,3)$-code and $\mathbf{0} \in C$. If all maximal faces of the Boolean function $d_C(\mathbf{x})$ have dimension $(n-1)/2$, then the code C is linear.*

PROOF. Put $k = (n-1)/2$ and $m = (n+1)/2$. Suppose that C is a nonlinear code, i.e., $\mathbf{z} = \mathbf{x} + \mathbf{y} \notin C$ for certain $\mathbf{x}, \mathbf{y} \in C$. Then the vertex $\mathbf{z}$ lies in a maximal face Γ of the function $d_C(\mathbf{x})$. By hypotheses, the dimension of Γ is equal to k. Consider the face fibering $\{B_i\}_{i=0}^m$ such that $B_0 \in F(\Gamma)$ and $\mathbf{0} \in B_0$. From Lemma 3 it follows that the face Γ is contained in an odd layer and the vertices of the code C are in even layers.

Write each vertex $\mathbf{x} \in E^n$ in block form: $\mathbf{x} = (\mathbf{ab})$, where $\mathbf{b} = (b_1, \dots, b_k)$ is the projection of $\mathbf{x}$ to the face B_0, and $\mathbf{a} = (a_1, \dots, a_m)$ are the coordinates of $\mathbf{x}$ in the other m directions. In this representation

$$\begin{aligned} \mathbf{x} &= (\mathbf{a}_x \mathbf{b}_x), &&\text{where } w(\mathbf{a}_x) = 2p, && p \in \{0, 1, \dots, (n+1)/4\}; \\ \mathbf{y} &= (\mathbf{a}_y \mathbf{b}_y), &&\text{where } w(\mathbf{a}_y) = 2q, && q \in \{0, 1, \dots, (n+1)/4\}; \\ \mathbf{z} &= (\mathbf{a}_z \mathbf{b}_z), &&\text{where } w(\mathbf{a}_z) = 2r - 1, && r \in \{1, \dots, (n+1)/4\}; \end{aligned}$$

Since $\mathbf{x} + \mathbf{y} = \mathbf{z}$, we have $\mathbf{a}_x + \mathbf{a}_y = \mathbf{a}_z$. Hence, $w(\mathbf{a}_x + \mathbf{a}_y) = w(\mathbf{a}_z)$ is an odd number. On the other hand,

$$w(\mathbf{a}_x + \mathbf{a}_y) = w(\mathbf{a}_x) + w(\mathbf{a}_y) - 2w(\mathbf{a}_x \wedge \mathbf{a}_y) = 2(p + q - w(\mathbf{a}_x \wedge \mathbf{a}_y)),$$

i.e., it is always even, and $w(\mathbf{a}_z)$ is odd. This is a contradiction. Hence, our assumption is false. Therefore, the sum of every two code vertices also belongs to the code, i.e., the code is linear. Theorem 3 is proved.

REMARK. In particular, Lemma 2 and Theorem 3 imply the well-known fact that each perfect $(7,3)$-code is linear: here the dimension of all maximal faces of code complemention is equal to $(n-1)/2$, since $\log(n+1) = (n-1)/2 = 3$, and so the code is linear.

It is clear that $S_1^n(\mathbf{x}) \subset D_C$ and $S_2^n(\mathbf{x}) \subset D_C$ for any $\mathbf{x} \in C$.

For a linear code H the following statement (see [5]) is valid: the sphere of radius 2 with center an arbitrary code vertex is covered by those and only those maximal faces of the function d_H that cover the sphere of radius 1 with center the same vertex. It turns out that this property is characteristic of a linear code, i.e., the following theorem is valid.

Theorem 4. *Assume that C is a perfect $(n,3)$-code, $\mathbf{0} \in C$, and the sphere of radius 2 with center an arbitrary code vertex is covered by those and only those maximal faces of the function d_C that cover the sphere of radius 1 with the same center. Then C is a linear code.*

PROOF. Let Γ be a maximal face of the function $d_C(\mathbf{x})$. Put $k = \dim \Gamma$ and $m = n - k$. Consider a face fibering $\{A_i\}_{i=0}^m$ such that $A_0 = \Gamma$; $(v_0, \dots, v_m)$ is the spectrum with respect to this fibering.

First, we show that $v_2 = 0$. Suppose that the layer A_2 contains a code vertex $\mathbf{x}$. Then the sphere of radius 2 with center $\mathbf{x}$ has nonempty intersection with the face $\Gamma = A_0$: $\Gamma \cap S_2^n(\mathbf{x}) \neq \varnothing$, although the sphere of radius 1 with the same center is disjoint from this face, i.e., $\Gamma \cap S_1^n(\mathbf{x}) = \varnothing$. This contradicts the hypothesis. Hence, $v_0 = v_2 = 0$, and from Lemma 3 we obtain $\dim \Gamma = k = (n-1)/2$.

This is true for every maximal face of the function $d_C(\mathbf{x})$. Hence, the dimensions of all maximal faces of the function $d_C(\mathbf{x})$ are equal to $(n-1)/2$ and, by Theorem 3, the code C is linear. Theorem 4 is proved.

The next corollary is straightforward from Theorems 3 and 4.

Corollary. *For nonlinear perfect $(n,3)$-codes the following properties are characteristic:*

(1) *the dimension of at least one maximal face containing no code vertices is strictly less than $(n-1)/2$;*

(2) *there exists at least one code vertex* $\mathbf{x}$ *at distance* 2 *from some maximal face containing no code vertices.*

Acknowledgments. The author is grateful to S. V. Avgustinovich for formulating the problem and making some valuable remarks.

References

1. H. S. Shapiro and D. L. Slotnick (1959) On the mathematical theory of error correcting codes, *IBM J. Res. Develop.* **3**, No. 1, 25–34.
2. F. J. MacWilliams and N. J. A. Sloane (1977) *The Theory of Error-Correcting Codes.* Vol. 1 and 2, North-Holland Publishing Co., Amsterdam, New York, and Oxford.
3. P. Delsarte (1972) Bounds for unrestricted codes by linear programming, *Philips Res. Reports* **27**, pp. 272–289.
4. A. K. Pulatov (1976) On the structure of close-packed $(n,3)$-codes (in Russian), *Metody Diskret. Anal.* **29**, 53–60.
5. A. M. Romanov (1983) Estimate of the length of the shortest disjunctive normal form for the negation of the characteristic function of the Hamming code (in Russian), *Metody Diskret. Anal.* **39**, 88–97.
6. I. G. Petrovskiĭ (1964) *Lectures on the Theory of Ordinary Differential Equations*, Nauka, Moscow.

Novosibirsk State University
ul. Pirogova, 2
Novosibirsk 630090
RUSSIA

TRANSLATED BY E. BODIN

DISTRIBUTIVE COLORING OF GRAPH VERTICES

V. G. Vizing

We investigate a special type of vertex colorings of graphs which are called distributive. Connection with the graph isomorphism problem is revealed.

1. Introduction. Basics

Let G be a simple graph [1, 2] with vertex set $V(G)$ and edge set $E(G)$. By a *coloring* of the vertices of G we mean a single-valued mapping Φ: $V(G) \to C$, where C is a set whose elements are called *colors*. If $\Phi(v) = c$ $(v \in V(G), c \in C)$, then it is said that the vertex v has color c. Two vertices v_1 and v_2 are called *co-colored* if $\Phi(v_1) = \Phi(v_2)$. Vertex subsets of the same cardinality $V_1 \subseteq V(G)$ and $V_2 \subseteq V(G)$ are called *co-colored* if for any color c the number of vertices colored c in V_1 is equal to the number of vertices colored c in V_2.

The subset of all vertices of G adjacent to a vertex v is called the *neighborhood* of v and is denoted by $N(v)$.

DEFINITION 1. A coloring of the vertices of a graph G is called *distributive* if the neighborhoods of any two vertices of the same color are co-colored. A distributive vertex coloring of a graph is called *minimal.*

In this paper we give an algorithm of polynomial complexity for finding a minimal distributive vertex coloring of a graph. We start with a few examples.

1. A coloring of all vertices of a regular graph with a single color is a minimal distributive coloring.

2. It is known [1, p. 44] that the vertex set of a graph G is partitioned into orbits of the automorphism group of this graph. Obviously, any coloring, in which the vertices are co-colored if and only if they belong to the same orbit, is distributive (but not necessarily minimal).

3. A vertex coloring of a graph in which all vertices are colored distinctly is distributive.

A graph on n vertices is called *uniquely colorable* if in a minimal distributive coloring of its vertices there are n colors. For $2 \leq n \leq 5$ there is no uniquely colorable graphs.

For each $n \geq 6$, it is possible to construct a uniquely colorable n-vertex graph (see Fig. 1). (The dashed line depicts a simple path joining v_6 to v_n.) The author possesses no complete description for all uniquely colorable graphs.

A. D. Korshunov (ed.), *Operations Research and Discrete Analysis*, 311–319.

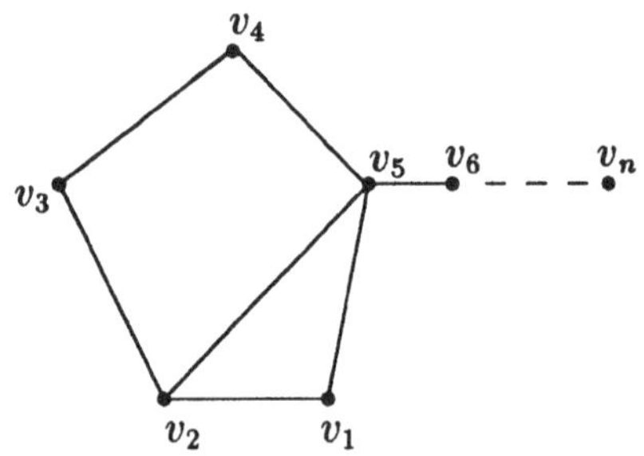

Fig. 1

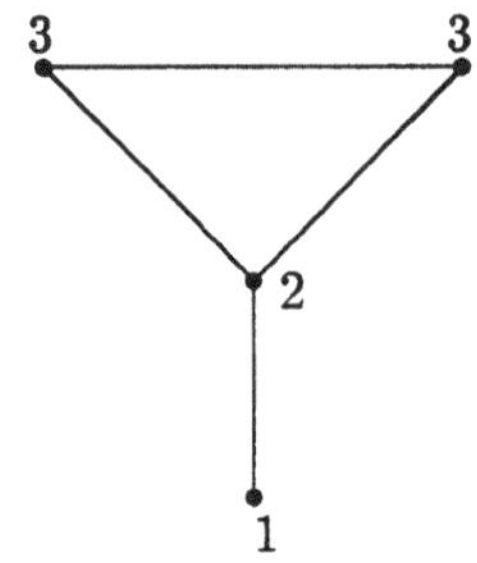

Fig. 2

4. Let G be a graph with distributively colored vertices. Delete from G all vertices of some color together with the incident edges. The remaining subgraph is colored distributively. However, the coloring of such subgraph can fail to be minimal distributive even if the coloring of G was minimal. For example, Fig. 2 depicts a minimal distributive coloring. Deleting the vertex colored 1 along with the incident edge, we obtain a distributive vertex coloring of the remaining graph which is not minimal.

2. Minimal Distributive Coloring

Define the concept of k-similar vertices of a graph and of k-similar subsets of vertices.

DEFINITION 2. Any two vertices of graph G are assumed to be 0-similar. Any two vertex subsets of the same cardinality are assumed to be 0-similar.

Let k be an integer, and let the concepts of $(k-1)$-similar vertices and $(k-1)$-similar subsets have been defined. Vertices x and y are called *k-similar* if the subsets $N(x)$ and $N(y)$ are $(k-1)$-similar. Subsets $X, Y \subseteq V(G)$ are called *k-similar* if there may be established a 1–1 correspondence between them under which the corresponding elements are k-similar.

It is easy to see that the relation of k-similarity is an equivalence.

Lemma 1. *Let k be an integer, then the following are $(k-1)$-similar:*
(a) *k-similar vertices;*
(b) *k-similar subsets of vertices.*

PROOF. The lemma is proved by induction on k. For $k = 1$ statements (a) and (b) are obvious. Suppose $k \geq 2$. Let x and y be k-similar vertices. By definition, $N(x)$ and $N(y)$ are $(k-1)$-similar. By the inductive hypothesis, $N(X)$ and $N(y)$ are $(k-2)$-similar. Hence, x and y are $(k-1)$-similar.

Now let X and Y be k-similar subsets of vertices. It means that the elements of X and those of Y may be put into a 1–1 correspondence in which any two corresponding vertices are k-similar. It follows from above that k-similar vertices are $(k-1)$-similar. Hence, X and Y are $(k-1)$-similar. Lemma 1 is proved.

It follows from Lemma 1 that every equivalence class consisting of k-similar vertices of a graph belongs to only one equivalence class that consists of $(k-1)$-similar vertices ($k \geq 1$). Therefore, with k increasing, the number of equivalence classes for the relation of k-similarity of vertices does not decrease. Let us explore the situation when this number remains constant in more detail.

Lemma 2. *Let an integer k be such that in G each class of $(k-1)$-similar vertices coincides with one of the classes of k-similar vertices. Then every two $(k-1)$-similar vertices in G are m-similar for any integer $m \geq 0$.*

PROOF. For $m \leq k$ the statement follows from Lemma 1. Therefore, we have to prove Lemma 2 for $m = k+1$.

Let the vertices x and y be $(k-1)$-similar. Then by assumption x and y are k-similar. Thus, the sets $N(x)$ and $N(y)$ are $(k-1)$-similar. Since any $(k-1)$-similar vertices are k-similar, the sets $N(x)$ and $N(y)$ are k-similar. This and Definition 2 imply that vertices x and y are $(k+1)$-similar. Lemma 2 is proved.

DEFINITION 3. Two vertices in $V(G)$ are called *similar* if they are k-similar for any $k \geq 0$. Two subsets from $V(G)$ are called *similar* if they are k-similar for any $k \geq 0$.

Lemma 3. *In every n-vertex graph the following are similar:*
(a) *$(n-1)$-similar vertices;*
(b) *$(n-1)$-similar subsets of vertices.*

PROOF. Statement (b) follows immediately from (a). Let us prove (a). Let a_k, $0 \leq k \leq n-1$, be the number of classes of k-similar vertices in an n-vertex graph G. By Lemma 1, we have $1 = a_0 \leq a_1 \leq a_2 \leq \cdots \leq a_{n-1}$. If for i such that $0 \leq i \leq n-2$, it turns out that $a_i = a_{i+1}$, then by Lemma 2, i-similar vertices are similar. But if $1 = a_0 < a_1 < a_2 < \cdots < a_{n-1}$, then $a_{n-1} = n$, i.e. each class of $(n-1)$-similar vertices consists of a vertex which is similar to itself. Lemma 3 is proved.

Corollary 1. *The neighborhoods of any two similar vertices are similar.*

Indeed, in an n-vertex graph similar vertices are n-similar. Therefore, their neighborhoods are $(n-1)$-similar.

Theorem 1. *Any vertices co-colored under a distributive coloring are similar.*

PROOF. It is sufficient to prove that any co-colored vertices are k-similar for any $k \geq 0$. We use induction on k. For $k = 0$ the statement is obvious. Assume it true for some $k \geq 0$. Let x be y co-colored in a distributive coloring of G. As the neighborhoods of any two co-colored vertices are co-colored, by the inductive hypothesis these neighborhoods are k-similar. Hence, the vertices x and y are $(k+1)$-similar. Theorem 1 is proved.

Theorem 2. *A vertex coloring of a graph in which vertices are co-colored if and only if they are similar is a minimal distributive coloring.*

PROOF. Corollary 1 implies that such a coloring is distributive. By Theorem 1, this coloring is a minimal distributive coloring. Theorem 2 is proved.

Theorems 1 and 2 imply the following statement:

Corollary 2. *If vertices x and y are co-colored in a distributive vertex coloring of a graph G, then x and y are also co-colored in a minimal distributive vertex coloring.*

Thus, there is a unique, up to the swapping of colors, minimal distributive vertex coloring of a graph.

Let us describe an algorithm for partitioning the vertex set of a graph into classes of similar subsets. Let G be an n-vertex graph.

1. Put all vertices of G to the class of the 0-similar. Go to Subsection 2.

2. Let a partition of the vertex set of G into classes of k-similar vertices be constructed, $k \geq 0$. Denote these classes by $V_1, \dots, V_s$. If $|V_i| = 1$ for all i, $1 \leq i \leq s$, then $V_1, \dots, V_s$ are the classes of similar vertices, and the algorithm stops. Otherwise, go to Subsection 3.

3. Every class V_i with $|V_i| \geq 2$ is partitioned into subclasses of $(k+1)$-similar vertices. To this end, attribute to each vertex $x \in V_i$ a tuple $(a_1, \dots, a_s)$, where a_j is the number of vertices in V_j which are adjacent to x $(1 \leq j \leq s)$. The vertices in V_i with the same tuple are related to the same class of $(k+1)$-similar vertices. If V_i is partitioned into 2 or more classes of $(k+1)$-similar vertices, we say that V_i has been split. If no class has been split, then $V_1, \dots, V_s$ are the classes of similar vertices and the algorithm stops. Otherwise, we put $k := k+1$ and proceed to Subsection 2.

By Lemma 3, the algorithm proceeds to Subsection 2 at most n times. Therefore, the algorithm always stops. In the end, the vertex set of G is partitioned into classes of similarity. To obtain a minimal distributive coloring, each class of similar vertices must be put in 1–1 correspondence with a color and all vertices of this class must be colored with this color.

Let us show that the algorithm requires at most $O(n^3)$ operations.

Denote by s_j the number of classes of j-similar vertices (j is an integer, $j \geq 0$). Clearly, the sequence $s_0, s_1, s_2, \dots$ is monotonically nondecreasing. If $s_{k-1} = s_k$, then by Lemma 2, $s_{k-1} = s_m$ for each $m \geq k-1$. Let l be the minimal subscript such that $s_l = s_{l+1}$. By Lemma 3, we have $l \leq n-1$. If $l = 0$, i.e., G is a regular graph, then to construct the minimal distributive coloring of G, $O(n^3)$ operations will suffice. Assume $0 < l \leq n - 1$. Then $1 = s_0 < s_1 < \cdots < s_l$. Let $k < l$, and let $V_1, V_2, \ldots, V_{s_k}$ be the classes of k-similar vertices. To determine whether two vertices in V_i are $(k+1)$-similar, we must compare the tuples that are assigned to them in Subsection 3 of the algorithm. The comparison requires $O(s_k) \leq O(n)$ operations. Let us estimate the number of comparisons required to construct the classes of $(k+1)$-similar vertices. Taking a vertex from V_i and comparing it with $|V_i| - 1$ vertices of this class, we determine one class of $(k+1)$-similar vertices. Thus, in $n - |V_i|$ comparisons, we shall determine s_k classes of $(k+1)$-similar vertices. To construct the remaining $(s_{k+1} - s_k)$ classes, it obviously requires less than $(s_{k+1} - s_k)n$ comparisons.

Hence, the passage from classes of k-similar to those of $(k+1)$-similar vertices takes less than $(s_{k+1} - s_k + 1)n$ comparisons and hence at most $(s_{k+1} - s_k + 1)O(n^2)$ operations. Further, to discover that l-similar vertices are similar, it requires either $(n - s_l)$ comparisons or $O(n)$ operations (by Subsection 2 of the algorithm); in both cases $O(n^2)$ operations will suffice. Thus, the total number of operations may be bounded above by $O(n^2)((s_1 - s_0 + 1) + (s_2 - s_1 + 1) + \cdots + (s_l - s_{l-1} + 1) + 1) = O(n^2)(s_l - s_0 + l + 1) = O(n^3)$.

It worth observing that during each execution of Subsection 3, each vertex must be assigned tuples as described above which requires at most $O(n^2)$ operations. As Subsection 3 is executed at most n times, to construct the tuples it requires $O(n^3)$ operations.

Thus, the complexity of the algorithm is at most $O(n^3)$.

3. Similar Graphs

DEFINITION 4. Let G and H be two n-vertex graphs such that $V(G) \cap V(H) = \varnothing$. Graphs G and H are called *similar* if the subsets of vertices $V(G)$ and $V(H)$ are similar in the graph $G \cup H$.

Since in any minimal distributive coloring co-colored subsets of vertices are similar and similar are co-colored, the following is true: $V(G) \cap V(H) = \varnothing$.

Theorem 3. *Two graphs G and H, where $V(G) \cap V(H) = \varnothing$, are similar if and only if in a minimal distributive vertex coloring of the graph $G \cup H$, the vertex subsets $V(G)$ and $V(H)$ are co-colored.*

It follows, to verify the similarity of two graphs one must construct a minimal distributive coloring of $G \cup H$ and verify that $V(G)$ and $V(H)$ are co-colored. Theorem 1 and Corollary 2 show that any (not necessarily minimal) distributive coloring of $G \cup H$ will do such that $V(G)$ and $V(H)$ are co-colored.

Theorem 3 implies the following statement concerning uniquely colorable graphs (see Example 3 in Subsection 1).

Corollary 3. *Two uniquely colorable graphs are isomorphic if and only if they are similar.*

Now we turn to connected graphs.

Theorem 4. *Let G and H be connected graphs such that $|V(G)| = |V(H)|$, $V(G) \cap V(H) = \varnothing$, and let there be a distributive coloring $G \cup H$ in which vertices $x \in V(G)$ and $y \in V(H)$ are co-colored. Then the sets $V(G)$ and $V(H)$ are co-colored, i.e., the graphs G and H are similar.*

PROOF. Let a distributive coloring of $G \cup H$ use l colors. If $l = 1$, then the claim is obvious: both graphs are regular and have the same degree. Suppose that $l \geq 2$. Let us verify that each of the l colors appears both on the vertices of $V(G)$, and on those of $V(H)$. Let $X \subseteq V(G)$ be the set of those vertices having a co-colored vertex in $V(H)$, and let $Y \subseteq V(H)$ be the set of those vertices having a co-colored vertex in $V(G)$.

By the hypothesis of the theorem, $x \in X$ and $y \in Y$. Let us show that $X = V(G)$. Suppose the contrary. As G is connected, there is a vertex $w \in V(G) \setminus X$ adjacent to a vertex $z \in X$. The vertex z is co-colored with a vertex $u \in Y \subseteq V(H)$. By the distributivity of the coloring of $G \cup H$, the neighborhoods of z and u must be co-colored. This contradicts the fact that vertex w in the neighborhood of z does not have a co-colored vertex in $V(H)$. It is proved similarly that $Y = V(H)$.

In a distributive vertex coloring of the graph $G \cup H$, the subsets $V(G)$ and $V(H)$ are partitioned into classes $X_i \subseteq V(G)$ and $Y_i \subseteq V(H)$ of co-colored vertices $(1 \leq i \leq l)$. We shall assume that all vertices in classes X_i and Y_i are colored with i $(1 \leq i \leq l)$. Colors i and j $(i \neq j)$ are called *adjacent* if in the neighborhood of a vertex colored i there is a vertex colored j; the adjacency relation on colors is clearly symmetric. If colors i and j are adjacent, then $|X_i|/|Y_i| = |X_j|/|Y_j|$.

Indeed, let q_{ij} be the number of vertices colored j and lying in the neighborhood of a vertex colored i; in a similar way we define q_{ji}. Then $|X_i|q_{ij} = |X_j|q_{ji}$ and $|Y_i|q_{ij} = |Y_j|q_{ji}$. This implies the equality we need.

Let $|X_1|/|Y_1| = t$. Put $K = \{k \in \{1, 2, \ldots, l\} \mid |X_k|/|Y_k| = t\}$, $P = \{1, 2, \ldots, l\} \setminus K$. Then $P = \varnothing$. For, otherwise by the connectivity of the graph G there would be a color $p \in P$ adjacent to a color in K. But then there would hold $|X_p|/|Y_p| = t$, i.e., the color p would not belong to P. Hence, $|X_k|/|Y_k| = t$ for any $k \in \{1, 2, \ldots, l\}$. Thus,

$$|V(H)| = |V(G)| = \sum_{k=1}^{l} |X_k| = t \sum_{k=1}^{l} |Y_k| = tV(H).$$

It follows that $t = 1$, i.e., the sets $V(G)$ and $V(H)$ are co-colored and the graphs G and H are similar. Theorem 4 is proved.

Theorem 5. *Every two similar trees are isomorphic.*

PROOF. Let G and H be two similar n-vertex trees such that $V(G) \cap V(H) = \varnothing$. Let us construct a distributive vertex coloring of $G \cup H$ such that the sets $V(G)$ and $V(H)$ are co-colored (this can be done due to Theorem 3). By induction on n, we prove that there is an isomorphism of G to H which maps each vertex to a co-colored vertex.

For $n \leq 2$ this is obvious. Suppose that $n \geq 3$. Take any pendant vertex v in $V(G)$. Let $X \subseteq V(G)$, and let $Y \subseteq V(H)$ be the subset of all pendant vertices of $G \cup H$ which are co-colored with v. Evidently, $|X| = |Y|$.

Consider the subgraph of $G \cup H$ spanned by $(V(G) \setminus X) \cup (V(H) \setminus Y)$; the vertices of this subgraph are colored distributively, while the subsets $V(G) \setminus X$ and $V(H) \setminus Y$ are co-colored. By the inductive hypothesis, there is an isomorphism of the tree G_1 with vertices $V(G) \setminus X$ to the tree H_1 with vertices $V(H) \setminus Y$, which maps co-colored vertices to co-colored. Let Z and W be the subsets of the vertices of G_1 and H_1, respectively, each vertices of which in $G \cup H$ is adjacent to a vertex of $X \cup Y$. Since $G \cup H$ is colored distributively, any two vertices of $Z \cup W$ are co-colored, $|Z| = |W|$, and each vertex of $Z \cup W$ is adjacent to the same number q of vertices of $X \cup Y$.

Let $Z = \{z_1, z_2, \dots, z_r\}$, $w_i = \varphi(z_i)$, let X_i be a q-element subset of X, consisting of the vertices adjacent to z_i, and let Y_i be a q-element subset of Y, consisting of the vertices adjacent to w_i, where $1 \leq i \leq r$. Let us establish a 1–1 correspondence ψ between X and Y which maps the vertices of X_i to those of Y_i $(1 \leq i \leq r)$. Then the 1–1 mapping of $V(G)$ to $V(H)$, which coincides with φ on $V(G) \setminus X$ and coincides with ψ on X, is an isomorphism of the trees G and H. Theorem 5 is proved.

4. Similar Graphs

DEFINITION 5. Let G and H be n-vertex graphs, and let k be an integer not greater than n. The graphs G and H are called *edgewise k-resembling* if there is a 1–1 mapping φ of $V(G)$ to $V(H)$ such that, for any k-subset $X \subseteq V(G)$, the subset of G spanned by the vertices of X has the same number of edges as the subgraph of H spanned by the vertices of $\varphi(X)$. We shall also say that φ preserves edgewise k-resemblance.

Lemma 4. *Let G and H be n-vertex edgewise k-resembling graphs, and let φ be a mapping preserving edgewise k-resemblance. Then*

(a) *for $2 \leq k \leq n$ the graphs G and H have the same number of edges;*

(b) *for $2 \leq k \leq n-1$ the graphs G and H have the same sets of vertex degrees, where the degrees of $v \in V(G)$ and $\varphi(V) \in V(H)$ are equal;*

(c) *for* $2 \le k \le n-2$ *the graphs* G *and* H *are isomorphic.*

PROOF. (a) For $k = n$ the statement is evident. Suppose that $2 \le k \le n$. Let M be the total number of edges in the $\binom{n}{k}$ distinct k-vertex subgraphs of G. Since each edge of G belongs to $\binom{n-2}{k-2}$ such subgraphs, G has $M/\binom{n-2}{k-2}$ edges. As G and H are edgewise k-resembling, the total number of edges in the $\binom{n}{k}$ distinct k-vertex subgraphs of H also equals M, and H has $M/\binom{n-2}{k-2}$ edges.

(b) Let v be any vertex of G, and let $\varphi(v) \in V(H)$. Consider the $(n-1)$-vertex subgraphs of G and H spanned by $V(G) \setminus \{v\}$ and $V(H) \setminus \{\varphi(v)\}$. As $2 \le k \le n-1$, from (a) it follows that the same number of edges is both in these subgraphs and in the graphs G and H. Hence, the degree of the vertices v and $\varphi(v)$ in G and H is the same.

(c) Let us show that φ is an isomorphism of G to H. Let v_1 and v_2 be distinct vertices of G. Due to (b), each vertex v_i has the same degree in G, as the vertex $\varphi(v_i)$ does in H, where $i = 1, 2$. Consider the $(n-2)$-vertex subgraphs of G and H spanned by $V(G)\setminus\{v_1 \cup v_2\}$ and $V(H)\setminus\{\varphi(v_1)\cup\varphi(v_2)\}$. Since $2 \le k \le n-2$, these subgraphs are edgewise k-resembling. By (a), they have the same number of edges. It follows from this that vertices v_1 and v_2 in G are adjacent if and only if vertices $\varphi(v_1)$ and $\varphi(v_2)$ are adjacent in H. Lemma 4 is proved.

DEFINITION 6. Let G and H be n-vertex graphs, and let k be an integer not greater than n. The graphs G and H are called *distributively* k-resembling if there is a 1–1 mapping φ of $V(G)$ on $V(H)$ such that, for any k-subset $X \subseteq V(G)$, the subgraph of G on X is similar to the subgraph of H on $\varphi(X)$.

Obviously, isomorphic graphs are distributively k-resembling. Since similar graphs have the same number of edges, the distributive k-resemblance implies the edgewise k-resemblance. Therefore, it holds the following

Theorem 6. *For any* $n \ge 4$, *two* n-*vertex graphs are isomorphic if and only if they are distributively* k-*resembling for some* k, *where* $2 \le k \le n-2$.

Theorem 6 should not be treated as solving the isomorphism problem on graphs: the problem of determining distributive k-similarity $(2 \le k \le n-2)$ may turn out to be NP-hard.

Certainly, it is possible to recognize the distributive $(n-1)$-resemblance of two n-vertex graphs by means of a polynomial algorithm. But what does it give? Is, for example, the following simple statement correct: If for $n \ge 3$ two n-vertex graphs are distributively $(n-1)$-resembling, then they are similar?

References

1. V. A. Emelichev, O. I. Mel'nikov, V. I. Sarvanov, and R. I. Tyshkevich (1990) *Lectures on Graph Theory* (in Russian), Nauka, Moscow.

2. A. A. Zykov (1987) *Fundamentals of Graph Theory* (in Russian), Nauka, Moscow.

Odessa State Academy
of Food Technologies
Kanatnaya, 112
Odessa 270039
UKRAINE

TRANSLATED BY O. V. BORODIN

DECIPHERING THRESHOLD FUNCTIONS OF k-VALUED LOGIC[†)]

N. Y. Zolotykh and V. N. Shevchenko

The problem of deciphering threshold functions of k-valued logic of n arguments is considered. A polynomial deciphering algorithm is proposed which, given n, uses at most $O(\log^n(k+1))$ appeals to the oracle.

Introduction

Consider the following two-person game. On the playing board $B(2,k)$ consisting of all points $(x,y) \in Z^2$ such that $0 \le x \le k-1$, $0 \le y \le k-1$, the first player chooses two distinct points and finds the equation $ax+by=c$ of the line through them. The second player must find the equation of this line by choosing points $(x_i, y_i) \in B(2,k)$ and asking the first player whether it is true that $ax_i + by_i \le c$. From the point of view of the second player, it is natural to find a strategy guaranteeing the answer with a small number of questions and an admissible time for computing the coordinates (x_i, y_i). Generalizing this situation, we introduce the following notation and notions [1, 2].

Let k and n be positive integers, $B(n,k)$ be the set of integral vectors $\mathbf{x} = (x_1, x_2, \ldots, x_n)$ such that $0 \le x_j \le k-1$ $(j = 1, 2, \ldots, n)$, $f(\mathbf{x})$ be a function mapping $B(n,k)$ into the set $\{0,1\}$, $M_0(f)$ and $M_1(f)$ be the sets of vectors $\mathbf{x} = (x_1, x_2, \ldots, x_n)$ such that $f(\mathbf{x})$ equals 0 and 1, respectively, i.e., $M_0(f) = \{\mathbf{x} \in B(n,k) \mid f(\mathbf{x}) = 0\}$, $M_1(f) = \{\mathbf{x} \in B(n,k) \mid f(\mathbf{x}) = 1\}$, and $N_i(f)$ be the set of the extreme points of the convex hull of $M_i(f)$ $(i = 0, 1)$.

A function $f(\mathbf{x})$ is a *threshold function* if there exist real numbers a_i $(i = 0, 1, \ldots, n)$ such that

$$M_0(f) = \left\{\mathbf{x} \in B(n,k) \mid \sum_{i=1}^{n} a_i x_i \le a_0\right\}. \tag{1}$$

The set of all threshold functions defined on $B(n,k)$ is denoted by $F(n,k)$.

[†)] This research was supported by the Russian Foundation for Basic Research (Grant 94–01–00491).

A. D. Korshunov (ed.), Operations Research and Discrete Analysis, 321–326.

Suppose that, to a function $f \in F(n,k)$, there corresponds some oracle that allows us to compute $f(\mathbf{x})$ given $\mathbf{x} \in B(n,k)$. We call such oracle the *M-oracle*. We understand the *M-deciphering* of the function $f \in F(n,k)$ to be a procedure for finding, with the help of the M-oracle, the numbers $a_0, a_1, \ldots, a_n$ for which (1) is satisfied.

An algorithm A for M-deciphering threshold functions is said to be *polynomial* if, for each function $f \in F(n,k)$, both the number $\tau(A)$ of appeals to the oracle and the number $\rho(A)$ of required operations are bounded above by some polynomials in n and $\log(k+1)$. (Throughout this paper, log denotes the base 2 logarithm.) In [2], it is proved that no polynomial algorithm exists for M-deciphering threshold functions. With a fixed dimension n, a polynomial algorithm is called *quasipolynomial*.

In [1], it is observed that to define a threshold function f, it suffices to define either the set $N_0(f)$ or the set $N_1(f)$, and it is shown that

$$|N_i(f)| \le (2\log(k+1))^n \quad (i = 0, 1).$$

In [3], a quasipolynomial algorithm is proposed for the construction of the convex hull of the set M of integral solutions to a system of linear inequalities based on the Lenstra algorithm [4] for finding a point $\mathbf{x} \in M$ and on the bound in [5] for the number of the extreme points and faces of the convex hull of M. These results are essentially used in the quasipolynomial algorithm A_1 for deciphering threshold functions which was developed in [2]. Later [6, 7], the upper bound of $\tau(A_1)$ is lowered on using refinement [3, 8, 9] of upper bounds on $|N|$; in [7], it has the form

$$\tau(A_1) = O((\log(k+1))^{n+\lfloor n/2\rfloor(n-1)}).$$

Note that no further decrease in the exponent seems possible with this approach since the bound on the number of extreme points obtained in [10] is best possible in order [11, 12].

We propose a quasipolynomial algorithm A_2 for deciphering $f \in F(n,k)$ such that $\tau(A_2) = O(\log^n(k+1))$.

To construct the algorithm, we use the results of [13], where some more informative oracle is associated with a threshold function. We call this oracle the *E-oracle*. It is defined as follows. Given any integers $a_0, a_1, \ldots, a_n$, the E-oracle of the function $f \in F(n,k)$ answers "yes" if $M_0(f)$ satisfies (1); otherwise, it answers "no" and produces a point $z = (z_1, z_2, \ldots, z_n)$ in $B(n,k)$ such that either $\sum_{i=1}^n a_i z_i > a_0$ and $f(z) = 0$ (positive counterexample) or $\sum_{i=1}^n a_i z_i \le a_0$ and

$f(z) = 1$ (negative counterexample). We understand the *E-deciphering* of a function f in the class $F(n, k)$ to be a sequence of appeals to the E-oracle of f which results in the answer "yes."

When constructing the algorithm A_2 in Section 2, we show that rather than appealing once to the E-oracle, it suffices to appeal to the M-oracle $O(\log^{n-1}(k+1))$ times.

1. Preliminary Results

We first reformulate the following result in [10].

Lemma 1. *Let P be a polyhedron defined by a system of m linear inequalities with integral coefficients whose absolute values do not exceed α, let $M(P)$ be the intersection of P with the integer lattice, and let $N(P)$ be the set of the extreme points of the convex hull of $M(P)$. Then, for any fixed m and n, it is true that $|N(P)| = O(\log \alpha^{n-1})$.*

Lemma 2. *Suppose the coefficients $a_0, a_1, \ldots, a_n$ with the length of the binary notation bounded above by a polynomial in $\log(k+1)$ are presented to the E-oracle. Then, in time polynomial in $\log(k+1)$, each such appeal to the E-oracle can be reduced to $O(\log^{n-1}(k+1))$ questions to the M-oracle (n is fixed).*

PROOF. First, using the algorithm in [3], find the set $N(a_0, a_1, \ldots, a_n)$ of vertices of the convex hull of the set

$$\left\{\mathbf{x} \in B(n,k) \mid \sum_{i=1}^{n} a_i x_i \leq a_0\right\}.$$

Next, with the help of the M-oracle, consecutively checking the value of f at each of these points, find a point $z \in N(a_0, a_1, \ldots, a_n)$ such that $f(z) = 1$ or find the absence of such points. Obviously, in the first case, z is a negative counterexample. In the second case, construct the set $N'(a_0, a_1, \ldots, a_n)$ of vertices of the convex hull of the set

$$\left\{\mathbf{x} \in B(n,k) \mid \sum_{i=1}^{n} a_i x_i > a_0\right\}$$

and appeal to the M-oracle at each point in $N'(a_0, a_1, \ldots, a_n)$ unless a point $z' \in N'(a_0, a_1, \ldots, a_n)$ is found such that $f(z') = 0$ or find the absence of such points. In the first case, z' is clearly a positive counterexample whereas, in the second case, as follows from [1],

$$M_0(f) = \left\{ \mathbf{x} \in B(n,k) \mid \sum_{i=1}^{n} a_i x_i \le a_0 \right\},$$

i.e., the function f is deciphered.

Thus, we have reduced one appeal to the E-oracle to a series of appeals to the M-oracle. We give an upper bound for the number of such appeals and the number of required arithmetic operations. In [1], it is shown that, given a fixed n, for each $f \in F(n,k)$, there exist integers $b_j (j = 0, 1, \ldots, n)$ such that $|b_j|$ is bounded by a polynomial in k and

$$M_0(f) = \left\{ \mathbf{x} \in B(n,k) \mid \sum_{i=1}^{n} b_i x_i \le b_0 \right\}.$$

It follows from this and Lemma 1 that, for given a fixed n,

$$|N(a_0, a_1, \ldots, a_n)| = |N(b_0, b_1, \ldots, b_n)| = O(\log^{n-1}(k+1)).$$

Similarly, it can be shown that $|N'(a_0, a_1, \ldots, a_n)| = O(\log^{n-1}(k+1))$. Therefore, the number of appeals to the M-oracle required for reducing one appeal to the E-oracle does not exceed $O(\log^{n-1}(k+1))$.

Now we give an upper bound for the number of arithmetic operations executed. To find points in the set $N(b_0, b_1, \ldots, b_n)$, we made use of the algorithm proposed in [3].

Given n, the running time of this algorithm is bounded by a polynomial in $\log(\alpha+1)$, where $\alpha = \max\{k, a_0, a_1, \ldots, a_n\}$. Since, for given n, the length of the binary notation of the coefficients $a_0, a_1, \ldots, a_n$ is bounded above by a polynomial in $\log(k+1)$, the number of operations required for reducing one appeal to the E-oracle to a series of appeals to the M-oracle is bounded above by a polynomial in $\log(k+1)$. Lemma 2 is proved.

2. Main Results

We provide some clarifications to the algorithm A_3 for E-deciphering threshold functions which was proposed in [13].

The algorithm A_3 uses an algorithm of finding a point in a convex body $W \subseteq \mathbb{R}^n$ defined by the *separated oracle* (see, for example, [14]). The separated oracle for W, for given a point $b \in \mathbb{R}^n$, answers "yes" if $b \in W$; otherwise, the oracle answers "no" and produces the coefficients of a hyperplane separating b from W.

Without loss of generality (see [13]), we may assume that $a_0 = 1$ in (1). Hence, to each threshold function f, there corresponds a convex body $W(f) = \{u \in \mathbb{R}^n \mid M_0(f) = \{\mathbf{x} \mid (u, x) \leq 1\}\}$ in the space $\mathbb{R}^n$.

Thus, the problem of E-deciphering a threshold function can be stated as a problem of finding a point in $W(f)$.

In [13], a way to construct the separated oracle for $W(f)$ based on the E-oracle for f is indicated. The ellipsoid algorithm [14, 15] can be used as an algorithm for finding a point in a convex body.

We state the result of [13] as a theorem.

Theorem 1 [13]. *There exists an algorithm A_3 for E-deciphering threshold functions such that, for given a fixed n, $\tau(A_3) = O(\log k)$ and both $\rho(A_3)$ and the length of the binary notation of each coefficient $a_0, a_1, \ldots, a_n$ under all the appeals to the E-oracle are bounded above by some polynomials in $\log(k+1)$.*

Theorem 1 and Lemma 2 imply the following

Theorem 2. *Given a fixed n, there exists a polynomial algorithm for M-deciphering threshold functions of k-valued logic of n arguments which uses at most $C_n \log^n(k+1)$ questions on the value of the function $f(\mathbf{x})$ at a point $\mathbf{x}$, where C_n is a constant depending only on n.*

References

1. V. N. Shevchenko (1985) On some functions of many-valued logic connected with integer programming (in Russian), *Metody Diskret. Anal.* **42**, 99–108.
2. V. N. Shevchenko (1987) On deciphering of threshold functions of many-valued logic (in Russian), in: *Combinatorial-Algebraic Methods and Their Applications*, Gor'kov. University, Gor'kiĭ, pp. 155–167.
3. V. N. Shevchenko (1985) The algebraic approach in integer programming (in Russian), *Kibernetika* (Kiev), No. 4, 36–41.
4. H. W. Lenstra, Jr. (1983) Integer programming with a fixed number of variables, *Math. Oper. Res.* **8**, No. 4, 538–548.
5. V. N. Shevchenko (1985) Convex polyhedral cones, comparison systems, and valid cuts in integer programming (in Russian), in: *Combinatorial-Algebraic*

Methods in Applied Mathematics, Gor'kov. University, Gor'kiĭ, pp. 109–119.
6. V. N. Shevchenko and S. I. Veselov (1989) Deciphering functions of many-valued logic (in Russian), *Theory and Program Realization of Discrete Optimization Methods*, Institute of Cybernetics, Kiev, pp. 30–34.
7. T. Hegedus (1994) Geometrical concept of learning and convex polytopes, *Proceedings of the 7th Annual Conference on Computational Learning Theory*, ACM Press, New Brunswick, New Jork.
8. V. N. Shevchenko (1985) Upper bounds for the number of extreme points in integer programming (in Russian), *Mathematical Problems in Cybernetics*. Vol. 4, Nauka, Moscow, pp. 65–72.
9. A. Yu. Chirkov and V. N. Shevchenko (1993) On the number of vertices in the convex hull of the intersection of a polyhedron with the integer lattice (in Russian), VINITI, No. 2165-B93, Moscow.
10. Cook W., Hartmann M., Kannan R., and McDiarmid C. (1992) On integer points in polyhedra, *Combinatorica* **12**, No. 1, 27–37.
11. S. I. Veselov (1984) A lower bound for the number of irreducible and boundary points for two problems of discrete programming (in Russian), VINITI, No. 619-B84, Moscow.
12. A. Yu. Chirkov (1994) A lower bound for the number of vertices in the convex hull of the intersection of a polyhedron with the integer lattice (in Russian), VINITI, No. 1361-B94, Moscow.
13. W. Maas and G. Turán (1991) How fast can a threshold gate learn? *IIG-Report Ser. Rep.* 321, Graz University of Technology.
14. A. Schrijver (1986) *Theory of Linear and Integer Programming*, John Wiley and Sons, Chichester.
15. L. G. Khachiyan (1979) A polynomial algorithm in linear programming (in Russian), *Dokl. Akad. Nauk SSSR* **244**, No. 5, 1093–1096.

Nizhniĭ Novgorod State University
Gagarin pr., 23
Nizhniĭ Novgorod 603600
RUSSIA

TRANSLATED BY S. I. SUSLOV

INDEX

Other *Mathematics and Its Applications* titles of interest:

P.H. Sellers: *Combinatorial Complexes. A Mathematical Theory of Algorithms.* 1979, 200 pp. ISBN 90-277-1000-7

P.M. Cohn: *Universal Algebra.* 1981, 432 pp. ISBN 90-277-1213- 1 (hb), ISBN 90-277-1254-9 (pb,

J. Mockor: *Groups of Divisibility.* 1983, 192 pp. ISBN 90-277-1539-4

A. Wwarynczyk: *Group Representations and Special Functions.* 1986, 704 pp. ISBN 90-277-2294-3 (pb), ISBN 90-277-1269-7 (hb)

I. Bucur: *Selected Topics in Algebra and its Interrelations with Logic, Number Theory and Algebraic Geometry.* 1984, 416 pp. ISBN 90-277-1671-4

H. Walther: *Ten Applications of Graph Theory.* 1985, 264 pp. ISBN 90-277-1599-8

L. Beran: *Orthomodular Lattices. Algebraic Approach.* 1985, 416 pp. ISBN 90-277-1715-X

A. Pazman: *Foundations of Optimum Experimental Design.* 1986, 248 pp. ISBN 90-277-1865-2

K. Wagner and G. Wechsung: *Computational Complexity.* 1986, 552 pp. ISBN 90-277-2146-7

A.N. Philippou, G.E. Bergum and A.F. Horodam (eds.): *Fibonacci Numbers and Their Applications.* 1986, 328 pp. ISBN 90-277-2234-X

C. Nastasescu and F. van Oystaeyen: *Dimensions of Ring Theory.* 1987, 372 pp. ISBN 90-277-2461-X

Shang-Ching Chou: *Mechanical Geometry Theorem Proving.* 1987, 376 pp. ISBN 90-277-2650-7

D. Przeworska-Rolewicz: *Algebraic Analysis.* 1988, 640 pp. ISBN 90-277-2443-1

C.T.J. Dodson: *Categories, Bundles and Spacetime Topology.* 1988, 264 pp. ISBN 90-277-2771-6

V.D. Goppa: *Geometry and Codes.* 1988, 168 pp. ISBN 90-277-2776-7

A.A. Markov and N.M. Nagorny: *The Theory of Algorithms.* 1988, 396 pp. ISBN 90-277-2773-2

E. Kratzel: *Lattice Points.* 1989, 322 pp. ISBN 90-277-2733-3

A.M.W. Glass and W.Ch. Holland (eds.): *Lattice-Ordered Groups. Advances and Techniques.* 1989, 400 pp. ISBN 0-7923-0116-1

N.E. Hurt: *Phase Retrieval and Zero Crossings: Mathematical Methods in Image Reconstruction.* 1989, 320 pp. ISBN 0-7923-0210-9

Du Dingzhu and Hu Guoding (eds.): *Combinatorics, Computing and Complexity.* 1989, 248 pp. ISBN 0-7923-0308-3

Other *Mathematics and Its Applications* titles of interest:

A.Ya. Helemskii: *The Homology of Banach and Topological Algebras.* 1989, 356 pp. ISBN 0-7923-0217-6

J. Martinez (ed.): *Ordered Algebraic Structures.* 1989, 304 pp. ISBN 0-7923-0489-6

V.I. Varshavsky: *Self-Timed Control of Concurrent Processes. The Design of Aperiodic Logical Circuits in Computers and Discrete Systems.* 1989, 428 pp. ISBN 0-7923-0525-6

E. Goles and S. Martinez: *Neural and Automata Networks. Dynamical Behavior and Applications.* 1990, 264 pp. ISBN 0-7923-0632-5

A. Crumeyrolle: *Orthogonal and Symplectic Clifford Algebras. Spinor Structures.* 1990, 364 pp. ISBN 0-7923-0541-8

S. Albeverio, Ph. Blanchard and D. Testard (eds.): *Stochastics, Algebra and Analysis in Classical and Quantum Dynamics.* 1990, 264 pp. ISBN 0-7923-0637-6

G. Karpilovsky: *Symmetric and G-Algebras. With Applications to Group Representations.* 1990, 384 pp. ISBN 0-7923-0761-5

J. Bosak: *Decomposition of Graphs.* 1990, 268 pp. ISBN 0-7923-0747-X

J. Adamek and V. Trnkova: *Automata and Algebras in Categories.* 1990, 488 pp. ISBN 0-7923-0010-6

A.B. Venkov: *Spectral Theory of Automorphic Functions and Its Applications.* 1991, 280 pp. ISBN 0-7923-0487-X

M.A. Tsfasman and S.G. Vladuts: *Algebraic Geometric Codes.* 1991, 668 pp. ISBN 0-7923-0727-5

H.J. Voss: *Cycles and Bridges in Graphs.* 1991, 288 pp. ISBN 0-7923-0899-9

V.K. Kharchenko: *Automorphisms and Derivations of Associative Rings.* 1991, 386 pp. ISBN 0-7923-1382-8

A.Yu. Olshanskii: *Geometry of Defining Relations in Groups.* 1991, 513 pp. ISBN 0-7923-1394-1

F. Brackx and D. Constales: *Computer Algebra with LISP and REDUCE. An Introduction to Computer-Aided Pure Mathematics.* 1992, 286 pp. ISBN 0-7923-1441-7

N.M. Korobov: *Exponential Sums and their Applications.* 1992, 210 pp. ISBN 0-7923-1647-9

D.G. Skordev: *Computability in Combinatory Spaces. An Algebraic Generalization of Abstract First Order Computability.* 1992, 320 pp. ISBN 0-7923-1576-6

E. Goles and S. Martinez: *Statistical Physics, Automata Networks and Dynamical Systems.* 1992, 208 pp. ISBN 0-7923-1595-2

Other *Mathematics and Its Applications* titles of interest:

M.A. Frumkin: *Systolic Computations*. 1992, 320 pp. ISBN 0-7923-1708-4

J. Alajbegovic and J. Mockor: *Approximation Theorems in Commutative Algebra*. 1992, 330 pp. ISBN 0-7923-1948-6

I.A. Faradzev, A.A. Ivanov, M.M. Klin and A.J. Woldar: *Investigations in Algebraic Theory of Combinatorial Objects*. 1993, 516 pp. ISBN 0-7923-1927-3

I.E. Shparlinski: *Computational and Algorithmic Problems in Finite Fields*. 1992, 266 pp. ISBN 0-7923-2057-3

P. Feinsilver and R. Schott: *Algebraic Structures and Operator Calculus*. Vol. I. Representations and Probability Theory. 1993, 224 pp. ISBN 0-7923-2116-2

A.G. Pinus: *Boolean Constructions in Universal Algebras*. 1993, 350 pp. ISBN 0-7923-2117-0

V.V. Alexandrov and N.D. Gorsky: *Image Representation and Processing. A Recursive Approach*. 1993, 200 pp. ISBN 0-7923-2136-7

L.A. Bokut' and G.P. Kukin: *Algorithmic and Combinatorial Algebra*. 1994, 384 pp. ISBN 0-7923-2313-0

Y. Bahturin: *Basic Structures of Modern Algebra*. 1993, 419 pp. ISBN 0-7923-2459-5

R. Krichevsky: *Universal Compression and Retrieval*. 1994, 219 pp. ISBN 0-7923-2672-5

A. Elduque and H.C. Myung: *Mutations of Alternative Algebras*. 1994, 226 pp. ISBN 0-7923-2735-7

E. Goles and S. Martínez (eds.): *Cellular Automata, Dynamical Systems and Neural Networks*. 1994, 189 pp. ISBN 0-7923-2772-1

A.G. Kusraev and S.S. Kutateladze: *Nonstandard Methods of Analysis*. 1994, 444 pp. ISBN 0-7923-2892-2

P. Feinsilver and R. Schott: *Algebraic Structures and Operator Calculus*. Vol. II. Special Functions and Computer Science. 1994, 148 pp. ISBN 0-7923-2921-X

V.M. Kopytov and N. Ya. Medvedev: *The Theory of Lattice-Ordered Groups*. 1994, 400 pp. ISBN 0-7923-3169-9

H. Inassaridze: *Algebraic K-Theory*. 1995, 438 pp. ISBN 0-7923-3185-0

C. Mortensen: *Inconsistent Mathematics*. 1995, 155 pp. ISBN 0-7923-3186-9

R. Abłamowicz and P. Lounesto (eds.): *Clifford Algebras and Spinor Structures*. A Special Volume Dedicated to the Memory of Albert Crumeyrolle (1919–1992). 1995, 421 pp. ISBN 0-7923-3366-7

W. Bosma and A. van der Poorten (eds.), *Computational Algebra and Number Theory*. 1995, 336 pp. ISBN 0-7923-3501-5

Other *Mathematics and Its Applications* titles of interest:

A.L. Rosenberg: *Noncommutative Algebraic Geometry and Representations of Quantized Algebras*. 1995, 316 pp. ISBN 0-7923-3575-9

L. Yanpei: *Embeddability in Graphs*. 1995, 400 pp. ISBN 0-7923-3648-8

B.S. Stechkin and V.I. Baranov: *Extremal Combinatorial Problems and Their Applications*. 1995, 205 pp. ISBN 0-7923-3631-3

Y. Fong, H.E. Bell, W.-F. Ke, G. Mason and G. Pilz (eds.): *Near-Rings and Near-Fields*. 1995, 278 pp. ISBN 0-7923-3635-6

A. Facchini and C. Menini (eds.): *Abelian Groups and Modules*. (Proceedings of the Padova Conference, Padova, Italy, June 23–July 1, 1994). 1995, 537 pp. ISBN 0-7923-3756-5

D. Dikranjan and W. Tholen: *Categorical Structure of Closure Operators*. With Applications to Topology, Algebra and Discrete Mathematics. 1995, 376 pp. ISBN 0-7923-3772-7

A.D. Korshunov (ed.): *Discrete Analysis and Operations Research*. 1996, 351 pp. ISBN 0-7923-3866-9

P. Feinsilver and R. Schott: *Algebraic Structures and Operator Calculus*. Vol. III: Representations of Lie Groups. 1996, 238 pp. ISBN 0-7923-3834-0

M. Gasca and C.A. Micchelli (eds.): *Total Positivity and Its Applications*. 1996, 528 pp. ISBN 0-7923-3924-X

W.D. Wallis (ed.): *Computational and Constructive Design Theory*. 1996, 368 pp. ISBN 0-7923-4015-9

F. Cacace and G. Lamperti: *Advanced Relational Programming*. 1996, 410 pp. ISBN 0-7923-4081-7

N.M. Martin and S. Pollard: *Closure Spaces and Logic*. 1996, 248 pp. ISBN 0-7923-4110-4

A.D. Korshunov (ed.): *Operations Research and Discrete Analysis*. 1997, 340 pp. ISBN 0-7923-4334-4

FSC
www.fsc.org
MIX
Papier | Fördert
gute Waldnutzung
FSC® C083411

Kolibri 360 GmbH
Ferdinand-Jühlke-Straße 7
99095 Erfurt, Deutschland
produktsicherheit@kolibri360.de